Ion Exchange and Solvent Extraction: Volume 23

Changing the Landscape in Solvent Extraction

Ion Exchange and Solvent Extraction Series

Series Editors
Arup K. SenGupta
Bruce A. Moyer

Founding Editors
Jacob A Marinsky
Yizhak Marcus

Contents of Recently Published Volumes

Browse contents of all other available volumes at https://www.crcpress.com/Ion-Exchange-and-Solvent-Extraction-Series/book-series/CRCIOEXSOEXT.

Ion Exchange and Solvent Extraction: Volume 23

Changing the Landscape in Solvent Extraction

Edited by

Bruce A. Moyer

CRC Press
Taylor & Francis Group
Boca Raton London New York

CRC Press is an imprint of the
Taylor & Francis Group, an **informa** business

CRC Press
Taylor & Francis Group
6000 Broken Sound Parkway NW, Suite 300
Boca Raton, FL 33487-2742

First issued in paperback 2022

© 2020 by Taylor & Francis Group, LLC

CRC Press is an imprint of Taylor & Francis Group, an Informa business
No claim to original U.S. Government works

ISBN: 978-1-03-240137-9 (pbk)
ISBN: 978-1-138-07920-5 (hbk)
ISBN: 978-1-315-11437-8 (ebk)

DOI: 10.1201/9781315114378

**Visit the Taylor & Francis Web site at
http://www.taylorandfrancis.com**

**and the CRC Press Web site at
http://www.crcpress.com**

Contents

Preface

My thematic vision and corresponding title for Volume 23 is *Changing the Landscape in Solvent Extraction*. Following my other Volumes 19 and 21, which respectively dealt with the specific topical areas of nuclear fuel recycle and supramolecular aspects of solvent extraction, my thought for Volume 23 was to capture new developments across the field more generally, some of which would be missed if I picked another very specific topic. And recently there have been quite a few important advances changing the landscape in solvent extraction! Note that the chosen title is not *The* Changing Landscape... but rather the active *Changing the Landscape....* The changes we are seeing are dramatic and deliberate, effected by an energetic new cadre of scientists and engineers, armed with powerful new tools and led in part by the contributors of this volume. Long-standing problems are being solved while entirely new directions are being struck, and the breadth of the impact is staggering. We now take for granted the fruits of solvent extraction in the form of electrical energy and the means to convey it; phosphors for energy-efficient lighting and displays; powerful magnets for motors and generators; high-performance alloys for construction, transportation, and aerospace; electrical devices for communications; computers of every size and description; petrochemicals feeding our thirst for fuels, chemicals, and polymers; pharmaceuticals that are solving global health problems; food products to feed the world; and analytical methods to quantify traces of almost any element in almost any conceivable matrix. Look anywhere on the periodic table, and it is difficult to find an element that has not in some way been touched by a solvent extraction technology in its recovery, purification, analysis, recycle, or academic study. Much of the foundational knowledge underlying this infusion of solvent extraction into our present high-tech economy can, in fact, be found in the early volumes of *Ion Exchange and Solvent Extraction*.

Perhaps the present state of solvent extraction and its changing landscape can be appreciated best by briefly reviewing where we have been. Growing out of the 1930s when analytical chemists began to experiment with simple ligands like acetylacetone and 8-hydroxyquinoline, the field of solvent extraction gained considerable momentum as a vehicle for studying and applying coordination chemistry as later driven by pressing needs in the nuclear industry and opportunities to recover metals in hydrometallurgy. A generation of scientists and engineers in the 1950s and 1960s opened up the field to enormous technological success, defining the underlying principles, introducing a basic palette of new extractants, developing efficient liquid–liquid contacting techniques, and implementing numerous large-scale applications worth hundreds of billions of dollars in today's world economy.

By the 1970s and 1980s, when I entered the field, solvent extraction had established itself so firmly in hydrometallurgy and nuclear applications that it came to be regarded as a mature separation technique. While this was a sure sign of success, it also came with a downside. Many practitioners baldly considered that they had all the chemistry and equipment needed to accomplish any desired separation, and with a few notable exceptions such as the introduction of molecular recognition and innovation in dealing with urgent needs for nuclear waste cleanup, progress seemed incremental for a couple of decades. Had the field followed the classical S-shaped maturation curve and was doomed to stagnation? If the field indeed had reached a plateau, it may have been a sign that certain problems arising from the lack of understanding of solvation, solution structure, complicated speciation, aggregation, interfacial phenomena, fluid dynamics, transport, etc. had become intractable with the available tools of the day. All the low-hanging fruit had been picked. Extractants were (and still are) mostly simple mono- and bidentate ligands rendered hydrophobic by empirically chosen substituents. Selectivity using such extractants is dictated by the principles of coordination chemistry, involving notions such as hard–soft acid–base bonding and ligand-field stabilization, and sometimes it could be modulated by steric effects to useful purposes. Liquid–liquid contacting in the laboratory had largely remained for a century in the form of shaking experiments in vials and separatory funnels, using various mechanical techniques and with little ability to scale down to volumes smaller than a milliliter. Industry developed its standard mixer-settlers and pulse columns, cautiously scaled up with empirical correlations. We knew that, as amphiphilic molecules, extractants tended to aggregate, but for the most part it was not possible to understand much about the structure of the aggregates let alone predict or control their behavior. Most extractants exhibit annoying third-phase behavior, and until recently, we could only suspect they had something to do with microemulsions but couldn't push past qualitative explanations. A friend of mine, Peter Tasker, likes to quote engineers as generously saying that any diluent is acceptable as long as it's kerosene. We have indeed had a hard time breaking out of that paradigm. Fledgling computational power could not tackle until recently the complexity of solvent extraction. The liquid–liquid interface lay shrouded in mystery, even though it was widely appreciated that it held the key to kinetics and coalescence phenomena. While these generalizations are simplistic and unfair, especially to the brilliant exceptions, it is true that our ability to fundamentally understand, predict, and control molecular and supramolecular behavior in solvent extraction had been sorely limited in the last century.

However, the last few years have seen some awesome developments that give me hope! Key developments have been happening in designing new tailor-made extractants and implementing them in efficient adaptive processes. Microfluidics methods are re-introducing solvent extraction to analytical chemistry once again, providing a potent method for reducing the scale of extraction systems, studying liquid–liquid kinetics, and possibly even intensifying process applications. Powerful new X-ray methods have enabled an unprecedented understanding of the structure and dynamics of the liquid–liquid interface, as well as the supramolecular structure of solvent species. Better models of liquid–liquid dispersions now make it possible to design and scale-up solvent extraction equipment more reliably than ever before. Ionic liquids have revealed fascinating, unexpected behavior in liquid–liquid extraction, and the rich possibilities for creating new chemistry for extraction seem endless. Computational techniques have been growing increasingly fast and accurate, allowing a new way to visualize the complex interactions in the solvent phase that have eluded chemical understanding for over half a century.

Recent advances have been changing the landscape in both the chemistry and engineering side of solvent extraction, creating new directions as well as deepening our understanding of structure and dynamics of liquid–liquid systems from the molecular- to nano- to meso- to bulk-scale. Some of these transformational changes are captured in this volume. Have we now solved all the problems? Is a new plateau looming? Judging by the accelerating pace of the developments outlined herein, I am quite sure that the breakthroughs occurring in structural techniques, computational methods, ligand design, equipment design, neoteric solvents, and miniaturization are only the start of major upswing in gaining new understanding and finding new applications that we could never have imagined a few decades ago. I can hardly wait to see what happens next!

Bruce A. Moyer
Oak Ridge National Laboratory,
Oak Ridge,
Tennessee

Acknowledgments

Multiple thanks are in order for the completion of this volume of *Ion Exchange and Solvent Extraction*. Two sponsors provided my major support: the U.S. Department of Energy, Office of Science, Basic Energy Sciences, Chemical Sciences, Geosciences, and Biosciences Division; and the Critical Materials Institute, an Energy Innovation Hub funded by the U.S. Department of Energy, Office of Energy Efficiency and Renewable Energy, Advanced Manufacturing Office. Both of these sponsors enable me to continue my career in separation science and technology, which in turn affords me the opportunity to muse about the direction of this essential technical field and to organize a book around the theme of changing the landscape of solvent extraction. In the same vein, I thank my supervisors at Oak Ridge National Laboratory for their encouragement. Even though this activity takes place in my "free time," editing this book series and, in particular, this volume has been a major commitment, and their enthusiasm has been energizing. The authors of the chapters especially deserve my thanks. They have done the lion's share of the work on this volume, providing the real substance for the theme of changing the landscape in solvent extraction. They have also been highly responsive to my many requests and graciously accommodated even rather involved revisions and additions toward a more complete work. I also thank Danielle Zarfati and Barbara Knott of Taylor & Francis for their help and patience while waiting and waiting for this book to come together. Finally, my wife Lily has always been supportive of numerous evenings and weekend days spent in front of this computer screen.

Bruce A. Moyer
Oak Ridge National Laboratory,
Oak Ridge,
Tennessee

Editor

Bruce A. Moyer is a corporate fellow at the Oak Ridge National Laboratory (ORNL) and fellow of the American Association for the Advancement of Science. Specializing over a 39-year career in both fundamental and applied aspects of separation science and technology, he has especially focused on the chemistry of solvent extraction and ion exchange as applied to problems in nuclear waste treatment, nuclear fuel recycle, and recovery of energy-critical materials. He received his BS degree summa cum laude with chemistry honors from Duke University in 1974 and was inducted into Phi Beta Kappa. He earned his Ph.D. in inorganic chemistry from the University of North Carolina at Chapel Hill in 1979 under the direction of Prof. Thomas J. Meyer. His graduate work dealt with fundamental mechanisms of redox catalysis, proton-coupled electron transfer, and ruthenium oxo complexes. In 1979, he joined the staff at ORNL and solved a variety of problems in separations chemistry, always with an eye on incorporating principles of molecular recognition. His interests in fundamental aspects include thermodynamics, equilibrium modeling, extractant design, and interfacial phenomena from third-phase formation to phase disengagement to CRUD. He has investigated diverse extractant classes including amines, sulfoxides, sulfonic acids, carboxylic acids, crown ethers, calixarenes, and various anion receptors such as calixpyrroles, macrocyclic amides, and sulfonamides. Extracted ions of interest have included alkali metals, alkaline earths, various transition metals, actinides, lanthanides, and a variety of inorganic anions. His most successful application is the development of the Caustic-Side Solvent Extraction (CSSX) process, which has been operating successfully at the Savannah River Site, processing approximately 7 million gallons of high-level waste. Dr. Moyer has published more than 200 open-literature articles, book chapters, proceedings papers, and reports, and he has edited three books. His 12 patents range from solvent extraction of cesium for nuclear waste cleanup to supported liquid membrane systems and novel anion exchange resins. In addition to his duties as Group Leader, Chemical Separations, in the ORNL Chemical Sciences Division, Dr. Moyer has been leading three programs for the U.S. Department of Energy: Principles of Chemical Recognition and Transport in Extractive Separations (Office of Science), the Sigma Team for Advanced Actinide Recycle (Office of Nuclear Energy), and the Diversifying Supply Focus Area of the Critical Materials Institute, a USDOE Energy Innovation Hub (Office of Energy Efficiency and Renewable Energy). He has also provided leadership for the chemical development of CSSX process, which won the 2013 Secretary of Energy's Award. Dr. Moyer also serves as co-editor of the journal *Solvent Extraction and Ion Exchange* and has served on the board of editors of the journal *Hydrometallurgy*. In 2008, Dr. Moyer chaired the technical program of the 2008 International Solvent Extraction Conference (ISEC 2008) and was editor-in-chief of the proceedings. Subsequently, he served ISEC 2011 as a member of the Advisory Committee, program chair for Nuclear Separations, and co-editor of the Proceedings.

Dr. Moyer has received a number of awards: 2018 American Association for the Advancement of Science Fellow; 2017 R&D-100 Award for development of an aluminum-cerium alloy; 2013 Secretary of Energy's Award and 2011 Council of Chemical Research Collaboration Award, both for development and implementation of high-level salt-waste processing technology (team award); 2004 IR-100 Award in 2004 for a highly selective, regenerable perchlorate treatment system; UT-Battelle Technical Achievement Award in 2000 for Contributions to the Development of Novel Resin Regeneration Techniques; Lockheed Martin Research Corporation Achievement awards in 1999–Leadership Award, Development Award for a novel bifunctional anion exchange resin, and Development Award for a novel process for cesium separation from waste.

Contributors

Jochen Autschbach
Department of Chemistry
University at Buffalo, State University
of New York
Buffalo, NY

Enrique R. Batista
Theoretical Division
Los Alamos National Laboratory
Los Alamos, NM

Wei Bu
Center for Advanced Radiation Sources,
ChemMatCARS
University of Chicago
Chicago, IL

Alexandre Chagnes
Université de Lorraine, CNRS
The GeoRessources Laboratory
Nancy, France

Aurora E. Clark
Department of Chemistry
Washington State University
Pullman, WA

Mark L. Dietz
Department of Chemistry and
Biochemistry
University of Wisconsin
Milwaukee, WI

David A. Dixon
Department of Chemistry and
Biochemistry
University of Alabama
Tuscaloosa, AL

Cory A. Hawkins
Department of Chemistry
Tennessee Technological University
Cookeville, TN

Qing He
State Key Laboratory of Chemo/
Biosensing and Chemometrics
Hunan University
Changsha, China

David Leleu
Department of Chemical Engineering
University of Liège
Liège, Belgium

Xiaosong Li
Department of Chemistry
University of Washington
Seattle, WA

Zhu Liu
Department of Chemistry
Washington State University
Pullman, WA

Jason B. Love
EaStCHEM School of Chemistry
University of Edinburgh
Edinburgh, UK

Guangsheng Luo
The State Key Laboratory of Chemical
Engineering, Department of
Chemical Engineering
Tsinghua University
Beijing, China

Edward J. Maginn
Department of Chemical and
    Biomolecular Engineering
University of Notre Dame
Notre Dame, IN

Ernesto Martinez-Baez
Department of Chemistry
Washington State University
Pullman, WA

Manuel Miguirditchian
CEA Marcoule
Alternative Energies and Atomic
    Energy Commission
Bagnols-sur-Ceze, France

Ken Newcomb
Department of Chemical and
    Biomolecular Engineering
University of Notre Dame
Notre Dame, IN

Andreas Pfennig
Department of Chemical Engineering
University of Liège
Liège, Belgium

Mark L. Schlossman
Department of Physics
University of Illinois at Chicago
Chicago, IL

Michael J. Servis
Department of Chemistry
Washington State University
Pullman, WA

Jonathan L. Sessler
Department of Chemistry
University of Texas
Austin, TX

Torin Stetina
Department of Chemistry
University of Washington
Seattle, WA

Jing Su
Theoretical Division
Los Alamos National Laboratory
Los Alamos, NM

Gabriela I. Vargas-Zúñiga
Department of Chemistry
University of Texas
Austin, TX

Kai Wang
The State Key Laboratory of Chemical
    Engineering, Department of
    Chemical Engineering
Tsinghua University
Beijing, China

Andrew Wildman
Department of Chemistry
University of Washington
Seattle, WA

Ping Yang
Theoretical Division
Los Alamos National Laboratory
Los Alamos, NM

Fang Zhao
State Key Laboratory of Chemical
    Engineering
East China University of Science and
    Technology
Shanghai, China

1 New Insights into the Recovery of Strategic and Critical Metals by Solvent Extraction

The Effects of Chemistry and the Process on Performance

Jason B. Love, Manuel Miguirditchian,
and Alexandre Chagnes

CONTENTS

1.1 INTRODUCTION

The exploitation of nonthermal chemical separation processes, which would lower global energy use, emissions, and pollution, is particularly pertinent to the recovery of metals from their primary ores and secondary resources [1]. In principle, the separation of metals by hydrometallurgical processes such as solvent extraction (SX, Figure 1.1) can achieve these outcomes due to ambient-temperature operation and the maintenance of materials balance through reagent recycling [2]. However, significant challenges remain, not least with attaining high levels of separation between metals, but also with issues such as reagent stability, complexity, cost, safety, and recyclability [3, 4].

The optimization of SX processes is one of the challenges for treating new unconventional primary and secondary resources, especially polymetallic and low-grade resources. Although the ease of contacting two liquid phases is one of the key advantages of SX, the physicochemistry involved is very complex due to the existence of associations at molecular and supramolecular levels, nonideality, and the presence of various reactions in the aqueous and organic phases as well as in the liquid–liquid interface, which governs the kinetics of extraction (Figure 1.2) [5].

The performance of solvent extraction processes mainly relies on the properties of extractant molecules for the selectivity needed for increasingly challenging problems. Surprisingly, there are relatively few extracting agents available on the market, and these molecules were mostly found more than 15 to 20 years ago in spite of great efforts to develop new ones. Industry representatives repeatedly cite the high costs of chemical development and long times for testing and regulatory approval of new reagents as impediments. Developing faster and better approaches to designing new extractants, therefore, appears of great importance toward building efficient, low-cost, and sustainable processes capable of extracting metals from new complex

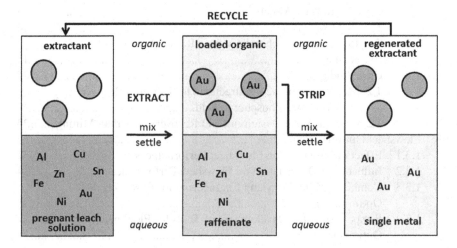

FIGURE 1.1 Schematic describing the separation of metals using solvent extraction. Circles in the organic phase represent extractant molecules or aggregates.

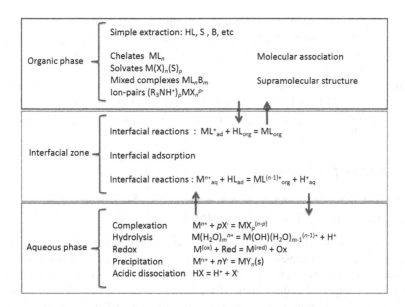

FIGURE 1.2 Main reactions involved in the liquid–liquid extractions of metals. Here, L is the conjugate anion of an acidic organic extractant, HL, often a chelant; B is a neutral coordinating extractant; X is an anion co-extracted with the metal M from the aqueous phase.

resources. However, the design of new extracting agents cannot be performed by considering only their chemical structures, because interactions involving extractants strongly influence extractant properties. The molecular environment around the extractant, including diluent, phase modifier, other extractant molecules, etc., must thus be first understood and then taken into account during extraction solvent design. For this goal, modeling tools such as DFT (density functional theory), semi-empirical, and QSPR (quantitative structure–property relationship) calculations are increasingly playing an important role [6–8]. Nevertheless, the degree of reliability of the predictions is still limited, and in the present state of the art, these techniques are likely more useful for optimization within a given family of extractants than to build *in silico* new reagents. The molecular modeling techniques provide binding energies between target metals and given ligands, as well as optimized chemical structures of the formed complexes. Thus, in principle, the information that can be deduced from the molecular modeling computations is richer than that provided by QSPR methods. Modeling tools are an asset to understand liquid–liquid extraction phenomena. The comprehension of the physicochemistry involved during liquid–liquid extraction is therefore particularly important to optimize solvent extraction processes. It paves the way of the development of simulation process tools which are of great importance for optimizing processes.

Likewise, the optimization of the flowsheets is another way to improve the performances of solvent extraction processes in terms of extraction efficiency and selectivity. The classical McCabe–Thiele approach is usually used to choose the number of contactors to implement in the process. Such an approach is only based on engineering process calculations. Undoubtedly, the development of new approaches combining both the physicochemistry of extraction solvent and the engineering process calculations would help the engineer to optimize its process while using a minimum of expensive experiments.

This chapter gives an overview of the new insights into the recovery of strategic and critical metals by solvent extraction. Particular attention is paid to explaining how it is possible to improve existing processes by playing both on the chemistry and the flowsheet design.

1.2 EXTRACTANT DESIGN

1.2.1 THE ROLE OF COORDINATION CHEMISTRY AND SUPRAMOLECULAR CHEMISTRY ON THE DESIGN OF NEW EXTRACTANTS

The development of new reagents for solvent extraction processes requires an understanding of the fundamental solution chemistry of metals and their compounds in both aqueous and organic phases. Ultimately, however, the thermodynamic stability of the metal compound in the organic phase must be greater than that in the aqueous phase for the equilibrium to favor extraction. These facets have been analyzed in terms of the mechanism of extraction, namely how the metal is transferred from the aqueous to the organic phase, as a cation (M^{n+}), anion (MX_m^{n-}), or salt (MX_m) [9]. This analysis requires combinations of expertise in coordination and supramolecular chemistry in particular, along with a myriad of experimental and computational characterization methods, to gain a proper understanding of the solvent extraction process. This part of the chapter will provide an overview on the role of coordination and supramolecular chemistry on solvent extraction processes, and will highlight the different types of extraction, the compounds formed, their characterization, and the impact on the design of new reagents. In keeping with the classifications of extractants described above, new advances will be described according to the metal species that are present in the organic phase: a discrete metal complex, a metal complex that has second-sphere (or outer-sphere) interactions (e.g., hydrogen bonding), ion pairs or clusters of ions, and reverse micelles (Figure 1.3).

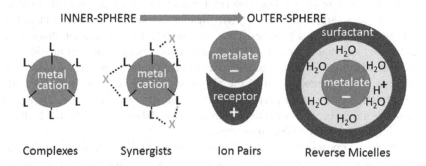

FIGURE 1.3 Modes of action generating charge-neutral assemblies in the water-immiscible phase in solvent extraction (L describes a neutral or anionic ligand).

1.2.1.1 Complex Formation

The formation of metal complexes that are soluble in the organic phase in a SX system is perhaps the most common, and best understood, mode of action, and can be described by coordination chemistry, namely the bonding between a metal cation and a ligand (sometimes known as first-, primary-, or inner-sphere bonding). The formation of discrete metal complexes can occur in a variety of ways, but primarily by acid–base chemistry (pH swing, Eq. 1.1) or neutral salt formation (solvation, Eq. 1.2).

$$M_{(aq)}^{n+} + nHL_{(org)} \rightleftharpoons \left[M\left(L\right)_n \right]_{(org)} + nH_{(aq)}^{+} \tag{1.1}$$

$$M_{(aq)}^{n+} + nX_{(aq)}^{-} + mL_{(org)} \rightleftharpoons \left[MX_n\left(L\right)_m \right]_{(org)} \tag{1.2}$$

One of the most studied systems that makes use of acid–base chemistry is the solvent extraction of copper using phenolic aldoximes and ketoximes (Figure 1.4). It is estimated that 20% of the global copper supply is delivered using these types of reagents in solvent extraction processes [10]. This area has been extensively reviewed, so only recent advances in the chemical understanding of the mode of action are highlighted here [11].

It was reported previously that the X-substituent of the phenolic oxime buttressed the hydrogen-bonding interaction between the phenolic oxygen atom and the adjacent oxime hydrogen atom, and that the nature of the X-substituent was a dominant effect on the strength of solvent extraction of Cu (Figure 1.4) [12, 13]. This is perhaps surprising, as the bond-dissociation energies of metal–ligand bonds are at least one order of magnitude greater than those of hydrogen bonds. Recently, however, this was studied further by analyzing similar Cu complexes by EPR (Cu^{2+} is d^9 with a single unpaired electron) and ^{1}H ENDOR (electron nuclear double resonance) spectroscopy, DFT calculations, and X-ray crystallography [14]. In this case, the ^{1}H ENDOR spectra are dominated by coupling of the unpaired electron to the azomethine and oximic protons, which provides information on the Cu$\cdots$H distances for the various complexes and gives an indication of the strength of the hydrogen bond and Cu–N/O bonds. For X=aminomethyl, this represents a lengthening of the Cu–N/O bonds and hence a decrease in the bond strength. However, a strong buttressing effect is seen by X-ray crystallography, and the strength of extraction of Cu by this ligand is anomalous. As such, the deprotonation energies of the pro-ligands, the binding

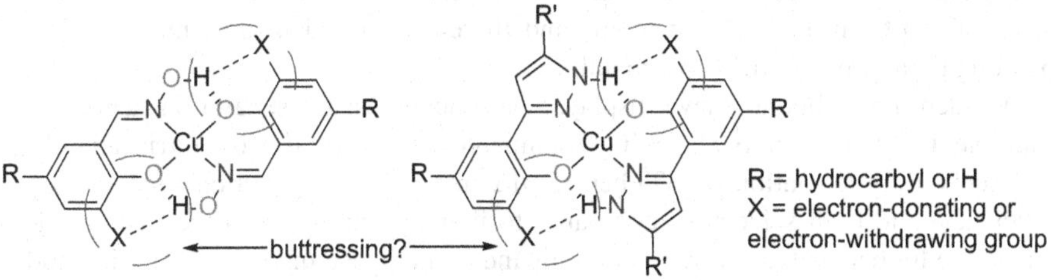

FIGURE 1.4 Buttressing of hydrogen-bonding interactions in *pseudo*-macrocyclic Cu complexes of phenolic aldoximes (left) and phenolic pyrazoles (right).

energies of the Cu complexes, and the formation energies of the Cu complexes were evaluated computationally. From these data, it is seen that the strength of Cu extraction is related to a combination of factors and not only hydrogen-bond buttressing. For example, for the $X = Br$ ligand, the ease of deprotonation due to the added EWG and the strong buttressing effect of the *ortho*-Br compensate for the weaker binding energy of the ligand to Cu that arises from the reduced basicity of the NO^- donor atoms. Overall, this provides sound chemical reasoning on the strength of solvent extraction exhibited by the phenolic oximes reagents.

Phenolic pyrazoles (Figure 1.4, right) have also been evaluated as extractants for Cu and the structures of the resulting complexes studied by X-ray diffraction and DFT calculations [15]. As with the phenolic oximes above, the presence of groups with the ability to buttress the phenolic oxygen–oxime hydrogen bond were shown to be the strongest extractants, although the ease of deprotonation of the pro-ligand is again a significant factor. Also, the pyrazole N–H donor is less accessible to the *ortho*-substituent compared to the equivalent in the phenolic oximes, and only for *ortho*-nitro substitution is strong buttressing of the hydrogen bonding seen, with otherwise minimal bonding seen from natural bond order analysis.

Tripodal polyamines based on the tris(2-aminomethyl)amine (tren) platform have potential in solvent extraction, as the NN_3 donor set can be tuned by N-substituent variation. This is exemplified by a series of tren ligands with aromatic N-substituents, which were used in the solvent extraction of transition metals at buffered pH 7.4 [16]. These ligands showed selectivity for Ag(I) and Zn(II) over Co(II), Ni(II), and Cd(II), which was attributed to a combination of lipohilicity of the ligand and the stability constants of the extracted complexes.

The separation, or partitioning, of lanthanides (Ln) from actinides (An) is a significant issue in nuclear waste remediation, as the lanthanides are efficient neutron absorbers and so inhibit the fission of recycled nuclear fuels and hence the transmutation of the highly radioactive trans-actinides. As the 5f orbitals have a greater radial extension compared to the 4f orbitals, it is thought that the actinide–ligand bond has a higher covalent contribution than the analogous lanthanide–ligand bond [17, 18]. This feature would provide a basis for Ln/An separation through the use of ligands that favor covalent interactions over ionic interactions. While the significance of aromatic nitrogen heterocycles in the separation of actinides from lanthanides is long understood, the chemical understanding of the extraction mechanisms is being increasingly studied [19].

Ligands such as triazinyl bipyridines and phenanthrolines (Figure 1.5) have an array of "soft" nitrogen donor atoms and therefore should favor a more covalent bonding than ionic bonding to a metal.

As such, these ligands have displayed significant An/Ln separation factors; for example, BTPhen exhibits $D_{Am} > 1000$ compared with $D_{Eu} < 10$ from nitric acid [20]. The locked conformation of BTPhen results in very fast extraction kinetics, high efficiency, and high selectivity compared to its bipy analog, with the fast kinetics due to the higher surface activity of the ligand at the phase interface [21]. The mode of extraction follows the metal salt route (Eq. 1.2) with the formation of the 2:1 ligand:metal complex $[Eu(NO_3)(BTphen)_2][NO_3]_2$, as verified by X-ray crystallography. Solution ^1H NMR (Nuclear Magnetic Resonance) dipolar paramagnetic shift

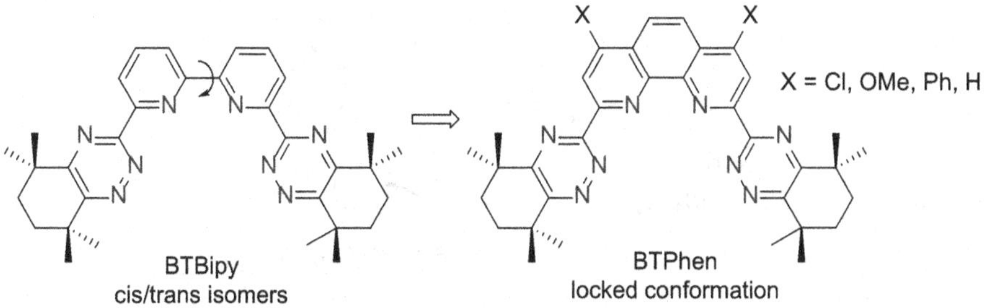

FIGURE 1.5 Triazinyl bipyridine and phenanthroline ligands for the separation of lanthanides from actinides.

analysis of the Yb(III) perchlorate analog showed the presence of a 2:1 ligand:metal complex. Good correlation between experimental and computed peak shifts is also seen in the presence of nitrate, but the solution structure could not be determined with accuracy. Clearly, the nitrate anions must be associated with the cationic complex to ensure solubility in the organic phase, and MD calculations of the Yb(III) analog showed that the structure is not coordinatively saturated.

The effect of electronic variation of BTPhen on the extraction of An/Ln was studied (Figure 1.4, right) and showed order-of-magnitude differences in distribution ratios while maintaining a high separation factor ($SF_{Am/Eu}$ *ca.* 110) [22]. The extraction mechanism was studied for Cm(III) by time-resolved laser-induced fluorescence spectroscopy (TRLFS). At low ligand concentrations (10^{-7} M), the 2:1 L:M complex was seen to form slowly (over 11 days), with the primary species in the solution being solvated Cm(III) and the 1:1 complex Cm(BTPhen)$^{3+}$. However, at ligand concentrations more representative of an extraction process (10^{-2} M), only the 2:1 complex Cm(BTPhen)$_2$$^{3+}$ is observed. The effect of HNO$_3$ concentration on the extraction profile was studied, showing that the basicity of the ligand is an important factor. Distribution ratios were seen to decrease on increasing the basicity of the ligand (e.g., X = OMe, pKa = 3.9), and the opposite was seen for less basic ligands (e.g., X = Cl, pKa = 1.7), indicating that it is the neutral ligand that transports the metal from the aqueous to the organic phase.

A raft of ligands containing combinations of pyridine and triazene groups with a variety of substituents have been reported and their modes of action studied [19]. The general extraction mechanism is by metal salt formation, and the use of lipophilic anions such as 2-bromohexanoic acid is seen to greatly enhance the extraction properties. More recently, the coordination properties of the zwitterionic nitrogen heterocyclic ligand HN$_4$bipy toward lanthanides and actinides was evaluated (Figure 1.6) [23].

In this case, HN$_4$bipy contains an ionizable hydrogen atom, so the ligand coordinates to the Lewis acidic metal as the anion N$_4$bipy$^-$, resulting in a stronger, electrostatic interaction. The Sm(III) complex was studied by X-ray crystallography, showing the 2:1 L:M complex Sm(OH)(OH$_2$)$_2$(N$_4$bipy)$_2$ in the solid state. The formation of Cu(III) and Eu(III) complexes of HN$_4$bipy were analyzed by TRLFS which showed 2:1 and 3:1 L:Cm complexes and 1:1 and 3:1 L:Eu complexes in solution.

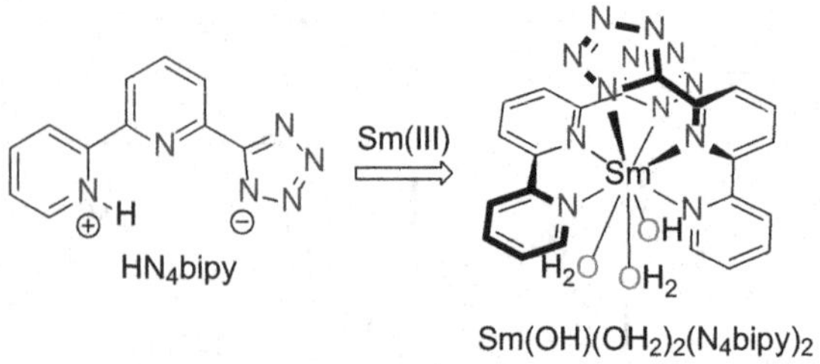

FIGURE 1.6 Coordination of the zwitterionic HN$_4$bipy to Sm(III) to form a 2:1 ligand:metal complex.

Determination of the stability constants of these complexes indicated a separation factor between Cu(III) and Eu(III) of *ca.* 500, an order of magnitude greater than the analogous alkylated ligands, RN$_4$bipy. As expected, the presence of 2-bromohexanoic acid only weakly competes with HN$_4$bipy, although more lipophilic versions of HN$_4$bipy would be required for its use in solvent extraction processes.

An alternative method for the separation of actinides from lanthanides is through the non-selective extraction of Ln and An into the organic phase (e.g., using the diglycolamide *N,N,N',N'*-tetraoctyldiglycolamide, TODGA) followed by selective back-transfer of the An into the aqueous phase through the use of hydrophilic ligands, for example, as shown in the TALSPEAK process [24], ALSEP process [25], and more recently the *i*-SANEX process [26, 27].

Recently, tetra-sulfonated versions of the above BTPhen ligands have been developed and exploited in this context (Figure 1.7, left) [28].

The addition of TS-BTPhen to the 0.5 M nitric acid aqueous phase in the extraction of Eu(III) and Am(III) by TODGA resulted in a separation factor of 616, compared to that of 3.5 in the absence of TS-BTPhen, significantly higher than the polyaminocarboxylate ligands used in the TALSPEAK process. Interestingly, the selectivity of TS-BTPhen for Am over Eu is similar to that seen for the hydrophobic analog BTPhen, which suggests that the coordination chemistry of this ligand class is preserved in both aqueous and organic phases.

As an alternative to TS-BTPhen, which suffers from radiolytic damage, the CHON-compliant, water-soluble bis(triazine)phenanthroline ligands have been

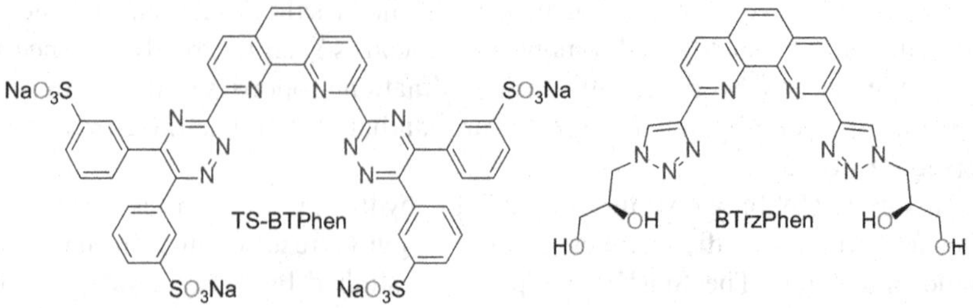

FIGURE 1.7 Hydrophilic ligands for the selective back-extraction of actinides into an aqueous phase.

developed and evaluated as selective Am(III) back-extractants (Figure 1.7, right) [29, 30]. Ligands such as BTrzPhen are able to differentiate between An and Ln, with separation factors of 36–47 favoring the An at low ligand concentrations. The solid-state structure of the Eu complex showed a 2:1 L:M ratio, similar to those seen for other An/Ln complexes extracted into an organic or aqueous phase.

As mentioned above, the diglycolamide TODGA (Figure 1.8) and its variants are used as neutral solvent extractants for Ln/An extraction and separation, often in combination with phosphorus acids.

As such, the role of these compounds in the mechanism of extraction has been studied and is driven fundamentally by the formation of the metal salt $LnX_3(L)_n$ (Eq. 1.2). While the solid-state structures determined by X-ray diffraction show a ligand:metal ratio of 3:1, with outer-sphere anions, the structures of these complexes in solution has only recently been determined. The Eu(III) complex $[Eu(TODGA)_3]$ $[BiCl_4]_3$ was analyzed by EXAFS (extended X-ray absorption fine structure) spectroscopy, and it was found that the cation $[Eu(TODGA)_3]^{3+}$ is present in solution, with no anion coordinated to Eu due to the rigidity of the chelating O_9-donor set [31]. This analysis parallels the structures determined in the solid state and implies that the variation in extraction seen across the Ln series is likely defined by the differences in the hydrophobicity due to outer-sphere effects of different charge-balancing anions; this may well suggest that aggregation or reverse-micelle formation is operative in these systems (see later).

The subtle energetic variations that underpin lanthanide extraction by diglycolamides were investigated by EXAFS spectroscopy and DFT calculations. As with the previous examples, EXAFS shows that the $[Ln(TODGA)_3]^{3+}$ structure is maintained in solution across the lanthanide series. On this basis, DFT calculations on a series of diglycolamides and related ligands showed that an interplay between steric strain and coordination energies gives rise to the nonlinear trend in lanthanide cation complexation; this shows the importance of strain energies to the design of the chelating ligands [32]. More recently, EXAFS and MD/DFT calculations on lanthanide diglycolamide extraction have shown that chloride or nitrate anions reside in clefts derived from the lipophilic arms of the diglycolamide ligands, through outer-sphere electrostatic and nonclassical C–H hydrogen-bonding interactions [33]. Also, subsequent work has shown that the co-extraction of water modifies the location of the

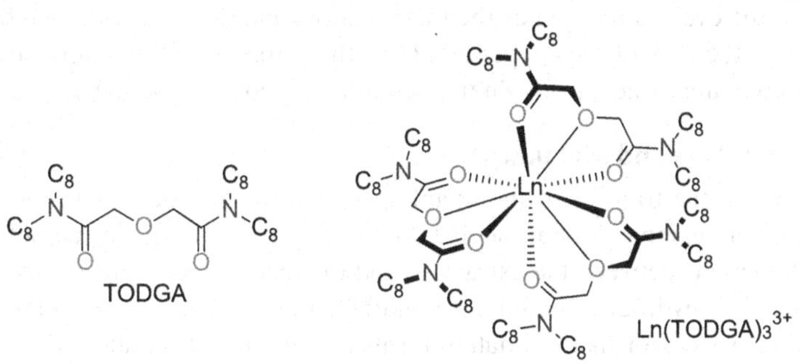

FIGURE 1.8 Lanthanide complexes of the diglycolamide TODGA.

anions within the structure which subsequently effects the separation of the light vs. heavy rare-earth elements [34]. Similar microhydration effects have been proposed recently for the mode of action of solvent extraction of lanthanides using ammonium ionic liquids [35].

The speciation of Eu and Nd complexes formed in an organic phase comprising the C-functionalized malonamide DMDOHEMA (see Figure 1.12) and the phosphoric acid D2EHPA was analyzed using ESI-MS, TRLFS, NMR, EXAFS, and DFT calculations in order to gain insight into synergism in Ln recovery by solvent extraction [36]. It was found that the order of addition of the individual extractants was important to the speciation. The phosphoric acid D2EHPA undergoes acid–base chemistry with the Ln cation to produce $M(L)_3(HL)_3$ complexes in the organic phase, and the subsequent addition of the malonamide causes the loss of neutral D2EHPA ligands (HL) forming a more lipophilic mixed-ligand complex. However, starting from malonamide-loaded organic phase, the Ln-containing species has attendant water and nitrate anions which are replaced on the addition of D2EHPA. As with the examples above, it is clear that while metal complexes of a defined coordination sphere are formed, the outer-sphere interactions are more difficult to probe, and so whether the extraction occurs by a molecular or micellar mechanism remains ill-defined.

EXAFS was used to probe solution speciation in the extraction of Nd(III) from ethylene glycol (EG) into dodecane using the phosphine oxide Cyanex 923. In this case, the mechanism of extraction was shown to be through the formation of the salt $Nd(L)_3(NO_3)_3(EG)$, and the use of this extraction procedure allowed the separation of heavy and light rare-earth elements [37].

While the fate of the anion in many of the above structures is unknown, namely in the inner- or outer-coordination sphere, or in a reverse-micelle, ligands that incorporate both metal cation and anion binding sites are able to transport metal salts as well-defined molecules from the aqueous into the organic phase. This approach requires an appreciation of both coordination and supramolecular chemistry in the design of metal–ligand and anion-receptor sites.

A series of N-donor ligands, such as macrocyclic cyclen, with attendant urea anion binding sites, has been prepared (Figure 1.9) and are shown to selectively solvent extract copper as the salts $[Cu(L)(SO_4)]$ in preference to Co(II), Ni(II), and Zn(II), albeit without the loading capacity for an efficient SX process [38]. In the solid state, the cyclen coordinates the metal cation, and the urea hydrogen bonds to the anion. In the case of the copper complex, the sulfate interacts with both the Cu center and the urea, whereas for Zn the two nitrate anions are bound separately.

1.2.1.2 Ion Pairs and Metalates

In many cases, the transfer of the metal species from the aqueous to the organic phase occurs as an ion pair or anion that does not form a coordination complex. That is, the interactions between the extractant and the metal species are electrostatic or supramolecular (hydrogen bonding, π-bonding); in this case, the extractant is best described as a receptor that ultimately forms a host–guest assembly on interaction with the metal species.

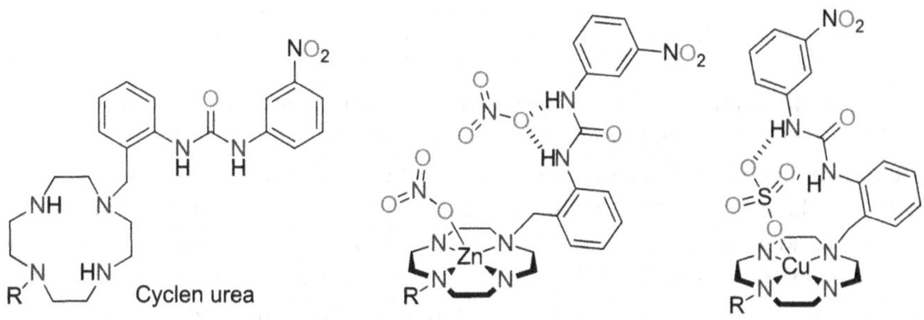

FIGURE 1.9 Ditopic ligands for simultaneous metal cation and anion binding.

Calix[4]pyrroles can be viewed as ditopic receptors, as they can bind metal cations and anions simultaneously and have been used as extractants for cesium halides [39], but unlike those described above (Figure 1.8), no coordinate bond is formed between the metal cation and a donor atom, with only hydrogen bonding, electrostatic, and ionic Cs-arene π interactions present (Figure 1.10) [38].

The simple octamethyl calix[4]pyrrole binds CsCl and CsBr in a 1:1:1 ligand:metal:halide stoichiometry on extraction into nitrobenzene from aqueous solution by an ion-pairing mechanism; in contrast, $CsNO_3$ behaves as a fully dissociated extraction system. Cation binding is absent without anion binding, as the strong interaction between the anion and receptor promotes organization of the receptor into a cone conformation of appropriate cavity size for the Cs cation, as seen by X-ray crystallography. A similar calix[4]pyrrole receptor with *meso*-hexyl substituents has also been shown to act as a receptor for CsBr [40]. Combined solution EXAFS, DFT, and MD studies on this system indicate that the toluene solvent interacts strongly with the Cs^+ cation, capping the coordination sphere through a cation-π-interaction, providing evidence that supposed "non-interacting" solvents may be more active in the structural chemistry than first envisaged (Figure 1.10, right).

The solvent extraction of anionic metalates from acidic media such as HCl using organic, protonatable bases is effective at achieving separation of base and precious metals (Eq. 1.3). This is perhaps surprising as the chemical separation is achieved through the formation of supposedly weak supramolecular interactions such as hydrogen bonds, and therefore little control over the host–guest assemblies formed in the organic phase would be expected. However, recent work has shown that control of these outer-sphere interactions can have a profound effect on the thermodynamic stability of the host–guest assembly, so effecting separation in a solvent extraction system.

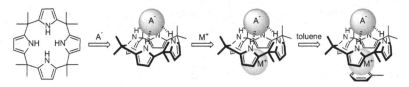

FIGURE 1.10 Calix[4]pyrroles as receptors for simple metal salts such as cesium halides.

$$MX_{y(aq)}^{n-} + nH_{(aq)}^{+} + nL_{(org)} \rightleftharpoons \left[(HL)_n MX_y \right]_{(org)} \qquad (1.3)$$

Pyridine-amides have been shown to be effective extractants for base metals such as Zn(II) and Co(II) [41, 42]. In these systems, the number of amido-substituents proximate to the pyridine is important to the selectivity of metalate over halide. On protonation of the pyridine, the resulting receptor is stabilized by an internal bifurcated hydrogen bond which provides a diffuse array of N–H and C–H bonds for interaction with the metalate (Figure 1.11). This "soft" set of hydrogen bonds interacts preferentially with the "soft" chlorometalate in preference to the "hard" chloride which is in excess in the aqueous phase.

The solid-state structure shows the formation of at least four N–H and C–H hydrogen bonds per receptor and the metalate $ZnCl_4^{2-}$. DFT calculations support the solid-state structure and show that the formation constant for the Zn assembly is greater than that for the Co assembly. This feature, along with the formation constant for $ZnCl_4^{2-}$ being higher than that for the Co analog $CoCl_4^{2-}$ results in favorable separation.

Other, more-simple receptor designs that make use of the hydrogen-bond chelate effect shown above have been reported (Figure 1.11, right) [43]. Intrinsic to this design is the feature that, on protonation, a stable, six-membered *pseudo*-chelate is formed. These receptors were found to show high selectivity for $ZnCl_4^{2-}$ over $FeCl_4^{-}$, an *anti*-Hofmeister bias, and Cl$^-$ which was rationalized using hybrid DFT calculations and is dependent on the number of N–H hydrogen bonds and the ease of protonation of the amido–amine through the formation of a proton-chelate structure.

Both amido–amine and amido–pyridine compounds have been found to act as receptors for perrhenic acid, facilitating the liquid–liquid extraction of perrhenate into the organic phase. In this case, higher-order clusters are present in solution, with ESI-MS (Electrospray Ionization-Mass spectrometry) showing ions representative of $(HL)_2(ReO_4)_2$ host–guest assemblies [44]. This is supported in the solid-state X-ray crystal structure in which two protonated pyridine diamide receptors are organized such that two perrhenate oxoanions are encapsulated by an array of polarized N–H and C–H hydrogen bonds. Also of note here is the propensity of the neutral amido–amine to decompose under strongly acidic conditions, which implies that these particular compounds act as synergistic extractants, namely a primary amide RC(O) NH$_2$ coupled with an ammonium salt R'$_2$NH$_2^+$.

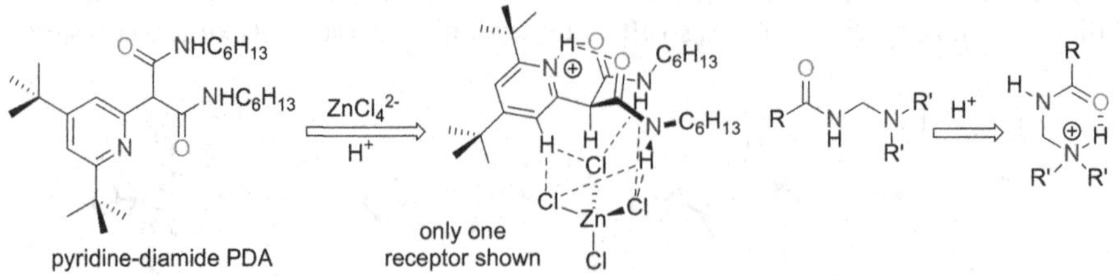

FIGURE 1.11 The formation of internal hydrogen bonds on protonation of a pyridine/amine-diamide resulting in receptors suited for metalate binding.

Mono-, bi-, and tripodal receptors derived from tris(aminoethyl)amine (tren) or tris(aminopropyl)amine (trpn) have been compared in the extraction of the precious chlorometalate $PtCl_6^{2-}$ [45, 46]. These reagents were designed to recognize the outer-coordination sphere of $PtCl_6^{2-}$ through the triangular faces of the octahedron, although ultimately they formed 2:1 receptor:chlorometalate assemblies in which the proton is encapsulated by a receptor at the axial amino nitrogen. Higher extraction of $PtCl_6^{2-}$ is seen with urea-substituents compared to amide and sufonamides, and tripodal amide receptors are more effective than bi- or mono-amides. In some cases, the formation of a proton-chelate similar to those described above results in very high selectivity for $PtCl_6^{2-}$ over chloride, templating of the receptor and providing positively polarized N–H and C–H hydrogen-bonding arrays.

Recently, attempts were made to exploit amido–amine and amido–pyridine (e.g., PDA, Figure 1.11) compounds in the extraction of lanthanides as their chlorometalates [47]. Previous reports on the extraction of lanthanides by ionic liquids such as quaternary ammonium salts under high aqueous-salt concentrations have suggested the formation of lanthanide chlorido- or nitratometalates LnX_6^{3-} in the organic phase. While successful uptake of $PtCl_6^{2-}$ by PDA from 6 M HCl was seen, no uptake of lanthanides or of other 3– chlorometalates such as $IrCl_6^{3-}$ occurred. DFT calculations showed that the substitution of aquo ligands by chloride, nitrate, and sulfate ligands in the aqueous phase is unfavorable, instead forming outer-sphere assemblies such as $[La(OH_2)_9][Cl]_x$ in which the aquo ligands are retained in the inner sphere. This highlights the importance of the Hofmeister-like selectivity and the need to consider the stability of metal species in the aqueous phase, not only that of the metal species transported into the organic phase.

1.2.1.3 Reverse Micelles

The types of amphiphiles used in solvent extraction experiments can assemble into nanoscale structures allowing hydrophilic and hydrophobic environments to coexist. These nanoscale structures are very different from the discrete, molecular entities described above, and so require different techniques and strategies to understand their formation.

The transfer of $Eu(NO_3)_3$ from an aqueous phase into a water-poor, amphiphile-in-oil system (c-functionalized malonamide DMDOHEMA in heptane) was studied due to its nonclassical distribution behavior (Figure 1.12) [48]. Here, the mechanism of extraction was found to be a combination of coordination and supramolecular/

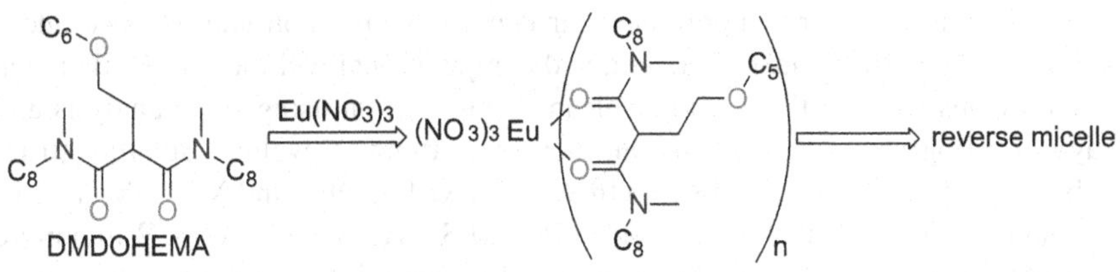

FIGURE 1.12 Sequential coordination and reverse-micelle formation in the extraction of $Eu(NO_3)_3$ by a malonamide amphiphile.

colloid chemistry in which structural evolutions in the organic phase change the properties of the solvent. As such, initial extraction is through the formation of metal salts $Eu(L)_3(NO_3)_3$ in the organic phase, as shown by EXAFS and TRLFS measurements. Subsequently, vapor pressure osmometry (VPO) and small/wide angle X-ray scattering (SAXS/WAXS) studies, combined with MD simulations showed the formation of reverse micelles, which draw water and the amphiphile into nanoscale domains. Further increasing the Eu(III) concentration results in further aggregation of the reverse micelles thought to be due to $Eu-NO_3$ interactions.

Tributyl phosphate (TBP) is an industrially relevant solvating extractant used to separate plutonium and uranium from nuclear fuel waste and is prone to form third phases under high acid and metal concentrations due to its amphiphilic nature and propensity to form reverse micelles [49]. While the study of these colloidal systems by X-ray and neutron scattering techniques is increasingly common [50], pulsed-field-gradient NMR spectroscopy has only recently been used to study the shape and size of metal-containing aggregates [51]. This diffusion NMR method is complementary to small-angle scattering techniques and was used to study the interactions between TBP aggregates containing uranium or zirconium cations and HNO_3 in a dodecane diluent. The aggregate sizes were found to be similar in size to those evaluated by diffraction methods, but in contrast a repulsive interaction between the aggregates was discovered, suggesting that this feature should be included in models to improve diffraction data simulations.

Structural insight into the multinuclear speciation of Ce(IV) in the TBP/dodecane solvent extraction system was recently reported [52]. A combination of XANES, EXAFS, and SAXS analysis indicated that tetranuclear Ce(IV)-oxo cores in reverse-micelle structures of ca. 6 Å diameters are formed, solvated by TBP. At low cerium concentrations (<0.14 M), these reverse micelles are randomly and homogeneously dispersed with some short-range interactions, whereas at high Ce concentration (1.5 M) a third phase is formed comprising correlated, long-range, percolated micellar aggregates.

1.2.1.4 Conclusion

It is becoming increasingly clear that an appreciation of the fundamental chemistry that underpins solvent extraction is necessary to make significant advances in the design and development of new processes that can deal with more economically and environmentally efficient metal recovery from a diversity of primary and secondary sources. Understanding the modes of action of known ligands, receptors, and solvating agents will help in the design of new extractants and requires knowledge of coordination chemistry, supramolecular chemistry, and colloidal science and a vast array of spectroscopic, diffraction, and computational techniques. While much of the previous work in this area has gleaned information using slope analysis and X-ray crystallography, there is now an increasing trend to evaluate solution structure by mass spectrometry, NMR spectroscopy, TRLFS, DFT and MD calculations, and solution diffraction techniques such as EXAFS, SAXS, and SANS. While more "sporting," these techniques are better able to describe solution structure and so provide a better reflection of the mechanism of extraction.

1.2.2 THE ROLE OF PHYSICOCHEMISTRY FOR A RATIONAL DESIGN OF NEW EXTRACTANTS

The physicochemistry involved in liquid–liquid extraction processes is complex because of the presence of an interphase between the aqueous solution and the organic solution, the non-ideality of the aqueous and organic phases, and the numerous chemical equilibria that can occur in each of these two phases (hydrolysis, precipitation, complexation, aggregation, etc.).

In order to precisely describe the phenomena of metal transfer from the aqueous phase to the organic phase (or from the organic phase to the aqueous phase in the case of a back-extraction), it is important to finely characterize the equilibria taking place in each of these phases and to determine the apparent thermodynamic constants associated with these equilibria. The comprehension of the physicochemistry of solvent extraction and a fine knowledge of the speciation in aqueous and organic phases will bring useful information to design new extractants and to model metal extraction in order to build tools which can be used to optimize solvent extraction processes (Figure 1.13).

Physicochemical models could be advantageously combined with engineering models in order to develop smart processes, namely processes that can adapt to the nature of the composition of the feed solution so that the process always works under optimized conditions. The smart processing approach is particularly relevant for treating resources like spent materials or tailings for which the composition can change drastically over time and the location.

By way of illustration, the development of physicochemical models for the recovery of uranium from concentrated phosphoric acid by a synergistic mixture of bis-(2-ethylhexyl)phosphoric acid (D2EHPA) and tri-n-octylphosphine (TOPO) is

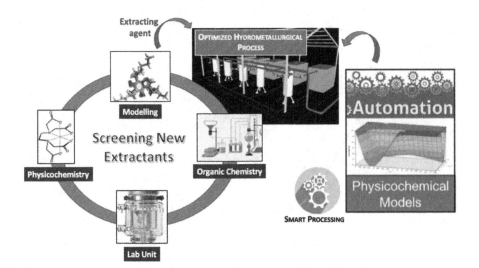

FIGURE 1.13 Smart processing approach.

presented below. Such a physicochemical model could be particularly interesting in order to anticipate changes of the formulation of the extraction solvent (for instance, because of the difference of solubility of D2EHPA and TOPO in the aqueous phase which leads to potential deformulation).

Although new understanding of uranium extraction and U/Fe selectivity with D2EHPA/TOPO has been gained recently [53–57], the mechanisms of uranium-selective extraction with these synergistic systems remain poorly understood. Empirical research has thus been mostly performed to propose better synergistic mixtures by changing the cation exchanger and/or the neutral-donor ligand in the reference D2EHPA/TOPO solvent.

Many different cation exchangers, belonging to phosphoric, phosphonic, or even phosphinic acids, have thus been tested in combination with TOPO for uranium(VI) extraction from WPA. Among them, dibutyl dithiophosphoric acid (HDBDTPA), bis(2-ethylhexyl)dithiophosphoric acid (D2EHDTPA) [58], dioctyl (DOPPA) [59] and dinonyl phenyl phosphoric acid (DNPPA) [60], bis(2-ethylhexyl) phosphinic acid (B2EHPA) [61], and (2-ethylhexyl)phosphonic acid mono-2-ethylhexyl ester (PC88A) [62] were used in mixtures with TOPO (Figure 1.14).

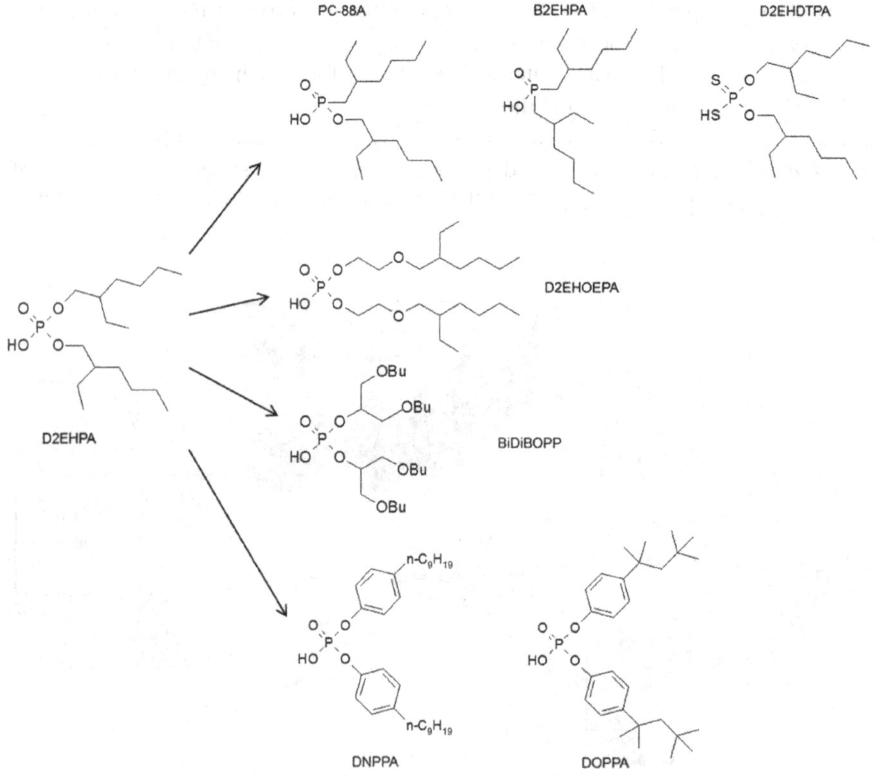

FIGURE 1.14 Molecular structure of different cationic exchangers tested in combination with TOPO for uranium extraction from WPA.

The extraction efficiency of uranium increases logically with the acidity of the cation exchanger in the following order: phosphinic acid < phosphonic acid < phosphoric acid < thiophosphoric acid. Derivatives with an ether function on the hydrophobic part of the extractant (D2EHOEPA and BiDiBOPP) were also synthesized and tested to show that the presence of additional oxygen atoms increases the uranium distribution ratio compared to D2EHPA [63].

On the other hand, several neutral-donor ligands such as tri-*n*-butylphosphate (TBP), di-*n*-butyl butyl phosphonate (DBBP) [60, 64, 65], or di-*n*-hexyl-methoxyoctylphosphine oxide (di-*n*-HMOPO) [56] were tested in combination with D2EHPA to substitute TOPO in the reference mixture. As with cation exchangers, uranium extraction increases with the basicity of the neutral ligand (phosphate < phosphonate < phosphine oxide) and with the introduction of an ether group on the phosphine oxide ligand (Figure 1.15).

Other synergistic mixtures containing a cation exchanger such as dibutyldithiophosphoric acid (DBDTPA), bis(2-ethylhexyl) dithiophosphoric acid (D2EHDTPA), bis(1,3-dibutoxyprop-2-yl)phosphoric acid (BiDiBOPP), dinonylphenyl phosphoric acid (DNPPA), or (2-ethylhexyl) phosphonic acid mono-2-ethylhexyl ester (PC88A) and a neutral synergistic molecule such as di-*n*-hexyl-methoxyoctylphosphine oxide (di-*n*-HMOPO), dibutyl butyl phosphonate (DBBP), tri-*n*-butyl phosphate (TBP), or octyl(phenyl)-*N,N*-diisobutylcarbamoyl methyl phosphine oxide (CMPO) were also tested for uranium extraction from WPA [63].

The development of a physicochemical model is very useful to understand the physicochemistry of extraction, the extraction equilibria, and to evaluate the speciation in the organic phase. For instance, Chagnes et al. developed a physicochemical model to describe uranium(VI) extraction by D2EHPA/TOPO, BiDiBOP/

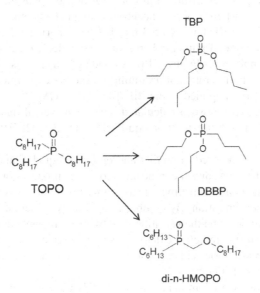

FIGURE 1.15 Molecular structure of different neutral-donor ligands tested in combination with D2EHPA for uranium extraction from WPA.

di-*n*-HMOPO, or other derivatives from 5.3 M phosphoric acid [64]. In this model, the dimerization of the acidic cationic extractant HL and the association between HL and the solvating agent S were taken into account by considering the formation of $(HL)_2$, $(HL)_2S$, $(HL)_2S_2$, and a supramolecular complex $(HL)_5S$:

$$2\overline{HL} \rightleftharpoons \overline{(HL)_2} \tag{1.3}$$

$$\overline{(HL)_2} + \overline{S} \rightleftharpoons \overline{(HL)_2 S} \tag{1.4}$$

$$\overline{(HL)_2} + 2\overline{S} \rightleftharpoons \overline{(HL)_2 S_2} \tag{1.5}$$

$$\overline{HL} + 2\overline{(HL)_2} + \overline{S} \rightleftharpoons \overline{(HL)_5 S} \tag{1.6}$$

Furthermore, it was considered that uranium(VI) is extracted according to the following equilibria:

$$UO_2^{2+} + 2\overline{(HL)_2} \rightleftharpoons \overline{UO_2 (HL_2)_2} + 2H^+ \tag{1.7}$$

$$UO_2^{2+} + 2\overline{(HL)_2} + \overline{S} \rightleftharpoons \overline{UO_2 (HL_2)_2 S} + 2H^+ \tag{1.8}$$

$$UO_2^{2+} + \overline{(HL)_2} + \overline{S} \rightleftharpoons \overline{UO_2 (HL)_2 S} + 2H^+ \tag{1.9}$$

By implementing these equilibria in the physicochemical model and by considering no significant changes in uranium complexation by phosphate and hydrogen phosphates (since phosphoric acid concentration and pH are constant), the distribution ratios of uranium(VI) between 5.3 mol L^{-1} phosphoric acid and HL/S diluted in Isane IP 185 (a grade of aliphatic kerosene) were calculated as a function of the organic-phase composition (Figure 1.16). A good agreement was obtained between the calculated distribution ratios of uranium(VI) and the experimental ones.

This model has been applied to the BiDiBOPP-di-*n*-HMOPO system. It appears that a good agreement between calculated and experimental data is obtained by considering the same set of extraction equilibria as for the D2EHPA/TOPO system (Figure 1.16).

Such a model can be used to predict and optimize liquid–liquid extraction as a function of the extraction solvent composition for many mixtures of organophosphorus cation exchangers and solvating agents. However, it can work only at constant phosphoric acid concentration. By combining this physicochemical model with an equation of state to describe the variation of activity coefficients in phosphoric acid, it is possible to perform predictive calculations as a function of the extraction solvent composition and phosphoric acid concentration. Such a model was also developed by Chagnes et al. [66, 67] (Figure 1.17).

The existence of the predominant species $UO_2(HL)_2L_2S$ predicted by the physicochemical model was demonstrated by means of TRLFS. It is particularly interesting

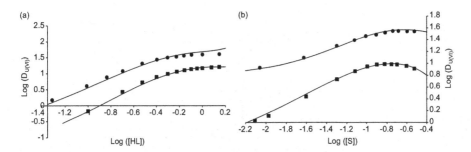

FIGURE 1.16 Logarithm of the distribution ratios of uranium(VI) between 5.3 M phosphoric acid and D2EHPA/TOPO (■) or BiDiBOPP/di-*n*-HMOPO (●) diluted in Isane IP 185 as a function of (a) the logarithm of initial D2EHPA or BiDiBOPP (HL) concentration at constant TOPO or di-*n*-HMOPO (S) concentration (0.125 M) and (b) the logarithm of initial TOPO or di-*n*-HMOPO concentration (S) at constant D2EHPA or BiDiBOPP (HL) concentration (0.5 M). Initial concentration of uranium = $1.43 \times 10^{-3} \times$ M, temperature = (25.0 ± 0.2)°C, phase volume ratio $V_o/V_a = 1$. ----: Calculated with the thermodynamic model.

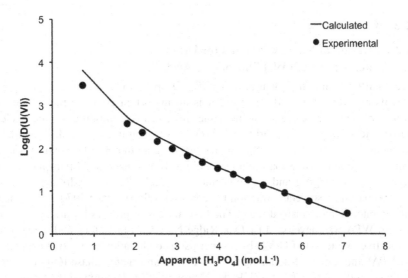

FIGURE 1.17 Experimental and calculated distribution ratios of uranium(VI) as a function of apparent concentration of phosphoric acid. Calculated line used the adjustable parameters of the thermodynamic model.

to highlight that the TOPO molecule in this species is located in the first solvation shell of uranium(VI), as it was also confirmed by DFT calculations (Figure 1.18).

The presence of such a species in which the cation exchanger and solvating agent are both in the first solvation shell of uranium(VI) paves the way to the development of bifunctional ligands as those recently developed by CEA (see paragraph 2.3.1 below as a case study).

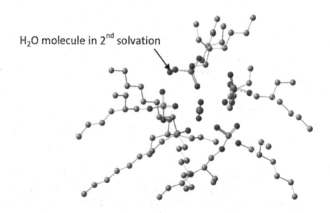

H₂O molecule in 2nd solvation

FIGURE 1.18 DFT calculation of the optimized geometry of $UO_2(HL)_2L_2(TOPO)H_2O$ (TOPO in the first solvation shell). Hydrogen atoms were removed from the figure for the sake of readability. Red atoms: oxygen. Orange atoms: phosphorus. Blue atoms: uranium. Gray atoms: carbon.

1.2.3 CASE STUDIES

1.2.3.1 Design of New Extractants for Uranium Recovery from Wet Phosphoric Acid

Since significant amounts of uranium (50–200 ppm) are contained in phosphate rocks, its recovery from industrial WPA is an important goal in order to decontaminate phosphoric acid and, at the same time, valorize uranium for the nuclear industry. According to the world's reserves, 4 Mt of uranium would be available from natural phosphate rocks [68], which constitutes an important secondary source for uranium production. Separation of uranium from concentrated phosphoric acid is, however, challenging and requires intensive R&D. Several hydrometallurgical processes based on solvent extraction have been developed since 1970, and some of them operated at an industrial scale. The first successful process for uranium recovery from WPA was developed by Oak Ridge National Laboratory and is based on the selective extraction of U(VI) by a synergistic combination of a cation exchanger (D2EHPA) and a neutral-donor ligand (TOPO) in an organic phase [69]. This process, later modified by Cogema (URPHOS process) [70, 71] allowed the production of hundreds of tons of uranium from WPA in the 1980s and 1990s. Uranium was sufficiently purified but required two extraction cycles to reach the specifications for nuclear-grade uranium.

Two other processes were also developed at an industrial scale and used phosphorus extractants to extract U(IV). The OPAP process [72] is based on a mixture of mono and dioctylphenyl phosphoric acid, while the OPPA process uses dioctyl-pyrophosphoric acid for uranium(IV) extraction [73, 74]. All these processes showed efficient extraction of U(IV) or U(VI) depending on the synergistic mixture, but different drawbacks such as insufficient uranium/impurities (iron in particular) selectivity, additional steps for uranium redox control, or degradation of the extractant under hydrolysis (for OPPA) were pointed out and limited their performances. These processes should thus be improved to be more competitive in order to be economically attractive for uranium production according to the current uranium price.

Improvement of these processes involves the design of new extracting molecules to bring higher affinity, selectivity, and robustness than the classical formulations used in the former industrial processes. Extensive research and development have been performed in this field to study new extractant molecules for the selective extraction of uranium from WPA [63, 75].

Several criteria must be taken into account to define efficient molecules capable of extracting uranium from concentrated phosphoric acid solutions. First, as phosphate anions are scarcely extracted into the organic phase, extraction by solvation (metal salt extraction) using a neutral ligand (i.e., uranium extraction by TBP from HNO_3) is not possible in this case. Moreover, with very few anionic complexes being formed between phosphate anions and uranium, anion exchange is also not suitable. Cationic exchange is thus the main extraction mode enabling uranium extraction from phosphoric acid. However, phosphate anions being strong complexing agents for uranium, a cation exchanger is generally not powerful enough by itself to compete with phosphate complexation in the aqueous phase. Another binding function is therefore required to help uranium transfer to the organic phase by decomplexing the aqueous uranyl cation from its bound phosphate ions and water molecules and completing the uranium coordination sphere. This effect is commonly explained by an additional solvation that would increase the uranium complex lipophilicity and its extractability in the organic phase.

Two different approaches can be therefore considered to tackle this issue: either using a combination of two molecules, a cation exchanger in addition to a new neutral-donor ligand, a so-called synergistic mixture (as already implemented in the Oak Ridge process) or designing new multifunctional ligands combining both cation exchanger and neutral-donor functions on the same molecular architecture.

As mentioned previously in this chapter, a large number of synergistic systems have been developed and tested for uranium extraction from phosphoric acid media. If improvements of uranium extraction efficiency compared to the Oak Ridge process were observed with some of these new synergistic combinations, these formulations remain insufficient to reach the desired selectivity for minimizing the impurity concentrations in the yellow cake (iron in particular) in only one extraction cycle. Besides, the process must operate at the optimum synergistic ratio of the mixture to reach the best performances while extraction efficiency and selectivity properties of the synergistic mixture are generally very sensitive to a slight deviation of this ratio. These reasons reinforced the need to develop new selective extractants instead of synergistic mixtures to fit industrial objectives.

This part of the chapter introduces the design of multifunctional ligands that combine both the cation exchanger (phosphoric, phosphonic acid) and the neutral-donor functions (phosphine oxide, amide) on the same molecular structure. The use of a single molecule combining the properties of the two functions of the synergistic mixture is interesting not only to simplify the process but also to potentially enhance the uranium distribution ratio and U/impurities selectivity. The idea is to pre-organize the two functions for uranium coordination in order to increase the complex stability by favoring the entropic contribution (the well-known "chelate effect").

Based on this approach, the first bifunctional extractant reported in the literature for uranium extraction from phosphoric acid was synthesized by Warshawsky et al. [76]. O-Methyldihexylphosphine oxide O′-hexyl-2-ethyl phosphoric acid (MDHPOH2EPA), depicted in Figure 1.19, includes a phosphoric acid as a cation exchanger while a phosphine oxide group acts as a neutral donor. A higher uranium extraction strength than obtained in the Oak Ridge system was obtained with this extractant, but U/Fe selectivity remains insufficient. Moreover, third-phase formation was observed under the extraction conditions with this system.

Pursuing this avenue, new bifunctional ligands, carrying a phosphonic acid group as cation exchanger and an amide group as a neutral-donor function, were designed by Turgis et al. [77]. Amidophosphonic acids, also named carbamoyl-alkylphosphonic acids, were first synthesized and tested toward uranium extraction from WPA. The influence of different modifications of the ligand structure (nature of the alkyl chains attached to the amide group, spacer length between the two functions, steric hindrance on the spacer, etc.) was studied through a structure–function approach and correlated to extraction efficiency and U/Fe selectivity. In solvent extraction, these structural modifications strongly influence the efficiency and selectivity of the extractant but also some important physicochemical parameters such as the ligand partitioning into the aqueous phase, the ligand solubility in aliphatic diluents, the metal complexes solubilities in the organic phase (limit toward third-phase formation, crud formation at the interface upon extraction or stripping, etc.), the organic-phase density and viscosity (impacting the phase settling and separation), or the ligand stability under hydrolysis. The best balance between the total number of carbons and the nature of hydrocarbon groups (linear or branched alkyl chains, short or long chains, aliphatic or aromatic groups, etc.) should be found to design the best ligand fulfilling the main criteria defined for process development.

It appears that uranium extraction was maximized with 2-ethylhexyl chains on the amide function and a methylene bridge between the amide group and the phosphonic acid, while U/Fe selectivity was enhanced by steric hindrance after alkylation of the methylene by a phenyl group. DEHCPBA (N,N-di-2-ethylhexyl-carbamoylbenzylphosphonic acid) exhibits higher distribution ratios than the D2EHPA-TOPO solvent, but the U/Fe separation factor remains still too low (<200), requiring additional efforts to define the best molecular design.

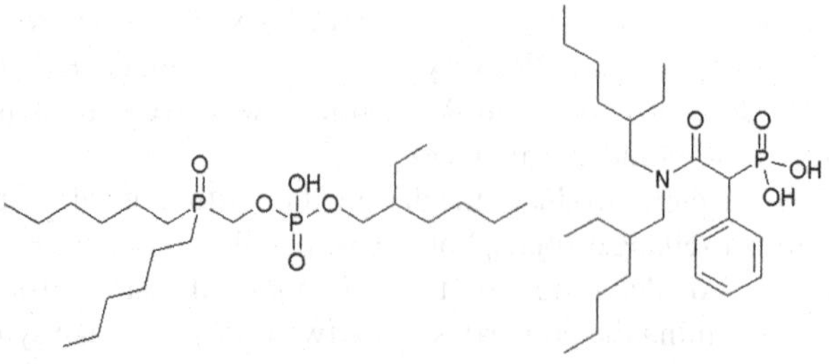

MDHPOH2EPA   DEHCPBA

FIGURE 1.19 Molecular structures of O-methyldihexylphosphine oxide O-hexyl-2-ethyl phosphoric acid and N,N-di-2-ethylhexyl-carbamoylbenzylphosphonic acid.

Based on the promising results obtained on DEHCPBA and carbamoylmethylphosphonic acids, amido phosphonate ligands were studied after monosaponification of the corresponding amido phosphonic acids. Keeping 2-ethylhexyl chains on the amide group and a methylene bridge between the two extracting functions, several amido phosphonate ligands were synthesized by changing the nature of the alkyl chain grafted on the spacer and on the phosphonate group. It appears that the presence of monosaponified phosphonate moiety in combination with the addition of an octyl pendant chain on the methylene bridge dramatically enhances the uranium extraction efficiency and the U(VI)/Fe(III) selectivity [78, 79].

The so-called DEHCNPB (butyl-1-[N,N-bis(2-ethylhexyl)carbamoylnonyl]phosphonic acid) reported in Figure 1.20 was put forward in regards to the outstanding results obtained for the selective extraction of uranium compared to the D2EHPA/TOPO synergistic solvent.

Indeed, the uranium distribution ratio and U(VI)/Fe(III) selectivity are more than 30 times and between 15 and 50 times higher than the reference D2EHPA/TOPO, respectively. Furthermore, DEHCNPB extracts uranium selectively from the other impurities (Mo, V, Al, ...) present in genuine industrial phosphoric acid with unequaled performance. This molecule was thus selected for process development [80, 81].

Other "autosynergic" bifunctional ligands have been more recently developed for uranium extraction by substituting the amide donor group of amido phosphonates by a phosphine oxide [82]. Keeping similar alkyl chains than those optimized in the case of DEHCNPB, new phosphine oxide-phosphonate ligands were synthesized and compared with DEHCNPB and the reference D2EHPA/TOPO mixture. Uranium was highly and selectively extracted by DEHNPB phosphine oxide phosphonate from a genuine industrial phosphoric acid solution with a better U/Fe

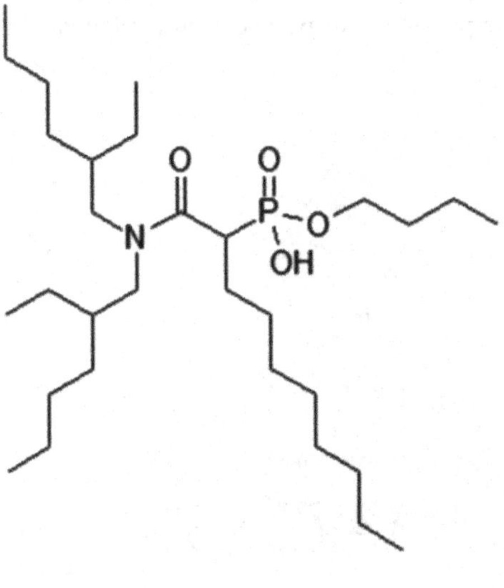

FIGURE 1.20 Molecular structure of butyl-1-[N,N-bis(2-ethylhexyl)carbamoylnonyl]phosphonic acid (DEHCNPB).

separation factor compared to D2EHPA/TOPO. The U/Fe selectivity remains nevertheless lower in comparison to the amido-derivative.

1.2.3.2 Chemical Design in Gold Recovery by Urban Mining

The case for the recycling of waste electronic and electrical equipment (WEEE) is compelling. In 2012, 49 million tons of WEEE were generated globally, primarily by the developed countries, containing at least 57 elements many of which are present in concentrations significantly higher than in their primary ore deposits and are listed as critical resources [83]. This is compounded by the exponential growth of the mobile electronics market coupled with the reduced lifetime of devices. While it is clear that prevention and reuse are preferable options in the e-waste management hierarchy and 3R policy (Figure 1.21), the recycling of WEEE would not only solve issues with landfill disposal and the leaching of hazardous elements, but also has enormous potential as a resource for valuable and critical materials through "urban mining" [84], potentially closing-the-loop and creating a circular economy. This has resulted in a series of global initiatives such as the Restriction of Hazardous Substances Directive (RoHS), the EU WEEE Directive 2012/19/EU, and Solving the E-waste Problem (StEP) [85, 86].

The recycling of WEEE using physical and chemical methods has been reviewed and is found to be complex, requiring dismantling, crushing, and classification of the materials followed by separation and refining steps by, for example, pyro-, hydro-, and/or biohydrometallurgical routes [83, 87]. While significant progress has been made in separating metallic from nonmetallic fractions and in finding economical uses for these latter waste products (e.g., fillers and composites), the recovery of individual metals remains an issue due to the difficulty in adapting technologies used to recover metals from primary ore feed streams for use in a WEEE feed stream. Such an issue is particularly true in pyrometallurgy, since many metals are lost in the slags, and it explains why more and more industries try implementing hydrometallurgy rather than pyrometallurgy. The eventual increase of the cost due to the

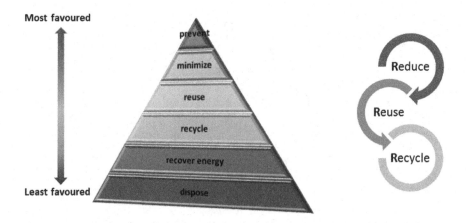

FIGURE 1.21   E-waste management hierarchy and 3R policy.

implementation of hydrometallurgy instead of pyrometallurgy may be paid by the recovery of many metals by hydrometallurgy in most cases.

The environmental impacts of WEEE treatment are also important to understand and have been evaluated by life-cycle analysis [88]. In this case, a hydrometallurgical process that was analyzed comprised two different leaching steps, in nitric acid and aqua regia, followed by electrodeposition for the recovery of Cu, Ag, and Au and adsorption steps for Ni and Sn recovery. The analysis found that the nitric acid leaching process has the highest environmental impact, contributing highly to eutrophication, acidification, human toxicity, global warming (from its synthesis), and abiotic depletion, with the adsorption steps proving inefficient and so also a high contributor. Furthermore, LCA evaluation of the recycling of smartphones shows it to have a net benefit in terms of carbon dioxide equivalents [89]. It is important to note here that improvements in the chemistry of leaching, separation, and refining steps that occur in a hydrometallurgical process would benefit the overall environmental (and economic) impact of WEEE recycling.

This part of the chapter will focus on recently reported research for the recovery of gold from WEEE; in particular, research that has used chemical understanding to inform on the mechanism of operation [90]. Even though gold is only present in small quantities in WEEE, for example ca. 0.25 g per smartphone, it represents the most value to recover [91]. WEEE is an important gold resource; it is estimated that 1 ton of smartphones contains 300 g of gold, which is a significantly more concentrated resource than the primary ore, of which one ton would yield 3–7 g [92]. Furthermore, it is also estimated that the recovery of gold from WEEE would decrease the environmental and mining footprint along with savings of 17,000 tons per ton in carbon dioxide emissions.

1.2.3.2.1 Leaching

As stated above, the leaching of precious metals such as gold from WEEE is environmentally impactful, as it necessitates the use of strong acids under oxidizing conditions, for example HNO_3 or aqua regia (1:3 HNO_3:HCl), or the use of extremely toxic chemicals such as cyanide/O_2 to dissolve gold by forming the Au(I) complex $Au(CN)_2^-$ [93]. Alternatively, leaching chemicals such as thiourea/Fe^{3+}, thiosulfate/O_2, and KI/I_2 are used, but their costs and environmental hazards are seen as prohibitive. Bio-oxidation and cyanidation are also possible methods for gold leaching [93]. The use of dithiobiuret reagents has proved beneficial for simultaneous leaching and extraction of gold from aqua regia and direct leaching into partially water-immiscible solvents such as CH_3CN in the presence of HCl/H_2O_2. Oxidative leaching using H_2O_2/H_2SO_4 is suitable for the leaching of Cu from waste printed circuit boards (WPCBs), providing both good Cu recovery and separation from the WPCB resin, but its potential for leaching other metals has not been explored [94]. Simple acid leaching in air has also been studied, and dilute HCl (1.0 M) was shown to lixiviate gold and other metals to a high degree from WPBCs provided suitable levels of the oxidant (O_2) is present and agitation occurs [95].

Very recently, a new, low-toxicity method for the leaching of gold from WPCBs using a synergistic mixture of N-bromosuccinamide (NBS) as an oxidant and

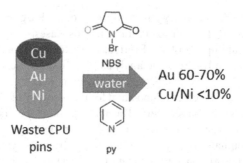

FIGURE 1.22 Selective leaching of gold using a synergistic mixture of NBS (70 mM) and py (100 mM) in water.

pyridine (py) as a ligand has been described (Figure 1.22) [96] Experimental variation of the reagents and conditions showed that gold dissolution was maximized (90%) using 10 mM NBS, 100 mM py, at pH 8.2 and 25°C, and contrasts to only 77% gold dissolution using KI/I_2; the cost of these latter reagents is 20 times greater.

The mechanism of gold dissolution was investigated, showing that the first step was oxidation of the gold surface by NBS at the relatively low potential of -0.854 V to form the Au(III) complex $AuBr_4^-$. This metalate then reacts with pyridine to form the neutral complex $AuBr_3(py)$ due to a high binding constant of 10^5–10^6 M. It is therefore clear that both NBS and py are required to maximize gold leaching (i.e., synergism occurs) and that the use of low concentrations of NBS and py are important to the efficacy of the system.

This procedure has proved effective in separating gold from other metals present in WEEE, with 90% gold recovery compared with 40% of other metals such as Ni, Sb, Zn, Mg, Cu, Sn, Al, and Fe. When leaching gold from the surface of Ni-coated Cu CPU pins, the selectivity increases due to the formation of a Ni_xBr_y passivating surface, which inhibits further Ni and Cu oxidation.

The dilute nature of the NBS and py reagents was found to mitigate against mammalian toxicity, showing 100% viability in mammalian models compared to 0% when dosed with more traditional reagents. Similar attenuation of toxicity was seen in aquatic creature models, making the NBS/py an attractive, environmentally benign alternative to conventional gold leaching methods.

1.2.3.2.2 Precipitation of Au(III) and Au(0)

Leaching metals from WEEE yields a pregnant leach solution (PLS) that generally contains a mixture of metals that require separation unless some separation process has operated at the leaching stage. Selectivity in separation is a key challenge to a hydrometallurgical metal recovery process, and a knowledge of coordination chemistry, supramolecular chemistry, and redox chemistry is important in designing a suitable system. One separation method is through precipitation and has been employed extensively in gold recovery, either carrying out separation as Au(III) or Au(I) complexes or by precipitation as Au(0) using carbonaceous- or bio-sorbents [91]. Gold deposition is also important in other fields where, for example, the size of the

Au(0) nanoparticles formed in TiO$_2$/Au materials is key to their efficacy as catalysts in a variety of chemical reactions [97].

Recently, the use of polyaniline reductants (conducting polymers) was studied for the recovery of gold from a PLS of WEEE [98], based on earlier chemistry that showed that the rate of Au(0) deposition was dependent on the surface area on the polyaniline and its intrinsic oxidation state [99]. In the former example, polyanilines were immobilized on supporting materials such as cotton fibers and proved effective at Au(0) deposition due to the electroless cycling of leucoemeraldine, emeraldine, and pernigranile during Au(III) reduction in aqueous HCl. Alternatively, glow discharge lamps have been used to reduce metals with positive standard reduction potentials, favoring separation of Au from Cu, Zn, and Fe [100]. In this case, the separation of Au(0) was facilitated through the use of poly(vinylpyrrolidone) which, under the aqueous conditions used, forms a film on the surface of the solvent in which Au(0) is embedded. Also, the use of reduced graphene oxide (GO) hydrogels to precipitate Au(0) from Au(III) solutions has been reported [101]. Here, GO was reduced by wild-type *Shenwanella oneidensis* MR-1 resulting in a biologically assembled GO hydrogel which could reduce Au(III) to Au(0) nanoparticles that become embedded in the hydrogel, with Au and Pd recovered preferentially from Cu, Zn, and Ni.

Perhaps one of the most exciting chemical advances in this area is the use of supramolecular chemistry to favor the self-assembly of gold precipitates for the selective isolation of gold from other metals. In one example, the addition of α-cyclodextrin (α-CD) to KAuBr$_4$ in water resulted in the spontaneous precipitation of a one-dimensional supramolecular complex with an extended chain superstructure $\{[K(OH_2)_6][AuBr_4]\subset(\alpha\text{-CD})_2\}_n$ (Figure 1.23) [102].

Single-crystal X-ray diffraction studies showed that a perfect match in molecular recognition between [AuBr$_4$]$^-$ and α-CD occurs, with axial orientation of the anion within the cyclodextrin cavity favoring a highly specific second-sphere electrostatic and hydrogen-bonding interaction between the [AuBr$_4$]$^-$ anion and the [K(OH$_2$)$_6$]$^+$ cation. In the structure, the [K(OH$_2$)$_6$]$^+$ cation is encapsulated by the secondary faces of two α-CDs, whereas in other adducts the cations are coordinated by OH groups and glucopyranosyl O atoms in the CDs. This feature explains the differences in precipitation yield when using β- or γ-cyclodextrins instead of α-cyclodextrin and when

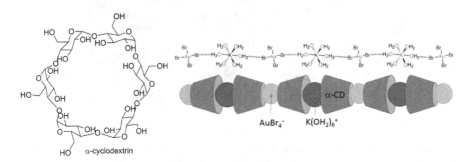

FIGURE 1.23 Spontaneous assembly of α-cycloαextrin (α-CD) wαth KAuBr$_4$ in water to form a one-dimensional extended-extended chain superstructure.

using different gold salts such as Na, Cs, or RbAuCl$_4$. Indeed, out of 24 combinations of salt:cyclodextrin, only three result in precipitation: α-cyclodextrin with either K, Cs, or RbAuCl$_4$.

Metal–organic frameworks (MOFs) are porous materials that can be constructed using increasingly well-defined synthetic methods that make use of metal centers as tectons for bridging ligands. While their physical and chemical properties have been exploited in, for example, gas absorption and catalysis, respectively, their ability to sequester metals selectively is more rare. Recently, however, MOFs that incorporate sulfur-donor atoms within the porous cavities have been prepared and exploited in gold recovery [103]. In this case, the MOF is constructed rationally using dinuclear copper(II) complexes of a chiral bis(L-methionine)oxalamide ligand as a tecton in which the *anti*-configuration adopted by the ligand (Figure 1.24a) favors chain growth and, on the addition of Ca(II) cations, causes curvature (Figure 1.24b) and the formation of the MOF. The MOF contains hexagonal channels of *ca.* 0.3 nm diameter that are decorated by L-methionine arms, which act as ligands for both Au(I) and Au(III). As such, soaking the MOF in water solutions containing AuCl$_3$ or AuCl results in gold uptake, forming the thioether complexes (RS)AuCl and (RS)AuCl$_3$ within the porous network. In the case of Au(I), the X-ray crystal structure shows the expected linear geometry for the (RS)AuCl complexes, along with short Au...Au distances of 3.04 Å indicative of aurophilic interactions (Figure 1.24c).

These materials are highly selective for gold over other metals in solution. Soaking the MOF in an equimolar mixture of AuCl$_3$, NiCl$_2$, CuCl$_2$, ZnCl$_2$, AlCl$_3$, and [Pd(NH$_3$)$_4$]Cl$_2$ showed rapid uptake of Au(III), maximized after 30 minutes, with no uptake of other metals observed. The recovered gold can be released from the MOF by soaked the Au-loaded material in a sulfur-containing solvent such as dimethylsulfide, thus recycling the MOF for further gold uptake.

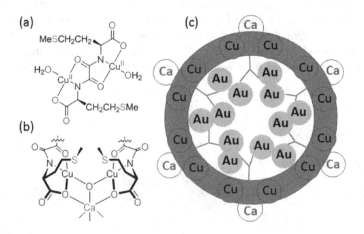

FIGURE 1.24 Methionine-decorated metal-organic frameworks for gold recovery: (a) dinuclear copper complex of the chiral bis(amino acid)oxalmide; (b) interaction of Ca^{2+} cations with dinuclear copper unit; (c) schematic of a single pore of the metal-organic framework structure in which the methionine groups coordinate to Au(I) centers within the pore.

1.2.3.2.3 Solvent Extraction of Au(III)

Solvent extraction (SX) is an attractive process for recovering metals from WEEE for, as long as the metals are leached, the PLS can be treated with a variety of reagents to separate and recover *all* of the metals sequentially. In principle, this would lead to complete recycling of the metals in WEEE with zero metal waste and, with recycling of the reagents used in SX processes, would provide excellent mass balance. If a single metal is targeted, SX offers significant environmental advantages over energy- and capital-intensive pyrometallurgical methods [3].

The solvent of extraction of gold from primary ores can be carried out from halide leach solutions using commercial reagents such as methyl isobutyl ketone (MIBK), dibutyl carbitol (DBC), or 2-ethylhexanol (2-EH) [104]. While the recovery of gold is efficient using these reagents, selectivity, safety, and mass balance issues are seen, and the chemistry that underpins their mode of action is poorly understood. As such, the development of new compounds that can selectively transport gold from aqueous to organic phases in a SX experiment remain desirable.

Several tertiary amides have been investigated as reagents for the solvent extraction of gold from HCl solutions (Figure 1.25).

The two tertiary amides DOAA and DOLA, which contain different chain-length substituents, were evaluated for gold transfer into 4:1 dodecane/ethyl hexanol from aqueous HCl in the presence of other metals including Pd(II), Pt(IV), Rh(III), Fe(II), Cu(II), Ni(II), and Zn(II) [105]. Good selectivity for Au was seen at lower HCl concentrations (<3.0 M), and the strength of extraction was higher for DOAA than DOLA. However, DOAA formed a third phase under these conditions, which was absent when using DOLA. Extractions using this latter reagent also allowed back-extraction using water (70%), whereas DOAA required the use of thiourea, thus affecting the overall mass balance. While these reagents show promise in gold extraction, the mode of action of extraction was not evaluated.

In a more recent study, a larger series of tertiary amides was evaluated for gold recovery from HCl solutions [106]. As with the above systems, extractions were carried out from mixtures of metals, Au(III), Pt(IV), Pd(II), Rh(III), Cu(II), and Ni(II),

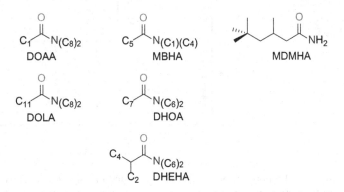

FIGURE 1.25 Tertiary and primary amides used in gold recovery by solvent extraction (C_n represents a straight-chain hydrocarbon substituent).

into 4:1 dodecane/2-ethylhexanol, and similar performance was seen with rapid phase transfer, quantitative (for MBHA, DHOA, and DHEHA), and dependent on the length and symmetry of the hydrocarbon substituents, with unsymmetrical substituents (MBHA) better than symmetrical substituents (DHOA, DHEHA). In this case, the analysis of the slopes of linear log D vs. log[L] plots ([L] = concentration of extractant) indicated that two reagent molecules were required for every Au transferred, suggesting a basic stoichiometry of $HAuCl_4(amide)_2$ as the extracted species.

While the use of tertiary and secondary amides in metal recovery by solvent extraction is well advanced, primary amides have been little studied in this area [107], perhaps due to the premise that the presence of multiple N–H groups would encourage extensive hydrogen bonding and so limit solubility in the organic phase. However, the use of amidoamines in base metal solvent extraction and from this the recognition that controlling hydrogen-bonding interactions can aid selectivity in the transfer of a metalate from the aqueous to organic phase [90] resulted in a new study on the use of primary amides in the solvent extraction of gold [108]. It was found that the transfer of Au(III) from aqueous HCl into a 0.1 M primary amide MDMHA solution in toluene (Figure 1.25) was rapid and quantitative, and unusually with straightforward back-extraction occurring using water. This reagent is also highly selective, extracting Au(III) from a mixture of Fe(III), Cu(II), Zn(II), Sn(II), and Ni(II) at concentrations similar to those found in a typical smartphone (Figure 1.26). Compared to MIBK, DBC, and 2-EH, MDMHA shows increased selectivity, and unlike the commercial reagents can be used in a diluent.

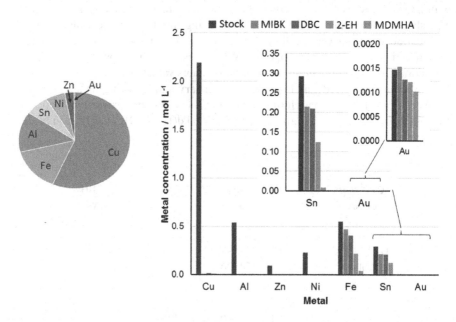

FIGURE 1.26 Left: Proportions of metals typically found in a smartphone device. Right: Comparison of the metals recovered by solvent extraction from a smartphone WEEE feed in 1.0 M HCl using the neat commercial reagents MIBK, DBC, and 2-EH and a 0.1 M toluene solution of the primary amide MDMHA.

Importantly, the chemistry that underpins the mode of extraction was studied in detail by a variety of analytical and spectroscopic techniques. Analysis of the slopes of linear log D vs. log[L] plots suggested that at a minimum a 1:2 Au:L stoichiometry was operating, namely the formation of $HAuCl_4(MDMHA)_2$ similar to that described above (Figure 1.27).

However, decreasing the gold concentration in the aqueous phase led to non-integer slopes (2.5), which implies that the speciation in the organic phase is variable. Karl–Fischer analysis of the organic phase showed that water is not transferred with gold, as no increase in the water concentration is seen on increased gold loading.

Analysis of the Au-loaded organic phase by EXAFS spectroscopy showed that $AuCl_4^-$ is present and that there were no close contacts between individual $AuCl_4^-$ complexes (Figure 1.27). This may imply that simple $HAuCl_4(MDMHA)_2$ ion pairs are present, but the analysis of the organic phase by ESI-MS showed a preference for cluster formation. In the mass spectrum (Figure 1.29), the dominant gold species have a generic formula of $[H(MDMHA)_2(AuCl_4)_n(H\text{-}MDMHA)_{n\text{-}1}]$ (where $n = 1\text{-}4$); that is, a basic 1:2 $HAuCl_4(MDMHA)_2$ complex exists, but higher-order clusters are favored and represented by the further addition of $HAuCl_4(MDMHA)$ units.

Further insight into the chemical speciation was provided by DFT and MD calculations (Figure 1.27). From these, it was found that aggregation into Au:amide clusters occurs readily in a solvated toluene box, with 4:10 $HAuCl_4/MDMHA$ calculations showing a variety of structural motifs in which hydrogen-bonded amides are able to bridge discrete $AuCl_4^-$ centers through electrostatic and hydrogen-bonding to defined edges of the square-planar gold cation. As such, the combined analytical

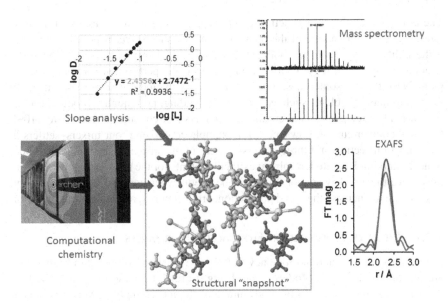

FIGURE 1.27 Combined analytical and spectroscopic techniques to determine the chemical speciation in the solvent extraction of gold from aqueous HCl using a toluene solution of MDMHA.

and spectroscopic analysis provides clear evidence for the spontaneous formation of supramolecular clusters facilitated by the simple primary amide, MDMHA.

1.2.3.2.4   Conclusion

The recovery of gold from WEEE is an important starting point for the complete recycling of these complex materials due to its economic value and relatively straightforward separation from other metals. However, one of the key technological advances that remains to be solved is its efficient, nontoxic, and potentially selective leaching from WEEE components but, as described above, some progress is being made in this area. The combination of efficient leaching with selective precipitation or solvent extraction techniques would result in a powerful process that could not only recover gold, but also other valuable or toxic metals from WEEE, so moving toward the closed-loop WEEE economy desired by society. As described above, chemists have made significant headway into understanding the mode of action of hydrometallurgical processes by studying the structures of the chemical compounds in aqueous, organic, or precipitated phases. The structural detail identified from these coordination and/or supramolecular chemical studies may point toward new designs of extractants that make use of recurring motifs to provide enhanced and efficient separation. Even so, while significant advances have been made in hydrometallurgical separation, their integration into a whole process needs to be considered, which relies upon continued input from, and collaboration between, chemists, biologists, engineers, economists, and the industries involved.

1.3   FLOWSHEET OPTIMIZATION

The efficiency of solvent extraction plants depends on various factors such as flow rates, concentration, nature of the feed solution, flowsheets, *etc.* Especially powerful is the ability to implement scrubbing stages to effect higher product purity. The modification of any one of these factors strongly affects the performance of an extraction plant. Nevertheless, sometimes, it may be necessary to modify one of these factors to face up to many problems such as low solvent loading, poor quality product, formation of cruds, precipitates, and emulsions, radiolysis or chemical stresses, etc. [109]. Classical counter-current flowsheets, for example comprising four mixers–settlers in the extraction section and three mixers–settlers in the stripping section, constitute a typical setup implemented in solvent extraction plants for the recovery of metals from ores. Nevertheless, they are not always the best setup when configured in this way, and it is possible to increase the metal production, product purity, and concentration factor.

1.3.1   EFFECT OF FLOWRATES ON FLOWSHEET PERFORMANCES

For instance, the extraction efficiency of the two flowsheets reported in Figure 1.28 are drastically different. Modeling can bring useful information to select the best flowsheet and optimize it to take advantage of the increasingly powerful emergent chemistry.

Flowsheet 4_3 is a classical counter-current flowsheet, whereas the other flowsheet, quoted 22_11*, is an unconventional flowsheet with two extraction-stripping

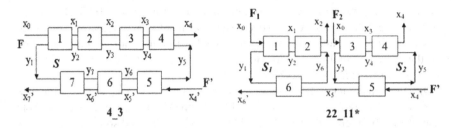

FIGURE 1.28 Classical counter-current and unconventional two-feed flowsheets, 4_3 and 22_11*, respectively. x_i and y_i represent the metal concentration in the aqueous and organics phases of the mixer-settler i, respectively. S denotes the solvent flow rate, F, F_1, and F_2 are leach solution flowrates, respectively, and F' is the stripping flow rate. More information about the chemistry of uranium extraction by Alamine 336 is detailed in refs [101–103].

loops. Each loop corresponds to two mixers–settlers in the extraction stage and one mixer-settler in the stripping stage. Chagnes et al. investigated the influence of the flowsheet for uranium extraction [110].

The influence of the solvent flow rate (S), the feed solution flow rate (F), and the stripping solution flow rate (F') on the residual fraction (defined as the uranium concentration in the raffinate (x_4) over its concentration in the inlet feed solution, $f=x_4/x_0$) for the classical flowsheet (4_3) and the unconventional flowsheet (22_11*) is displayed in Figure 1.29.

The residual fraction decreases when the solvent flow rate increases and remains constant after a threshold value of S equal to 29 m³ h⁻¹ for both flowsheets. The classical counter-current flowsheet permits to reach a residual fraction close to zero for

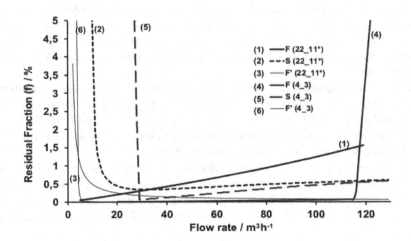

FIGURE 1.29 Influence of solvent flow rate (S), feed flow rate (F) and stripping flow rate (F') on the residual fraction (f) for the flowsheets 4_3 and 22_11*. The operating conditions for 4_3 and 22_11* are: $S=29$ m³ h⁻¹, $F=115$ m³ h⁻¹, $F'=4.8$ m³ h⁻¹, initial uranium concentration $x_0=1677$ mg L⁻¹, and $S=29$ m³ h⁻¹, $F=90$ m³ h⁻¹, $F'=4.8$ m³ h⁻¹, initial uranium concentration $x_0=400$ mg L⁻¹, respectively.

solvent flow rates higher than 29 m³ h⁻¹, whereas f remains close to 0.5% at the same range of solvent flow rates. A sharp decrease of the residual fraction is also observed when the stripping solution flow rate increases but the drop in f is more important with the 4_3 configuration: f is close to zero with the 4_3 flowsheet and f = 1.1% with the 22_11* flowsheet at F' = 5.5 m³ h⁻¹.

The classical counter-current flowsheet ensures that you reach a feed solution flow rate of 115 m³ h⁻¹ without a significant increase of the residual fraction. No threshold value of f is observed with the 22_11* flowsheet when the feed solution flow rate is lower than 130 m³ h⁻¹. The sharp increase of f at solvent flow rates higher than 115 m³ h⁻¹ can be explained by the saturation of the solvent, which cannot then extract more uranium, as there are not enough mixers–settlers to treat the feed solution in the stripping stage at a high S/F' ratio. On the other hand, no saturation phenomenon is observed with the 22_11* flowsheet even if the residual fraction with 22_11* remains higher than with 4_3 before the threshold value (F = 115 m³ h⁻¹). The 22_11* flowsheet does not enhance the extraction efficiency, but it permits the avoidance of the saturation of the organic phase above 130 m³ h⁻¹.

1.3.2 INFLUENCE OF DEGRADATION ON FLOWSHEET PERFORMANCE

Solvent extraction processes may undergo stresses such as radiolysis or chemical degradation. For example, chemical degradation of trioctylamine and tridecanol in n-dodecane as a diluent may occur due to the presence of strong oxidant metal ions in the feed solution such as vanadium(V) [111]. In the case of vanadium(V), the degradation occurs as follows: (i) Vanadium(V) is extracted by Alamine 336 from aqueous phase to the organic phase; (ii) the strong oxidation power of V(V) is responsible of the oxidation of the modifier (tridecanol) to carboxylic acid according to a radical mechanism; (iii) the preceding radicals react quickly with Alamine 336 to form dioctylamine (DOA) as a degradation product [111]. Furthermore, the degradation of Alamine 336 to dioctylamine is responsible for a drop of extraction efficiency, as dioctylamine is a less efficient extractant for uranium(VI) than is trioctylamine.

By including the physicochemistry of degradation into the flowsheet simulation, it is possible to investigate the sturdiness of candidate flowsheets against chemical degradation of the solvent. Figure 1.30 displays the sturdiness of the 4_3 and 22_11* flowsheets against the degradation progress, that is, DOA molar fraction in the Alamine 336–tridecanol–n-dodecane system [112].

There is no influence of the chemical degradation on the extraction efficiency for the classical counter-current flowsheet as long as the molar fraction of DOA is lower than 0.1. At higher DOA molar fractions, the residual fraction increases linearly. The use of DOA instead of Alamine 336 is responsible for an extraction loss equal to 8.8%.

Therefore, the implementation of the 22_11* flowsheet in a solvent extraction process for the recovery of uranium(VI) from sulfuric acid media permits the enhancement of the resistance of the solvent extraction process toward chemical degradation. In fact, a slow increase of residual fraction is observed when the molar fraction of DOA in organic phase increases (Figure 1.3). The use of a second loop in the

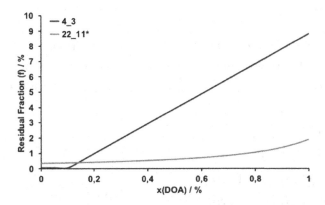

FIGURE 1.30 Influence of the dioctylamine molar fraction (x(DOA)) on the residual fraction (f) for the flowsheets 4_3 and 22_11*. The operating conditions for 4_3 and 22_11* are: solvent flow rate $S=29$ m^3 h^{-1}, feed flow rate $F=115$ m^3 h^{-1}, initial uranium concentration $x_0=1677$ mg L^{-1}, and stripping flow rate $F'=4.80$ m^3 h^{-1}, feed flow rate $F=90$ m^3 h^{-1}, initial uranium concentration $x_0=400$ mg L^{-1}, respectively.

extraction process permits to increase strongly the resistance of the process against degradation. Works are in progress for testing other configurations to understand the origin of the enhancement of the flowsheet sturdiness. It may be supposed that the effect of the solvent extraction degradation may be counterbalanced by the increase of the extraction capacity of unconventional flowsheets.

1.3.3 COMBINING CHEMISTRY AND ENGINEERING FOR FLOWSHEET OPTIMIZATION

Likewise, modeling tools for optimizing flowsheets and operation conditions have been used by CEA in order to optimize a flowsheet for extracting uranium(VI) from WPA by using the new bifunctional molecules DEHCNPB (see above). A chemical model was elaborated for uranium and iron extraction by DEHCNPB in order to precisely simulate the behavior of these elements in the SX process whatever the concentrations of uranium, phosphoric acid, or DEHCNPB concentration are. Based on experimental extraction isotherms and, when it was possible, on speciation studies of uranium and iron both in phosphoric acid solutions and in the organic phase, extraction equilibria were proposed, and extraction constants were adjusted to fit the experimental values. First, uranium and iron speciation in 5 M H$_3$PO$_4$ were calculated by selecting the main equilibria formed in aqueous solution among the different equilibria reported in the literature [113, 114]. Miguirditchian et al. [115] considered four equilibria to describe uranium speciation in phosphoric media, while two equilibria were sufficient for iron (Table 1.1).

Then, based on the speciation results obtained in the organic phase, cation exchange for uranium(VI) by DEHCNPB was considered in its proton–dimer form. Extraction of uranium by 2.5 dimers of DEHCNPB (i.e., two dimers and one monomer) along with the formation of a bimolecular uranium complex (2:4) allowed reaching the best fit of the experimental isotherm. The global form of these equilibria and the associated thermodynamic constants are given by Eqs. 1.10 and 1.11.

TABLE 1.1

Equilibria Considered in Aqueous Phase to Calculate Uranium(VI) and Iron(III) Speciation in Phosphoric Acid

Equilibrium	$\log(\beta)$ at $I=0$ and 25°C
$H_3PO_4 \rightleftharpoons H_2PO_4^- + H^+$	-2.14
$H_2PO_4^- \rightleftharpoons HPO_4^{2-} + H^+$	-7.21
$HPO_4^{2-} \rightleftharpoons PO_4^{3-} + H^+$	-12.35
$UO_2^{2+} + H_3PO_4 \rightleftharpoons UO_2(H_2PO_4)^+ + H^+$	1.5
$UO_2^{2+} + H_3PO_4 \rightleftharpoons UO_2(H_3PO_4)^{2+}$	1.3
$UO_2^{2+} + 2H_3PO_4 \rightleftharpoons UO_2(H_2PO_4)_2 + 2H^+$	1.3
$UO_2^{2+} + 2H_3PO_4 \rightleftharpoons UO_2(H_2PO_4)(H_3PO_4)^+ + H^+$	2.3
$Fe^{3+} + 2H_2PO_4^- + H_3PO_4 \rightleftharpoons FeH(H_2PO_4)_3^+$	8.3
$Fe^{3+} + 3H_2PO_4^- \rightleftharpoons Fe(H_2PO_4)_3$	9.8

It is usually rigorous to take into account activity coefficients in the aqueous phase to correct deviations from ideality in process conditions. Different theories such as the "simple solution" concept [116, 117] or SIT (specific interaction theory) [118, 119] are, for instance, often used in nitric acid or sulfuric media to calculate activity coefficients. In the present model, variation of activity coefficients in phosphoric acid has not been taken into account. Activity coefficients in the organic phase were also considered constants in the operating uranium concentration range to simplify the modeling.

$$UO_2^{2+} + 2.5\left(\overline{HY}\right)_2 \rightleftharpoons \overline{UO_2Y_2(HY)_3} + 2H^+ \quad K_{U1} = \frac{\left[\overline{UO_2Y_2(HY)_3}\right]\cdot\left[H^+\right]^2}{\left[UO_2^{2+}\right]\cdot\left[\left(\overline{HY}\right)_2\right]^{2.5}} \quad (1.10)$$

$$2\,UO_2^{2+} + 2\left(\overline{HY}\right)_2 \rightleftharpoons \overline{\left((UO_2)_2\,Y_4\right)} + 4H^+ \quad K_{U2} = \frac{\left[\overline{\left((UO_2)_2\,Y_4\right)}\right]\cdot\left[H^+\right]^4}{\left[UO_2^{2+}\right]^2\cdot\left[\left(\overline{HY}\right)_2\right]^2} \quad (1.11)$$

Iron(III) extraction by DEHCNPB was also modeled by a cation-exchange mechanism involving a combination of a dimer and a monomer according to the following equilibrium:

$$Fe^{3+} + 1.5\left(\overline{HY}\right)_2 \rightleftharpoons \overline{FeY_3} + 3H^+ \quad K_{Fe} = \frac{\left[\overline{FeY_3}\right]\cdot\left[H^+\right]^3}{\left[Fe^{3+}\right]\cdot\left[\left(\overline{HY}\right)_2\right]^{1.5}} \quad (1.12)$$

Extraction constants were optimized by minimizing the differences between experimental and calculated organic concentrations after application of the mass action law on each equilibrium. Mass balances were resolved in both phases in order to determine the free concentrations. The comparison between experimental and calculated data for the isotherm of uranium(VI) extraction is depicted in Figure 1.31. A very good agreement between experimental and calculated data is reached with an average deviation lower than 5%.

The chemical model was then implemented into the PAREX simulation code [120] to calculate a flowsheet in order to perform a counter-current laboratory run (Figure 1.32).

Taking into account the objectives for uranium recovery and purification, an optimized flowsheet including three stages for uranium extraction, three stages for iron scrubbing, one stage for water scrubbing and four stages for uranium stripping was proposed. This flowsheet was tested in laboratory scale mixer-settlers from a genuine industrial phosphoric acid solution. More than 95% of uranium was recovered

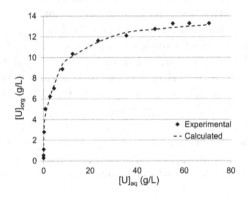

FIGURE 1.31 Comparison of experimental and calculated isotherms of uranium(VI) extraction with 0.1 M DEHCNPB/TPH and 5 M H_3PO_4.

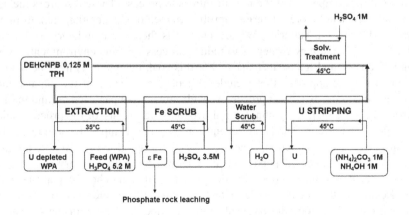

FIGURE 1.32 Flowsheet of the continuous test performed in mixer-settlers with DEHCNPB.

and very well decontaminated from iron and other impurities, confirming the high selectivity of DEHCNPB for uranium from WPA and the high potential of this new molecule for future industrial development.

Therefore, simulation tools based on the physicochemistry of solvent extraction are very useful to model and optimize processes. A deep understanding of the equilibria taking place in solvent extraction is the key to imagine new efficient processes.

1.3.4 PERSPECTIVE IN THE DEVELOPMENT OF TOOLS FOR FLOWSHEET OPTIMIZATION

In order to reduce OPEX, the process must be able to adapt to raw material composition and to anticipate issues. This is especially true for recycling processes, as waste composition can dramatically vary throughout time and depending on the origin of the wastes. Indeed, waste composition changes depending on technology progress, the country's economic health, location, etc. The implementation of engineering and chemical codes is the key to developing smart processes capable of adapting to the operating conditions and the flowsheet automatically so that the process always operates under optimized conditions. The tools presented in the previous part can be viewed as preliminary blocks for implementation in smart processes. However, the development of such models requires the acquisition of thermodynamic data, which can be time-consuming. Nevertheless, the time spent for developing the models and the associated database will be undoubtedly compensated by OPEX gain resulting from their use in hydrometallurgical processes.

The impacts of this approach on process performance can be drastically improved by combining physicochemical and engineering models with artificial intelligence (machine learning, deep-learning, big data) and predictive maintenance technologies promoted by advanced manufacturing approach [121].

1.4 CONCLUSION

The extraction and valorization of metals is presently more difficult than in the past because of the complexity of the new resources to process. These resources include poor and polymetallic ores (laterite-saprolite, pyrochlore, pegmatite, etc.), tailings, e-wastes (lithium-ion batteries, WEEE, etc.). It is therefore mandatory to develop new approaches and new strategies to reduce process cost and environmental footprint while keeping high performances. For this goal, many efforts must be paid to develop a holistic approach that includes a geochemical cycle of the metals, concentration methods, extraction processes, environmental impact studies, and market analysis. In particular, the technical and economic feasibility of the process relies on the flowsheet design. The latter depends on resources composition and technologies involved in the extraction-separation stages. In the case of solvent extraction processes, performances depend both on engineering and chemistry. It is therefore of great importance to optimize these engineering and chemistry at the same time. Regarding the chemistry, the extractants are at the center of solvent extraction processes. Extractant design needs to have a fine description of the speciation in the leach solution and metal–ligand interaction and coordination. Obviously, the extractant molecule is not alone in the solvent, and its environment must be taken into account as well (diluent effect, molecular, and supramolecular association). The physicochemistry is complex and modeling tools are especially adapted to investigate the influence of the chemical structure of a molecule on these properties. Regarding flowsheet design, the combination of physicochemical and engineering models is the key to develop disruptive and optimized flowsheets. Thanks to this integrated approach, the flowsheet can adapt to changes in compositions or external constraints, and, therefore, operate under optimal conditions. The next generations of hydrometallurgical processes should also rely on the artificial intelligence and predictive maintenance technologies promoted by the advanced manufacturing approach, since these technologies will likely lead to smart processing that will be able to adapt to changes in operation conditions and to immediately fix the main issues. This approach will likely drastically reduce operating costs, maintenance costs, energy consumption, and environmental impact by keeping the process optimized at any time.

REFERENCES

1. Sholl, D. S.; Lively, R. P. 2016. *Nature* 532, 435–437.
2. Tasker, P. A.; Plieger, P. G.; West, L. C. 2005. *Comprehensive Coordination Chemistry II*, eds. J. A. McCleverty, T. J. Meyer, Elsevier Ltd, Oxford, ch. 9.17, 759–808.
3. Izatt, R. M.; Izatt, S. R.; Bruening, R. L.; Izatt, N. E.; Moyer, B. A. 2014. *Chemical Society Reviews* 43, 2451–2475.
4. Xie, F. Z.; Zhang, T. A.; Dreisinger, D.; Doyle, F. 2014. *Minerals Engineering* 56, 10–28.
5. Chagnes, A. 2015. Fundamentals in electrochemistry and hydrometallurgy. In *Lithium Process Chemistry: Resources, Extractions, Batteries and Recycling*, eds A. Chagnes, J. Swiatowska, Elsevier, Amsterdam, 41–80.
6. Chagnes, A.; Moncomble, A.; Cote, G. 2013. *Solvent Extraction and Ion Exchange* 31, 499–518.
7. Prestianni, A.; Joubert, L.; Chagnes, A.; Cote, G.; Ohnet, M. N.; Rabbe, C.; Charbonnel, M. C.; Adamo, C. 2010. *The Journal of Physical Chemistry. A* 114, 10878–10884.
8. Prestianni, A.; Joubert, L.; Chagnes, A.; Cote, G.; Adamo, C. 2011. *Physical Chemistry Chemical Physics: PCCP* 13, 19371–19377.
9. Turkington, J. R.; Bailey, P. J.; Love, J. B.; Wilson, A. M.; Tasker, P. A. 2013. *Chemical Communications* 49, 1891–1899.
10. US Geological Survey; Doggett, M. D. Global mineral exploration and production – The impact of technology, https://pubs.usgs.gov/circ/2007/1294/reports/paper10.pdf (accessed July 5, 2018).
11. Wilson, A. M.; Bailey, P. J.; Tasker, P. A.; Turkington, J. R.; Grant, R. A.; Love, J. B. 2014. *Chemical Society Reviews* 43, 123–134.
12. Forgan, R. S.; Wood, P. A.; Campbell, J.; Henderson, D. K.; McAllister, F. E.; Parsons, S.; Pidcock, E.; Swart, R. M.; Tasker, P. A. 2007. *Chemical Communications*, 4940–4942.
13. Forgan, R. S.; Roach, B. D.; Wood, P. A.; White, F. J.; Campbell, J.; Henderson, D. K.; Kamenetzky, E.; McAllister, F. E.; Parsons, S.; Pidcock, E.; Richardson, P.; Swart, R. M.; Tasker, P. A. 2011. *Inorganic Chemistry* 50, 4515–4522.
14. Healy, M. R.; Carter, E.; Fallis, I. A.; Forgan, R. S.; Gordon, R. J.; Kamenetzky, E.; Love, J. B.; Morrison, C. A.; Murphy, D. M.; Tasker, P. A. 2015. *Inorganic Chemistry* 54, 8465–8473.

15. Healy, M. R.; Roebuck, J. W.; Doidge, E. D.; Emeleus, L. C.; Bailey, P. J.; Campbell, J.; Fischmann, A. J.; Love, J. B.; Morrison, C. A.; Sassi, T.; White, D. J.; Tasker, P. A. 2016. *Dalton Transactions* 45, 3055–3062.

16. Wenzel, M.; Hennersdorf, F.; Langer, M.; Gloe, K.; Antonioli, B.; Buschmann, H.-J.; Lindoy, L. F.; Bernhard, G.; Gloe, K.; Weigand, J. J. 2017. *Separation Science and Technology* 53 (8), 1273–1281.

17. Kaltsoyannis, N. 2013. *Inorganic Chemistry* 52, 3407–3413.

18. Choppin, G. R. 2002. *Journal of Alloys and Compounds* 344, 55–59.

19. Panak, P. J.; Geist, A. 2013. *Chemical Reviews* 113, 1199–1236.

20. Drew, M. G. B.; Foreman, M. R. S. J.; Hill, C.; Hudson, M. J.; Madic, C. 2005. *Inorganic Chemistry Communications* 8, 239–241.

21. Lewis, F. W.; Harwood, L. M.; Hudson, M. J.; Drew, M. G. B.; Desreux, J. F.; Vidick, G.; Bouslimani, N.; Modolo, G.; Wilden, A.; Sypula, M.; Vu, T. H.; Simonin, J. P. 2011. *Journal of the American Chemical Society* 133, 13093–13102.

22. Edwards, A. C.; Wagner, C.; Geist, A.; Burton, N. A.; Sharrad, C. A.; Adams, R. W.; Pritchard, R. G.; Panak, P. J.; Whitehead, R. C.; Harwood, L. M. 2016. *Dalton Transactions* 45, 18102–18112.

23. Kratsch, J.; Beele, B. B.; Koke, C.; Denecke, M. A.; Geist, A.; Panak, P. J.; Roesky, P. W. 2014. *Inorganic Chemistry* 53, 8949–8958.

24. Weaver, B.; Kappelmann, F. A. 1964. TALSPEAK, A new method of separating americium and curium from the lanthanides by extraction from an aqueous solution of an aminopolyacetic acid complex with a monoacetic organophosphate or phosphonate. ORNL-3559.

25. Gelis, A. V.; Lumetta, G. J. 2014. *Industrial and Engineering Chemistry Research* 53, 1624–1631.

26. Paiva, A. P.; Malik, P. 2004. *Journal of Radioanalytical and Nuclear Chemistry* 261, 485–496.

27. Geist, A.; Müllich, U.; Magnusson, D.; Kaden, P.; Modolo, G.; Wilden, A.; Zevaco, T. 2012. *Solvent Extraction and Ion Exchange* 30, 433–444.

28. Lewis, F. W.; Harwood, L. M.; Hudson, M. J.; Geist, A.; Kozhevnikov, V. N.; Distler, P.; John, J. J. 2015. *Chemical Science* 6, 4812–4821.

29. Edwards, A. C.; Mocilac, P.; Geist, A.; Harwood, L. M.; Sharrad, C. A.; Burton, N. A.; Whitehead, R. C.; Denecke, M. A. 2017. *Chemical Communications* 53, 5001–5004.

30. Malmbeck, R.; Magnusson, D.; Geist, A. 2017. *Journal of Radioanalytical and Nuclear Chemistry* 314, 2531–2538.

31. Antonio, M. R.; McAlister, D. R.; Horwitz, E. P. 2015. *Dalton Transactions* 44, 515–521.

32. Ellis, R. J.; Brigham, D. M.; Delmau, L.; Ivanov, A. S.; Williams, N. J.; Vo, M. N.; Reinhart, B.; Moyer, B. A.; Bryantsev, V. S. 2017. *Inorganic Chemistry* 56, 1152–1160.

33. Brigham, D. M.; Ivanov, A. S.; Moyer, B. A.; Delmau, L. H.; Bryantsev, V. S.; Ellis, R. J. 2017. *Journal of the American Chemical Society* 139, 17350–17358.

34. Baldwin, A. G.; Ivanov, A. S.; Williams, N. J.; Ellis, R. J.; Moyer, B. A.; Bryantsev, V. S.; Shafer, J. C. 2018. *ACS Central Science* 4, 739–747.

35. Hunter, J. P.; Dolezalova, S.; Ngwenya, B. T.; Morrison, C. A.; Love, J. B. 2018. *Metals* 8, 465.

36. Muller, J. M.; Berthon, C.; Couston, L.; Guillaumont, D.; Ellis, R. J.; Zorz, N.; Simonin, J.-P.; Berthon, L. 2017. *Hydrometallurgy* 169, 542–551.

37. Batchu, N. K.; Vander Hoogerstraete, T.; Banerjee, D.; Binnemans, K. 2017. *Separation and Purification Technology* 174, 544–553.

38. Carreira-Barral, I.; Mato-Iglesias, M.; De Blas, A.; Platas-Iglesias, C.; Tasker, P. A.; Esteban-Gómez, D. 2017. *Dalton Transactions* 46, 3192–3206.

39. Wintergerst, M. P.; Levitskaia, T. G.; Moyer, B. A.; Sessler, J. L.; Delmau, L. H. 2008. *Journal of the American Chemical Society* 130, 4129–4139.

40. Ellis, R. J.; Reinhart, B.; Williams, N. J.; Moyer, B. A.; Bryantsev, V. S. 2017. *Chemical Communications* 53, 5610–5613.

41. Ellis, R. J.; Chartres, J.; Henderson, D. K.; Cabot, R.; Richardson, P. R.; White, F. J.; Schröder, M.; Turkington, J. R.; Tasker, P. A.; Sole, K. C. 2012. *Chemistry – A European Journal* 18, 7715–7728.

42. Ellis, R. J.; Chartres, J.; Sole, K. C.; Simmance, T. G.; Tong, C. C.; White, F. J.; Schröder, M.; Tasker, P. A. 2009. *Chemical Communications*, 583–585.

43. Turkington, J. R.; Cocalia, V.; Kendall, K.; Morrison, C. A.; Richardson, P.; Sassi, T.; Tasker, P. A.; Bailey, P. J.; Sole, K. C. 2012. *Inorganic Chemistry* 51, 12805–12819.

44. Cokoja, M.; Markovits, I. I. E.; Anthofer, M. H.; Poplata, S.; Pöthig, A.; Morris, D. S.; Tasker, P. A.; Herrmann, W. A.; Kühn, F. E.; Love, J. B. 2015. *Chemical Communications* 51, 3399–3402.

45. Warr, R. J.; Bell, K. J.; Gadzhieva, A.; Cabot, R.; Ellis, R. J.; Chartres, J.; Henderson, D. K.; Lykourina, E.; Wilson, A. M.; Love, J. B.; Tasker, P. A.; Schröder, M. 2016. *Inorganic Chemistry* 55, 6247–6260.

46. Bell, K. J.; Westra, A. N.; Warr, R. J.; Chartres, J.; Ellis, R.; Tong, C. C.; Blake, A. J.; Tasker, P. A.; Schröder, M. 2008. *Angewandte Chemie* 47, 1745–1748.

47. Doidge, E. D.; Carson, I.; Love, J. B.; Morrison, C. A.; Tasker, P. A. 2016. *Solvent Extraction and Ion Exchange* 34, 579–593.

48. Ellis, R. J.; Meridiano, Y.; Muller, J.; Berthon, L.; Guilbaud, P.; Zorz, N.; Antonio, M. R.; Demars, T.; Zemb, T. 2014. *Chemistry – A European Journal* 20, 12796–12807.

49. Osseo-Asare, K. 1991. *Advances in Colloid and Interface Science* 37, 123–173.

50. Ferru, G.; Gomes Rodrigues, D.; Berthon, L.; Diat, O.; Bauduin, P.; Guilbaud, P. 2014. *Angewandte Chemie* 53, 5346–5350.

51. Baldwin, A. G.; Yang, Y.; Bridges, N. J.; Braley, J. C. 2016. *The Journal of Physical Chemistry. Part B* 120, 12184–12192.

52. Antonio, M. R.; Ellis, R. J.; Estes, S. L.; Bera, M. K. 2017. *Physical Chemistry Chemical Physics: PCCP* 19, 21304–21316.

53. Beltrami, D.; Chagnes, A.; Haddad, M.; Laureano, H.; Mokhtari, H.; Courtaud, B.; Jugé, S.; Cote, G. 2014. *Hydrometallurgy* 144–145, 207–214.

54. Beltrami, D.; Cote, G.; Mokhtari, H.; Courtaud, B.; Chagnes, A. 2012. *Hydrometallurgy* 129–130, 118–125.

55. Beltrami, D.; Mercier-Bion, F.; Cote, G.; Mokhtari, H.; Courtaud, B.; Simoni, E.; Chagnes, A. 2014. *Journal of Molecular Liquids* 190, 42–49.

56. Dourdain, S.; Hofmeister, I.; Pecheur, O.; Dufrêche, J. F.; Turgis, R.; Leydier, A.; Jestin, J.; Testard, F.; Pellet-Rostaing, S.; Zemb, T. 2012. *Langmuir: The ACS Journal of Surfaces and Colloids* 28, 11319–11328.

57. Pecheur, O.; Dourdain, S.; Guillaumont, D.; Rey, J.; Guilbaud, P.; Berthon, L.; Charbonnel, M. G.; Pellet-Rostaing, S.; Testard, F. 2016. *Journal of Physical Chemistry. Part B* 120, 2814–2823.

58. Fitoussi, R.; Musikas, C. 1980. *Separation Science and Technology* 15, 845–860.

59. Krea, M.; Khalaf, H. 2000. *Hydrometallurgy* 58, 215–225.

60. Singh, H.; Vijayalakshmi, R.; Mishra, S. L.; Gupta, C. K. 2001. *Hydrometallurgy* 59, 69–76.

61. Beltrami, D.; Chagnes, A.; Haddad, M.; Laureano, H.; Mokhtari, H.; Courtaud, B.; Juge, S.; Cote, G. 2013. *Science and Technology* 48, 480–486.

62. Singh, S. K.; Tripathi, S. C.; Singh, D. K. 2010. *Science and Technology* 45, 824.

63. Beltrami, D.; Cote, G.; Mokhtari, H.; Courtaud, B.; Moyer, B. A.; Chagnes, A. 2014. *Chemical Reviews* 114, 12002–12023.

64. Bunus, F. T. 1977. *Talanta* 24, 117–120.

65. Chen, H. M.; Chen, H. J.; Tsai, Y. M.; Lee, T. W.; Ting, G. 1987. *Industrial and Engineering Chemistry Research* 26, 621–627.

66. Dartiguelongue, A.; Provost, E.; Chagnes, A.; Cote, G.; Fürst, W. 2016. *Solvent Extraction and Ion Exchange* 34, 241–259.

67. Dartiguelongue, A.; Chagnes, A.; Provost, E.; Fürst, W.; Cote, G. 2016. *Hydrometallurgy* 165, 57–63.

68. Gabriel, S.; Baschwitz, A.; Mathonnière, G.; Eleouet, T.; Fizaine, F. 2013. *Annals of Nuclear Energy* 58, 213–220.

69. Hurst, F. J.; Crouse, D. J.; Brown, K. B. 1972. *Industrial and Engineering Chemistry Process Design and Development* 11, 122–128.

70. Francois, A.; Sialino, A. 1977. Patent NL 7806490 (FR 2396803, US 4238457), Azote Produits Chimiques.

71. Michel, P.; Ranger, G.; Corompt, P.; Bon, P. 1980. New process for the recovery of uranium from phosphoric acid. In *Proceedings of the International Solvent Extraction Conference ISEC 80*, Liege, Belgium, p 80.

72. Hurst, F. J.; Crouse, D. J. 1974. *Industrial and Engineering Chemistry Process Design and Development* 13, 286–291.

73. Khorfan, S. 1993. *Chemical Engineering and Processing: Process Intensification* 32, 273–276.

74. Mc-Cready, W. L.; Wethington, J. A.; Hurst, F. J. 1981. *Nuclear Technology* 53, 344–353.

75. Singh, D. K.; Mondal, S.; Chakravartty, J. K. 1966. *Solvent Extraction and Ion Exchange* 34, 201–225.

76. Warshawsky, A.; Kahana, N.; Arad-Yellin, R. 1989. *Hydrometallurgy* 23, 91–104.

77. Turgis, R.; Leydier, A.; Arrachart, G.; Burdet, F.; Dourdain, S.; Bernier, G.; Miguirditchian, M.; Pellet-Rostaing, S. 2014. *Solvent Extraction and Ion Exchange* 32, 478–491.

78. Arrachart, G.; Aychet, N.; Bernier, G.; Burdet, F.; Leydier, A.; Miguirditchian, M.; Pellet-Rostaing, S.; Plancque, G.; Turgis, R.; Zekri, E. 2013. Patent WO 2013/167516 (FR 2990207), AREVA Mines.

79. Turgis, R.; Leydier, A.; Arrachart, G.; Burdet, F.; Dourdain, S.; Bernier, G.; Miguirditchian, M.; Pellet-Rostaing, S. 2014. *Solvent Extraction and Ion Exchange* 32, 685–702.

80. Bernier, G.; Miguirditchian, M.; Pacary, V.; Bertrand, M.; Cames, B.; Hérès, X.; Mokhtari, H. 2014. New process for the selective extraction of uranium from phosphoric ores, ISEC, Wurzburg, Germany.

81. Miguirditchian, M.; Bernier, G.; Pacary, V.; Balaguer, C.; Sorel, C.; Berlemont, R.; Fries, B.; Bertrand, M.; Camès, B.; Leydier, A.; Turgis, R.; Arrachart, G.; Pellet-Rostaing, S.; Mokhtari, H. 2016. *Solvent Extraction and Ion Exchange* 34, 274–289.

82. Leydier, A.; Arrachart, G.; Turgis, R.; Bernier, G.; Marie, C.; Miguirditchian, M.; Pellet-Rostaing, S. 2017. *Hydrometallurgy* 171, 262–266.

83. Kaya, M. 2016. *Waste Management* 57, 64–90.

84. Cui, J.; Zhang, L. 2008. *Journal of Hazardous Materials* 158, 228–256.

85. Jadhao, P.; Chauhan, G.; Pant, K. K.; Nigam, K. D. P. 2016. *Waste Management* 57, 102–112.

86. Chagnes, A.; Cote, G.; Ekberg, C.; Nilson, M.; Retegan, T. 2016. *Recycling of Waste Electrical and Electronic Equipment: Research, Development and Policies*, Elsevier, 212 pages.

87. a) Chatterjee, A.; Abraham, J. 2017. *International Journal of Environmental Science and Technology* 14, 211–222; (b) Sun, Z.; Cao, H.; Xiao, Y.; Sietsma, J.; Jin, W.; Agterhuis, H.; Yang, Y. 2017. *ACS Sustainable Chemistry and Engineering* 5, 21–40; (c) Priya, A.; Hait, S. 2017. *Environmental Science and Pollution Research International* 24, 6989–7008; (d) Diaz, L. A.; Lister, T. E.; Parkman, J. A.; Clark, G. G. 2016. *Journal*

of Cleaner Production 125, 236–244; (e) Dodson, J. R.; Parker, H. L.; Muñoz García, A.; Hicken, A.; Asemave, K.; Farmer, T. J.; He, H.; Clark, J. H.; Hunt, A. J. 2015. *Green Chemistry* 17, 1951–1965.

88. Iannicelli-Zubiani, E. M.; Giani, M. I.; Recanati, F.; Dotelli, G.; Puricelli, S.; Cristiani, C. 2017. *Journal of Cleaner Production* 140, 1204–1216.

89. Baxter, J.; Lyng, K. A.; Askham, C.; Hanssen, O. J. 2016. *Waste Management* 57, 17–26.

90. a) Wilson, A. M.; Bailey, P. J.; Tasker, P. A.; Turkington, J. R.; Grant, R. A.; Love, J. B. 2014. *Chemical Society Reviews* 43, 123–134; (b) Turkington, J. R.; Bailey, P. J.; Love, J. B.; Wilson, A. M.; Tasker, P. A. 2013. *Chemical Communications* 49, 1891–1899.

91. Syed, S. 2012. *Hydrometallurgy* 115–116, 30–51.

92. Hagelüken, C.; Corti, C. W. 2010. *Gold Bulletin* 43, 209–220.

93. Karthikeyan, O. P.; Rajasekar, A.; Balasubramanian, R. 2015. *Critical Reviews in Environmental Science and Technology* 45, 1611–1643.

94. Ou, Z. J.; Li, J. 2016. *Waste Management* 57, 57–63.

95. Jadhav, U.; Hocheng, H. 2015. *Scientific Reports* 5, 14574.

96. Yue, C.; Sun, H.; Liu, W. J.; Guan, B.; Deng, X.; Zhang, X.; Yang, P. 2017. *Angewandte Chemie* 56, 9331–9335.

97. Zanella, R.; Giorgio, S.; Henry, C. R.; Louis, C. 2002. *The Journal of Physical Chemistry B* 106, 7634–7642.

98. Wu, Y.; Fang, Q.; Yi, X.; Liu, G.; Li, R.-W. 2017. *Progress in Natural Science: Materials International* 27 (4), 514–519.

99. Ting, Y. P.; Neoh, K. G.; Kang, E. T.; Tan, K. L. 1994. *Journal of Chemical Technology and Biotechnology* 59, 31–36.

100. Li, M.; Sun, Q.; Liu, C.-j. 2016. *ACS Sustainable Chemistry and Engineering* 4, 3255–3260.

101. He, Y.-R.; Cheng, Y.-Y.; Wang, W.; Yu, H. 2015. *Chemical Engineering Journal* 270, 476–484.

102. a) Liu, Z.; Frasconi, M.; Lei, J.; Brown, Z. J.; Zhu, Z.; Cao, D.; Iehl, J.; Liu, G.; Fahrenbach, A. C.; Botros, Y. Y.; Farha, O. K.; Hupp, J. T.; Mirkin, C. A.; Fraser Stoddart, J. 2013. *Nature Communications* 4, 1855; (b) Liu, Z.; Samanta, A.; Lei, J.; Sun, J.; Wang, Y.; Stoddart, J. F. 2016. *Journal of the American Chemical Society* 138, 11643–11653.

103. Mon, M.; Ferrando-Soria, J.; Grancha, T.; Fortea-Pérez, F. R.; Gascon, J.; Leyva-Pérez, A.; Armentano, D.; Pardo, E. 2016. *Journal of the American Chemical Society* 138, 7864–7867.

104. Grant, R. A.; Drake, V. A. 2002. Presented in part at the International Solvent Extraction Conference 2002.

105. Narita, H.; Tanaka, M.; Morisaku, K.; Abe, T. 2006. *Hydrometallurgy* 81, 153–158.

106. Mowafy, E. A.; Mohamed, D. 2016. *Separation and Purification Technology* 167, 146–153.

107. Preston, J. S.; du Preez, A. C. 1995. *Solvent Extraction and Ion Exchange* 13, 391–413.

108. Doidge, E. D.; Carson, I.; Tasker, P. A.; Ellis, R. J.; Morrison, C. A.; Love, J. B. 2016. *Angewandte Chemie* 55, 12436–12439.

109. Ritcey, G. M. 1996. Solvent extraction processing plants, problems, assessments, solutions. In *Proceedings of the ISEC'96*, Australia.

110. Collet, S.; Chagnes, A.; Courtaud, B.; Thiry, J.; Cote, G. 2009. *Journal of Chemical Technology and Biotechnology* 84, 1331–1337.

111. Chagnes, A.; Fossé, C.; Courtaud, B.; Thiry, J.; Cote, G. 2011. *Hydrometallurgy* 105, 328–333.

112. Collet, S.; Chagnes, A.; Courtaud, B.; Thiry, J.; Cote, G. 2009. *Journal of Chemical Technology and Biotechnology* 84, 1331–1337.

113. Ciavatta, L.; Iuliano, M. 1995. *Annali di chimica* 85, 235–255.

114. Markovic, M.; Pavkovic, N. 1983. *Inorganic Chemistry* 22, 978–982.
115. Miguirditchian, M.; Bernier, G.; Pacary, V.; Balaguer, C.; Sorel, C.; Berlemont, R.; Fries, B.; Bertrand, M.; Cames, B.; Leydier, A.; Turgis, R.; Arrachart, G.; Pellet-Rostaing, S.; Mokhtari, H. 2016. *Solvent Extraction and Ion Exchange* 2016, 34.
116. Mokili, B.; Poitrenaud, C. 1996. *Solvent Extraction and Ion Exchange* 14, 617–634.
117. Mokili, B.; Poitrenaud, C. 1995. *Solvent Extraction and Ion Exchange* 13, 731–754.
118. Brønsted, J. N. 1922. *Journal of the American Chemical Society* 44, 877–898.
119. Brønsted, J. N. 1922. *Journal of the American Chemical Society* 44, 938–948.
120. Sorel, C.; Montuir, M.; Balaguer, C.; Baron, P.; Dinh, B. 2011. The PAREX code: A powerful tool to model and simulate solvent extraction operations. In *Proceedings of the ISEC'11 Conference*, Santiago, Chile.
121. Iung, B.; Levrat, E. 2014. *Procedia CIRP* 22, 15–22.

2 Liquid–Liquid Separation by Supramolecular Systems

Gabriela I. Vargas-Zúñiga, Qing He,
and Jonathan L. Sessler

CONTENTS

2.1 INTRODUCTION

The goal of this chapter is to provide an overview of the most representative cation, anion, and ion-pair receptors that have been used as extractants under liquid–liquid extraction conditions in recent years. The emphasis is thus on one of the most historically important methods used to purify ions.

Ions are particles that exist as positively or negatively charged species—anions and cations. These particles can be monoatomic or polyatomic (e.g., Na^+, F^-, or NH_4^+, $[Pu(NO_3)_6]^{2-}$). Many of these ions play important roles in biological and industrial processes. For example, iodide and bicarbonate are involved in the biosynthesis of hormones and pH regulation, respectively.[1] Other entities of physiological interest, such as polypeptides, proteins, nucleic acids, or metal cations coordinated to enzymes, are complex ions that participate in essential life processes.[2] In metallurgy, metals are usually won from naturally occurring salts *via* what are often complex purification and processing sequences. Oxyanions, such as sulfate and nitrate, are components in acid rain, whereas phosphates and nitrates used in fertilizers can lead to the eutrophication of lakes and other waterways.[3] Actinide, lanthanide, and other metallic salts are important constituents of radioactive tank waste.[4] Zwitterions, such as amino acids, are charged species capable of existing as anions, cations, or net neutral entities depending on the pH of the medium.[2]

In solution, anions and cations are commonly found surrounded by solvent molecules. The underlying interaction between the ion and solvent molecules (i.e., solvation) is dependent on several factors that include the polarity of the solvent, temperature, the nature of the ions in question, and the amount or concentration of the charged species present in the solution. One of the best solvents in which ions establish interactions is water; here, depending on the charge of the ion, the neighboring water molecules can be oriented in two different limiting ways. For cations, the water molecules interact by orienting one or two of its oxygen lone pairs toward the positively charged particle, giving rise to the so-called first hydration shell.[2] The number of solvating water molecules depends on the cation in question. In contrast to cations, the interactions between water and anions involve predominantly hydrogen bonds. However, it is currently believed that there is no definite number of water molecules surrounding each given anion.[2,5]

The interactions formed between ions and water induce polarization of the water molecules surrounding the ion, resulting in the formation of second hydration shell in which the hydrogen-bonding interactions are stronger than the ones occurring in pure water. This second sphere of interaction is especially prevalent in the case of hydrated small cations, such as Na^+, K^+, Mg^+, some transition metals, lanthanides and actinides, multivalent anions such as CO_3^{2-}, SO_4^{2-}, and univalent anions such as F^- or $H_2PO_4^-$ that are capable of binding a large number water molecules either by accepting and/or donating hydrogen bonds.[2,5] Thus, the degree of hydration of an ion is determined by the affinity of the ion for water and, in mixed media, the interactions between water and the non-aqueous solvent.

Due to the wide involvement of ions in industry and biology, a considerable body of effort has been devoted to the problem of ion separation. Important separation techniques include *inter alia* selective crystallization, liquid–solid, and liquid–liquid extraction. Liquid–liquid extraction, in particular, is a time-honored practice traditionally used to separate charged species from aqueous environments. In this chapter, we review recent progress in the development of receptors that are effective as extractants for ions and ion pairs under liquid–liquid extraction conditions. Although complex thermodynamic and kinetic factors are involved in ion extraction, these are lightly touched on in this structure-focused chapter. Interested readers are recommended to consult what is a rather detailed literature involving this aspect of extraction-based separation processes.[2,5–7]

2.2 LIQUID–LIQUID ANION EXTRACTION BY SUPRAMOLECULAR SYSTEMS

This section is focused on anion extraction. The challenges associated with the design of receptors that can facilitate the transfer of anions from an aqueous to an organic environment reflect, in large measure, the intrinsic interactions between anions and water that must be overcome to have effective extraction.[8] The strong hydrogen-bonding interactions that highly charged anions establish with water is reflected in their hydration energies (e.g., $\Delta G_h = -1090$ and -306 KJ mol^{-1} for SO_4^{2-} and NO_3^-, respectively).[5] Not surprisingly, a more hydrophobic anion is expected to be extracted more easily into an organic phase. This bias toward extraction in the case of more hydrophobic anions is termed the Hofmeister bias and reflects a series by that name that roughly parallels the degree of hydration; cf. Figure 2.1.[7,8] Thus, anion-receptor interactions are often hampered by water itself, which can both hydrate the solute in question and function actively as a competitive guest. In addition to the difficulties associated with the strong hydrogen-bonding interactions with water that are typical of highly hydrated anions, many of these anions are susceptible to protonation at lower pH values. This latter feature can result in oftentimes complex protonation equilibria and cause full or partial neutralization of the negative charge on a target anion. This, in turn, can preclude strong binding to receptors that rely predominantly on Coulombic interactions.

Also complicating the underlying receptor–anion interactions is the fact that anions have larger radii than their isoelectronic cations (e.g., $r_{K^+} = 1.38\,\text{Å}$ vs. $r_{Cl^-} = 1.81\,\text{Å}$).[2] Many are characterized by nonspherical geometries, including tetrahedral, octahedral, trigonal planar, and linear. Finally, in the case where a neutral receptor is employed as the extractant, the participation of the counter-cation and the interactions that they may establish with the receptor can influence both the formation of an anion-receptor complex and its ability to undergo effective transfer into an organic phase. Proper design will lead to receptor systems that account for these considerations or, even in ideal cases, exploit them to maximize recognition, selectivity, and extraction efficiency.

2.2.1 LIQUID–LIQUID EXTRACTION BY CHARGED RECEPTORS

The use of receptors bearing positively charged groups for the recognition and extraction of anions is a widely employed strategy. It is attractive because the receptor can function as the counter-ion and allow formation of a neutral anion-receptor complex that is likely to be more inherently soluble in an organic phase. Recognition of oxoanions, such as phosphates, sulfates, and chromates, has benefited from this approach. These anions are of considerable interest due to the impact they can have on the environment. However, these particular solutes are highly hydrated. This makes their extractive removal from aqueous phases inherently challenging. However, as detailed below notable examples of success have been recorded using appropriately designed cationic receptor systems.

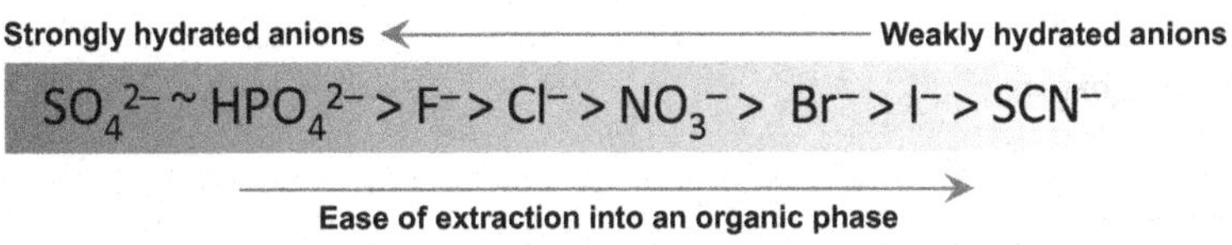

FIGURE 2.1 The Hofmeister series.

In an early report, Schmidtchen and coworkers described the extraction properties of the bis-guanidinium-based receptor **1**. This system was found effective for sulfate anions and biologically relevant molecules, such as ATP, over a wide pH range (Figure 2.2).[9] Studies with receptor **1**, involving an immiscible $CHCl_3$/H_2O (pH 8.7) solvent system with various sodium salts, revealed a higher degree of extraction for SO_4^{2-} (99.8%) than for HPO_4^{2-} (8.6% as its sodium salt).[9] Effective extraction (96.5%) from water into chloroform was also seen for the relatively hydrophobic anion, I^- (as its sodium salt).[9] In addition, receptor **1** allowed 98.3% of the ATP to be extracted.[9] In the case of ATP and sulfate, the extraction features of receptor **1** were found to be superior to those of the lipophilic quaternary ammonium salt methyl(*n*-octyl,*n*-decyl)ammonium chloride (Aliquat 336N™, abbreviated A336N) or *N,N*-bis(2-ethylhexyl)guanidine (LIX 79), both of which are classic phase-transfer catalysts.[10] The higher affinity and better extractability observed for sulfate over hydrogen phosphate was attributed to more effective $N-H^+$ hydrogen-bonding interactions between the target anion and the guanidinium groups of receptor **1** (four vs. two hydrogen bonds for SO_4^{2-} and HPO_4^{2-}, respectively).

Phase-transfer catalysts are often used in liquid–liquid extraction as a way to overcome the Hofmeister bias.[11,12] However, in the case of highly hydrated anions, such as sulfate and phosphate, their use is generally accompanied by the formation of a third phase. The term third phase in this context refers to a situation in which the amount of water in the organic layer at the critical point of phase splitting correlates with the hydration energies of the targeted anions. This can lead to difficult-to-manipulate emulsions. Nevertheless, the phase-transfer catalyst approach was used successfully in 2007 by Sessler, Moyer, and coworkers to effect the liquid–liquid extraction of the sulfate anion using calix[8]pyrrole **2** (Figure 2.3).[13,14] In the solid state, the protonated form of this receptor (2•2H)$^{2+}$ was found to form a 1:1 complex with a sulfate anion located within the center of this relatively large macrocycle and stabilized by hydrogen-bonding interactions involving the eight N–H pyrroles making up this particular expanded porphyrin system (Figure 2.3).[14]

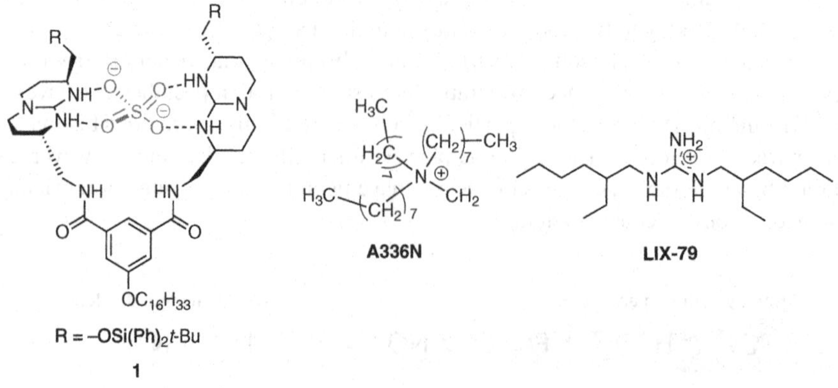

R = −OSi(Ph)$_2$*t*-Bu

1

A336N

LIX-79

FIGURE 2.2 Schematic representation of the 1:1 complex of sulfate with receptor **1** and structures of the monofuctional coextractants Aliquat 336N™ and LIX®79.

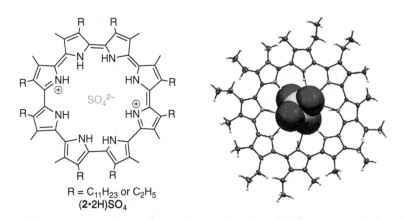

R = $C_{11}H_{23}$ or C_2H_5
(2·2H)SO₄

FIGURE 2.3 Cyclo[8]pyrrole sulfate complex, [(2·2H)SO₄] and the single-crystal X-ray structure of the sulfate complex of the parent form of **2**, which bears ethyl and methyl substituents on the β-pyrrolic positions. (Data from the Cambridge Crystallographic Data Centre [CCDC # 176189].[14])

An embodiment of receptor **2** bearing undecyl chains on eight of the β-pyrrolic positions was found to be more soluble in organic solvents than its less highly functionalized analogs. It was thus judged more suitable for use in liquid–liquid extraction applications. Initial extraction experiments revealed that the use of only deprotonated receptor $(2·2H)^{2+}$ as the extractant led to the transfer of SO_4^{2-} from the aqueous phase into toluene in the presence of an excess of the more hydrophobic and normally more extractable NO_3^- anion. However, the extraction was characterized by slow-exchange kinetics, associated with the formation of the 1:1 complex [(2·2H)SO₄]. This problem was attributed to poor interfacial activity arising from the hydrophobic nature of the macrocycle.[13] The use of Aliquat 336N™ as a phase-transfer catalyst resulted in enhanced exchange kinetics and gave a $K_{exch} = 1.1 \pm 0.4$.[13] After extraction, the sulfate anion could be stripped from the extractant by treating the complex [(2·2H)SO₄] with 0.1 mM trioctylamine in aqueous HNO_3 (0.1 mM), followed by consecutive washes with aqueous NaOH.[13] The extractant was recovered by contacting with a HNO_3 solution.

In 2017, Moyer and coworkers reported a simple guanidinium system that exhibits selective sulfate extraction properties.[15] The receptor in question, N,N'-bis(4-dodecyl-2-pyridyl)guanidinium chloride **3**, was found to extract sulfate (as $Na^{35}SO_4^{2-}$; used as radiotracer) from 10 mM aqueous NaCl solutions into a 1,2-dichloroethane organic phase via the formation of the complex [3·SO₄] as shown in Figure 2.4. Based on linear regression parameters obtained from the experimental data, a separation factor of $SF_{SO4/Cl} = 32$ [3·Cl] was calculated. The high value obtained reflects the relative ability of **3** to separate sulfate from chloride under conditions where the latter anion is present in a 100-fold excess.[15] The separation factor was found to be concentration-dependent due to the disparity of the charges involved in the ion exchange process (2– for sulfate vs. 1– for chloride). Compared with a control exchange system consisting of Aliquat 336™ (A336, chloride salt) tested under similar experimental conditions,

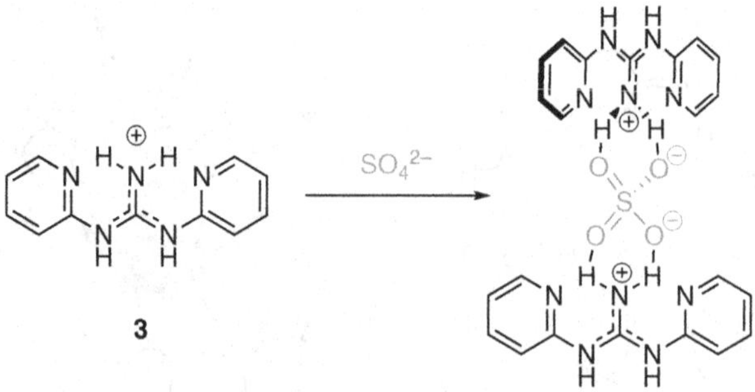

FIGURE 2.4   Schematic representation of the complex [3•SO₄].

the sulfate distribution in the organic phase using receptor **3** was superior by a factor of 55 ± 5 (allowing conditional K_{exch} values of 7.40 ± 0.04 and 5.66 ± 0.07 to be calculated for **3** and A336, respectively).[15] This superior selectivity was ascribed to the strong hydrogen-bonding donating ability of receptor **3** and its ability to stabilize strong Coulombic interactions with the charge-dense sulfate anion. Support for this contention came from a single-crystal X-ray diffraction analysis of the less lipophilic *N,N'*-bis(2-pyridyl)guanidinium receptor **3**, for which a 2:1 (receptor:anion) complex was observed.[15] Binding constants associated with the formation of the sulfate complex in the presence of less-well hydrated anions, such as NO_3^- and Cl^-, further revealed that the guanidinium-based receptor **3** exhibits selectivity for the sulfate anion (i.e., $SO_4^{2-} \gg NO_3^- > Cl^-$).

Metallic oxoanions, such as arsenate and dichromate, are highly toxic and carcinogenic and represent an environmental concern due to their high solubility in water. Arsenate (e.g., $H_2AsO_4^-/H_2AsO_4^{2-}$) is largely produced through anthropogenic activities (e.g., from pesticides, fossil combustion, or mining) and is readily accumulated in soils and groundwater.[16] Several methods to treat arsenate-containing wastewater are known, including precipitation, co-precipitation, ion exchange, ultrafiltration, reverse osmosis, and solvent extraction; in many instances, these methods are effective.[17,18] Dichromate salts (e.g., $K_2Cr_2O_7$) are commonly used as additives in cement, photographic screen printing, and for the treatment of wood.[19] Dissolved in water, dichromate is highly corrosive due to its powerful oxidizing properties. Dichromate is also harmful to living organisms. It easily diffuses through cell membranes producing hydroxyl radicals and other reactive oxygen species (ROS) that react adversely with DNA and proteins.[20,21] This chemistry is thought to underlie the carcinogenicity of Cr(VI) species.

In 2013, Bayrakci and coworkers reported the tetra-substituted calix[4]arene receptors **4** and **5** and their use in promoting the liquid–liquid extraction of the arsenate anion.[22] Receptors **4** and **5** bear pyridyl and furyl groups, respectively, and were found to be susceptible to protonation at low pH values. Extraction experiments at several pH values (3.5, 4.5, 5.5, and 7.0) using aqueous solutions containing 10^{-5} M NaHAsO₄⁻ in contact with 10^{-3} M solutions of 4 in 1,2-dichloroethane, revealed that the maximum extraction efficiency (81.6%; log $K_{exch} = 2.95 \pm 0.1$ at 25°C) was achieved at pH 3.5.[22] The ability to promote the extraction seen in the case of **4** was attributed to the presence of protonated pyridyl amine groups (cationic pyridinium sites) that can form complexes with the arsenate anions via a combination of electrostatic and hydrogen-bonding interactions involving the pyridine moieties and the oxygen atoms of the arsenate anion.

At pH 3.5, $H_2AsO_4^-$ (formally As(V)) is essentially the only arsenate species present in the aqueous medium. The use of this pH value is desirable in an operational sense since $H_2AsO_4^-$ has a lower hydration energy than the corresponding dianionic form, $H_2AsO_4^{2-}$, a species more important at higher pH. The ability to target $H_2AsO_4^-$ over $H_2AsO_4^{2-}$ thus facilitates the extraction process. Studies involving the extraction of arsenate (at 10^{-5} M) in the presence of other anions, including Cl^-, NO_3^-, and SO_4^{2-} (at 10^{-3} M), revealed that these anions did not interfere appreciably with the extraction of $H_2AsO_4^-$ into a 1,2-dichloroethane phase at pH 3.5. In fact, 78.8% of the $H_2AsO_4^-$ proved to be extracted by receptor **4** under these conditions.[22]

In the case of receptor **5**, a system bearing furyl groups, the maximum extraction percentage observed was significantly less (17.7%) than for receptor **4**.[22] This decrease in efficiency was attributed to the relatively more electron-donating nature of the furyl groups. Interference from other interactions was also invoked. In fact, extraction was found to be dependent on the binding of Na^+ ions within the calixarene cavity. Thus, it was suggested that the extraction process occurs via ion-pair recognition, rather than through pure anion complexation (Figure 2.5). To test this hypothesis, extraction experiments were carried out under similar conditions using the corresponding potassium arsenate salt. In this case, the maximum extraction percentage was found to drop to ~3–4%.[22] This was taken as a clear demonstration that receptor **5** functions as a ditopic extractant and that the extraction of arsenate is governed by the formation of an ion-pair complex with the receptor.

In a recent report, Yilmaz and coworkers reported the terpyridine-conjugated calix[4]arene receptor **6** (Figure 2.6).[23] Extraction of dichromate (from a 0.1 mM $Na_2Cr_2O_7$ aqueous solution) by receptor **6** (1 mM in dichloromethane) at different pH

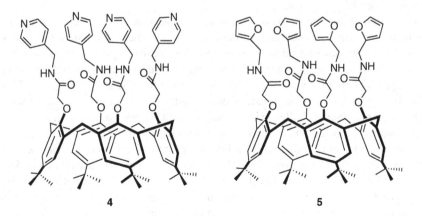

4   5

FIGURE 2.5 Structures of receptors **4** and **5**.

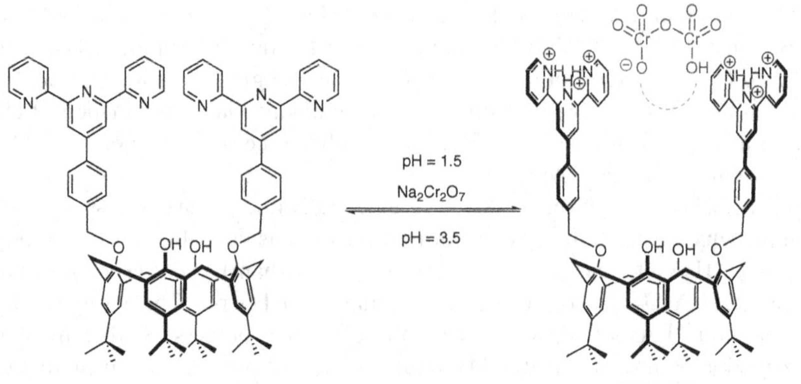

FIGURE 2.6 Schematic views of the terpyridine-conjugated calix[4]arene receptor **6** and its dichromate complex.

values (pH 1.5–4.5), was achieved by contacting the dichromate aqueous solution with the organic phase containing **6** and determining the concentration of dichromate extracted *via* the absorbance difference seen in the aqueous source phase.[23] The results revealed that the extraction efficiency is affected by the pH. For instance, at pH > 2.5, little extraction of dichromate was observed, but at a pH of 1.5 a maximum dichromate extraction efficiency of ~65% (as $HCr_2O_7^-$) was observed (log $K_{exch} = 6.36 \pm 0.5$ M^{-1}).[23] The extraction selectivity of **6** for the dichromate anion was analyzed by contacting aqueous solutions containing $HCr_2O_7^-$ and other potentially competitive anions, such as Cl$^-$, SO$_4^-$, and NO$_3^-$, with dichloromethane solutions containing this receptor. Extraction studies carried out at pH 1.5 revealed that the ability of receptor **6** to act as an extractant for the dichromate anion not affected by the presence of other anions.[23] This combination of selectivity and effectiveness was attributed to the strong electrostatic and hydrogen-bonding interactions provided by the protonated terpyridinium groups that combine to favor formation of a 1:1 complex with the $HCr_2O_7^-$ anion.

Extraction of fluoride from aqueous solutions has attracted researchers' attention due to its recognized physiological affects, particularly when administered in excess. For instance, fluoride can act as Ser/Thr phosphate inhibitor giving rise to an interruption in cell signaling processes.[24] The design of fluoride extractants faces some challenges. The first is the high hydrophilicity of the fluoride anion ($\Delta G_h = -465 \pm 6$ kJ mol^{-1}),[25] which makes the transfer from an aqueous to an organic phase difficult. The second recognized challenge is the basicity of the fluoride anion, which can lead to protonation of the anion and a reduced affinity for a putative anion receptor.

Gabbaï and Chiu studied the ability of receptor **7** to bind fluoride anions (Figure 2.7).[26] NMR spectroscopic (^{1}H, ^{11}B, and ^{19}F) analyses carried out in deuterated chloroform revealed that the fluoride anion interacts with both the methylene group bearing the borane subunit and the trimethylammonium moieties of receptor **7**. Interestingly, no evidence of complex formation with other halide anions was observed. Computational studies provided support for the contention that the

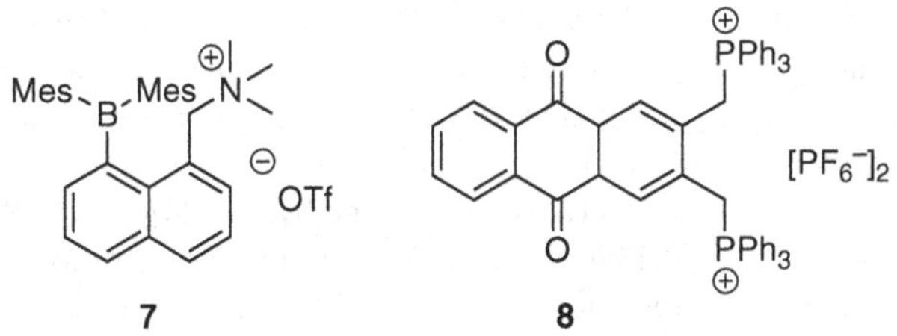

FIGURE 2.7 Cationic boronate-based receptor **7** reported by Chiu and Gabbaï,[26] and structure of the phosphonium salt-based receptor **8** reported by Das and coworkers.[27]

observed selectivity has its origins in a favorable combination of moderate hydrogen-bonding interactions with the acidic C–H borane methylene groups and electrostatic interactions involving the trimethylamine group. The fluoride extraction ability of **7** was tested using a two-solution setup consisting of $CDCl_3$ and D_2O layers with tetrabutylammonium fluoride (TBAF) being used as the fluoride anion source. Under these conditions, a yield of 82% of the complex [**7**•F] was seen in the organic phase.[26]

In 2011, Ganguly, Das, and coworkers reported that the phosphonium salt-based receptor **8** shown in Figure 2.7 displays an affinity for fluoride anions in acetonitrile.[27] This receptor was found to produce a colorimetric change upon exposure to fluoride anions that was not observed in the presence of other anions, *viz.* Cl^-, Br^-, I^-, HSO_4^-, NO_2^-, NO_3^-, AcO^-, ClO_4^-, IO_4^-, and $H_2PO_4^-$. ^1H and ^{31}P NMR spectroscopic studies revealed that the two acidic methylene ($-CH_2$) hydrogen atoms attached to the anthraquinone core of receptor **8** interact with fluoride anions giving rise to an overall 1:2 receptor:anion stoichiometry (i.e., formation of complex [**8**•2F$^-$]).[27] The fluoride extraction properties of receptor **8** were tested using water samples containing NaF that were taken from the Arabian Sea and the Sambhar lake in India. The results of these studies provided support for the conclusion that receptor **8** is able to extract F$^-$ from aqueous environments into an organic phase consisting of CH_2Cl_2 or $CHCl_3$ and do so with an efficiency of 99.3%, even at fluoride concentrations as low as 0.06 ppm.[27]

2.2.2 Liquid–Liquid Extraction by Neutral Receptors

Anion transfer from an aqueous environment to an organic solvent or solvent mixture is a process that benefits from charge neutrality. When electroneutral receptors are used, the design benefits from having receptors that possess appropriately oriented hydrogen-bond-donating groups able to interact with the target anion. In favorable cases, encapsulation of the negatively charged species allows it to overcome hydration effects, resulting in the extraction into the organic phase. In addition, the effect of the counter-cation can be significant, and the requirement of net charge neutrality means these ions are not necessarily spectators but can play an important role in regulating the extraction process. In operational terms, it is thus often useful to view anion extraction by a neutral receptor in terms of ion-pair recognition.

Calix[4]pyrroles and their derivatives have been used by our group to develop neutral extractants. These macrocycles consist of four pyrrole subunits linked by four fully substituted sp³ hybridized *meso*-carbon atoms. They are readily accessible via the acid-catalyzed condensation of pyrroles with ketones. The synthesis of the parent member of this class macrocycles, octamethylcalix[4]pyrrole **9**, is credited to Baeyer and dates back to 1886.[28] However, the anion-recognition properties of calix[4]pyrroles were only recognized by Sessler and coworkers in 1996.[29] A latter (2005) report, describing the binding of an ion pair of halide salts (CsX where X=F, Cl, Br), provided an inspiration to synthesize dedicated ion-pair receptors and extractants based on the calix[4]pyrrole framework.[30] As detailed below, some of these functionalized calix[4]pyrroles have been attached to polymeric backbones; others have been tested directly as anion extractants, and several have been further modified with the view to preparing selective ion-pair receptors.

In 2008, Akar, Bielawski, Sessler, and coworkers described the synthesis of the organic solvent soluble methyl methacrylate calix[4]pyrrole-containing polymer **10** (Figure 2.8).[31] Thermogravimetric and ¹H NMR spectral studies revealed that compound **10** was able to extract chloride and fluoride anions (as their tetrabutylammonium (TBA⁺) salts) into a CD_2Cl_2 phase. In this case, the extraction behavior of **10** followed the Hofmeister series in that less-well hydrated chloride anions were extracted into the deuterated dichloromethane phase more effectively than the fluoride anion.[31] Moreover, the incorporation of calix[4]pyrrole to a polymer backbone also resulted in an enhanced extraction of halide anions relative to what was seen using octamethylcalix[4]pyrrole alone.

In a latter report, Sessler, Aydogan, Lee, and coworkers synthesized gold nanoparticles decorated with a double-armed calix[4]pyrrole **11** and found that the resulting constructs were able to extract fluoride in preference over chloride (as their respective TBA⁺ salts) from D_2O into a CD_2Cl_2 phase.[32] In both cases, the use of a more hydrophobic counter-cation, such as TBA⁺ had a big influence on the extraction efficiency.

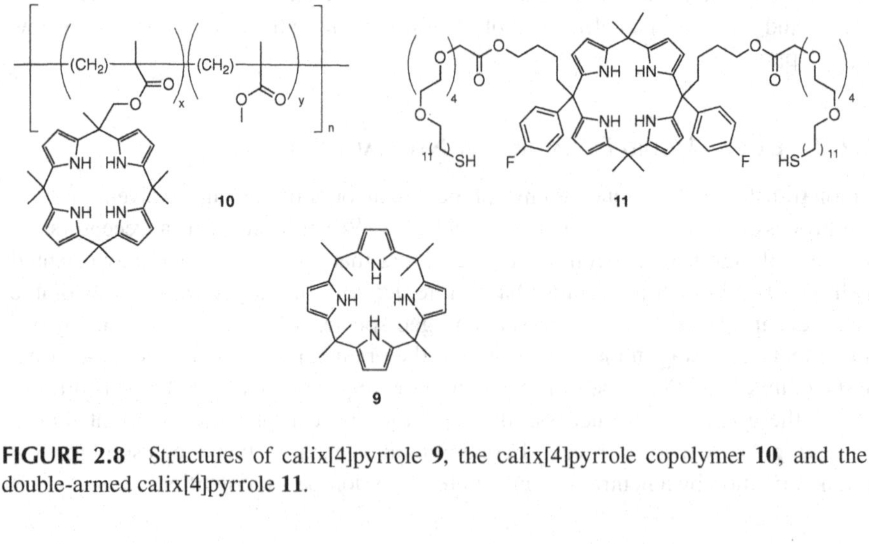

FIGURE 2.8 Structures of calix[4]pyrrole **9**, the calix[4]pyrrole copolymer **10**, and the double-armed calix[4]pyrrole **11**.

In 2010, Fowler, Moyer, Sessler, and coworkers reported that octamethylcalix[4] pyrrole **9** and its octafluorinated derivative could be used to facilitate the transfer of SO_4^{2-} into chloroform from an aqueous source phase.[33] The extraction was performed over a range of concentrations (0.1–10 mM) and involved contacting **9** dissolved in chloroform with a 0.1 mM aqueous Na_2SO_4 solution containing an excess of $NaNO_3$ (10 mM) in the presence of the anion exchange agent A336N. Under these conditions, receptor **9** acted to extract the sulfate anion in preference over the nitrate anion; a finding rationalized in terms of the mechanism shown in Figure 2.9.[33]

In order to assess the role of the alkyl ammonium exchanger, several symmetrical long-chain quaternary ammonium salts were tested. Under otherwise identical conditions, it was found that sulfate ion exchange into the organic phase is "turned off" when bulkier ammonium exchangers are used. In contrast, when A336N was utilized, a higher SO_4^{2-} anion extraction selectivity relative to NO_3^- was observed. This preference for methyl-containing alkylammonium counter-cations was explained in terms of the calixpyrrole complex (**9•**SO_4) being able to bind the A336 asymmetrical ammonium cation within the cup-like portion of the calix[4]pyrrole,[33] thereby forming an ion-pair complex.

The same research team later reported that the extraction of the sulfate anion from an aqueous source phase could be improved by using the strapped calix[4] pyrroles **12** and **13** (Figure 2.10).[34] It was found that extraction of SO_4^{2-} (aqueous, 0.1 mM) in the presence of excess NaCl (at 10 mM) was particularly effective when receptor **12** was used in conjunction with A336 (as the chloride salt). Improvements relative to receptor **9** and even the strapped receptor **13** were seen.[34] This enhanced capability, which is reflected by the higher distribution ratio observed for receptor **12** ($K_{D\ SO4} = [SO_4^{2-}]_{org}/[SO_4^{2-}]_{aq} = 0.013$ for receptor 12 vs. 0.00061 and 0.00055 for receptors **13** and **9**, respectively), was ascribed to the ability of **12** to encapsulate most effectively the sulfate anion within its cavity.[34]

Theoretical calculations have led to the suggestion that the sulfate anion will be complexed optimally within a tetrahedral cavity that provides 12 hydrogen-bond donors.[35,36] The tripodal hexaurea receptor **14**, shown in Figure 2.11, reported by Wu, Li, and coworkers was found to encapsulate the sulfate anion in such a near-idealized tetrahedral geometry via hydrogen-bonding interactions with the urea groups.[37] In solution, ^{1}H NMR spectroscopic studies performed in DMSO-d_6/0.5% D_2O revealed that the stoichiometries of the sulfate complexes stabilized by **14** were

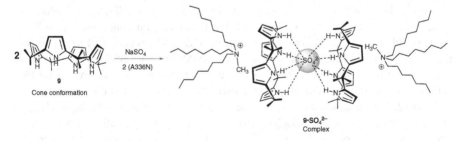

FIGURE 2.9 Proposed mechanism of sulfate complexation by calix[4]pyrrole **9** in synergistic combination with lipophilic quaternary ammonium nitrate, A336N.

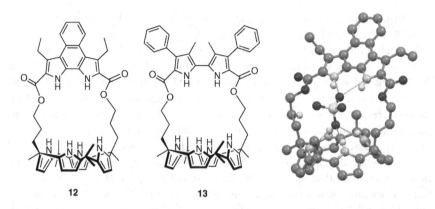

FIGURE 2.10 Strapped calix[4]pyrroles **12** and **13**, and the single-crystal structure of the sulfate complex of **12**. The solvent molecules (methanol, water, and chloromethane) and the tetramethylammonium counter-cations have been removed for clarity. (Data from the Cambridge Crystallographic Data Centre [CCDC # 1017884].[34])

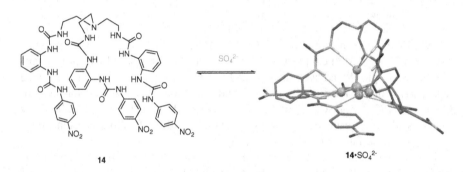

FIGURE 2.11 Tripodal hexaurea-based receptor **14** and its sulfate complex [**14**•SO$_4$]. The solvent molecules (dimethyl sulfoxide) and the TBA$^+$ counterions have been removed for clarity. (Data from the Cambridge Crystallographic Data Centre [CCDC # 784518].[37])

concentration-dependent. It was observed that as the relative concentration of the sulfate anion increased, a 1:1 complex is first formed, followed by formation of a 1:2 (SO$_4^{2-}$: **14**) complex. Extraction experiments using aqueous (TBA)$_2$SO$_4$ solutions containing an excess of NaNO$_3$ revealed that under certain conditions the sulfate anion could be extracted into the organic phase (CDCl$_3$) by receptor **14** nearly quantitatively.[37] Once extracted, the sulfate anion could be removed from the receptor via a back-extraction process involving the use of aqueous BaCl$_2$ to precipitate the sulfate anion as BaSO$_4$. This allowed the extractant to be recycled for reuse.

In 2012, Ghosh and coworkers reported the anion recognition and liquid–liquid extraction properties of the tris(2-aminoethyl)amine-based pentafluorophenyl substituted urea receptor **15** (Figure 2.12a).[38] ^{1}H NMR spectroscopic and isothermal calorimetry titration binding studies revealed that the nature of the interactions with

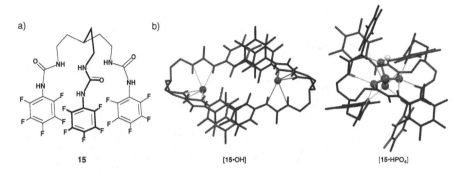

a)   b)

15   [15•OH]   [15•HPO₄]

FIGURE 2.12 (a) Structure of receptor **15** and (b) single crystals of the pseudocapsule assemblies 15•OH and 15•HPO₄. The TBA⁺ counterions are removed for clarity. (Data from the Cambridge Crystallographic Data Centre [CCDC # 861095 and 861094].[38])

negatively charged species depend on the charge of the targeted anion. In the presence of mononegative anions, such as F⁻, OH⁻, and $H_2PO_4^-$, receptor **15** forms 1:1 complexes in the form of pseudo-capsular assemblies. In contrast, exposure of **15** to dianionic species, such as CO_3^{2-}, SO_4^{2-}, and HPO_4^{2-}, results in the formation of 2:1 ([**15**:X²⁻]) complexes where the oxoanion is encapsulated by the receptor. These stoichiometries were found to persist in the solid state as inferred from single-crystal X-ray diffraction analyses of the complexes [15•OH] and [15•HPO₄] (Figure 2.12b).[38] Interestingly, the encapsulation of OH⁻ and CO_3^{2-} was not the result of an active addition of hydroxide and carbonate salts, respectively. Instead, complex [15•OH] was obtained by the treatment of **15** with TBACN under aerobic conditions, whereas the complex [15•CO₃] was isolated upon treating **15** with TBAOH without protection from the laboratory atmosphere. The authors proposed that [15•OH] comes about from the presence of hydroxide produced *in-situ* via cyanide anion-mediated proton abstraction involving first one of the acidic urea N–H protons then water from the atmosphere. Carbon dioxide trapping presumably explains formation of the carbonate anion complex, [15•CO₃].

The high solubility of complex [15•CO₃] in chloroform allowed the authors to study the sulfate solvent extraction properties of **15** via anion exchange. Exchange of carbonate for sulfate was found to be favored due to the combination of a higher association constant and reduced hydrophilicity in the case of sulfate relative to carbonate (i.e., $K_a = 4.68 \times 10^4$ M⁻¹ and $\Delta G_h = -1080$ kJ mol⁻¹ [5] vs. $K_a = 1.17 \times 10^4$ M⁻¹ and $\Delta G_h = -1315$ kJ mol⁻¹,[5] for SO_4^{2-} and CO_3^{2-}, respectively; K_a calculated from ITC analyses).[38] In this case, complex [15•CO₃] (0.1 mM in chloroform) was contacted with an aqueous solution of SO_4^{2-} (as its potassium salt at 0.1 mM). The dark pink color observed after addition of phenolphthalein was used to signal the presence of CO_3^{2-} in the aqueous solution as the result of anion exchange. Moreover, the isolation of the organic phase and evaporation yielded a colorless solid with 99% conversion, which was characterized as the sulfate anion complex by ¹H and ¹³C NMR spectroscopies, as well as powder and single-crystal X-ray diffraction analyses.

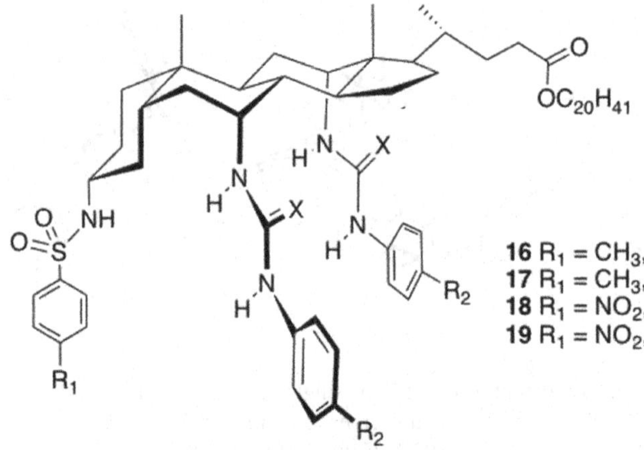

16 $R_1 = CH_3$, $R_2 = H$, X = O
17 $R_1 = CH_3$, $R_2 = NO_2$, X = O
18 $R_1 = NO_2$, $R_2 = NO_2$, X = O
19 $R_1 = NO_2$, $R_2 = NO_2$, X = S

FIGURE 2.13 Neutral cholapods **16–19** reported by Davis and coworkers.[39]

In in a separate experiment, $BaCl_2$ was added into the isolated chloroform solution to precipitate the sulfate extracted by receptor **15**. The formation of a precipitate corresponding to $BaSO_4$ was taken as evidence for the proposed exchange of sulfate for carbonate that allows the latter ion to be extracted into the chloroform phase. This conversion yield was supported by gravimetric studies of the barium sulfate salt, which was obtained in 96.7% yield by adding barium chloride to the contacted aqueous solution. This number is in agreement to what was calculated from the NMR spectroscopic data.[38]

Davis and coworkers reported the neutral cholapods **16–19** that contain two urea or thiourea moieties and other subunits appended to a lipophilic steroid skeleton and which as a consequence provide up to five hydrogen-bond-donor groups (Figure 2.13).[39] These receptors were studied as extractants for a range of monovalent anions that included Cl^-, Br^-, I^-, NO_3^-, AcO^-, ClO_4^-, and $EtSO_3^-$ (as their TBA^+ salts) using chloroform and water as the two phases. The ^1H NMR spectroscopy data of the organic phase (using 1,1,2,2-tetrachloroethane as an internal standard), obtained after contacting receptors **18** and **19** dissolved in deuterated chloroform with an aqueous solution of the anion in question, led the authors to suggest that both Cl^- and AcO^- ($K_a \approx 10^{11}\ M^{-1}$) were extracted into the organic phase more effectively than the other monoanions studied ($K_a \approx 10^6$–$10^9\ M^{-1}$).[39] The higher affinity observed for Cl^- and AcO^- was ascribed to the ability of **18** and **19** to form strong hydrogen bonds with these particular anions. The selectivity values calculated for receptors **16**, **18**, and **19** from competitive ^1H NMR spectroscopic titrations (i.e., Cl^- vs. $EtSO_3^-$; as their TBA^+ salts) using water-saturated $CDCl_3$ revealed separation factors ($SF_{Cl/SO3} = K_{aCl}/K_{aEtSO3}$) of 1.4, 3.1, and 16 for **16**, **18**, and **19**, respectively.

2.3 LIQUID–LIQUID CATION EXTRACTION BY SUPRAMOLECULAR SYSTEMS

Metal isolation and purification by liquid–liquid extraction is widely used in hydrometallurgy. It plays a role in the recycling of rare earth elements from strong magnets, phosphors, and electronics, and in the treatment and management of radioactive waste. In a very general sense, the recycling of these metals relies on contacting a highly acidic or basic aqueous solution of what are generally multivalent salts with a water-immiscible organic phase that contains an extractant. In some cases, the formal extractant is the organic solvent itself. In this section, our emphasis will be on receptor-based extractants of use (or potential use) in targeting radioactive cations.

As the result of the Cold War and attendant legacy radioactive waste, considerable focus has been placed on the reprocessing of low burn-up uranium and weapons-grade plutonium fission byproducts present in these so-called tank wastes. On the other hand, the reprocessing of spent nuclear fuel (SNF), generated from civil high burn-up fuel, has also received attention, albeit mostly in Europe.[40–43] In one day, an operating 1 GWe nuclear power plant consumes ca. 3.2 kg of actinides by fission reactions.[44] This results in the generation of almost the same amount of fission products (FPs) and about 1.2 kg of transuranic elements (TRUs) of which the major constituent is plutonium, along with several minor actinides (MA; Np, Am, and Cm).[44] The MA represent around 0.1% of the total heavy metal mass in the SNF. However, these particular cations usually dominate in terms of both the radiotoxicity and the heat load of the waste. Their presence can thus impact negatively on the environment and lead to reduced repository capacities. In addition, these cations are often accompanied by a mixture of lanthanides (Ln), transition metals, and other FPs. Besides U and Pu, other FPs, such as Ru, Rh, Pd, ^{99}Tc, ^{241}Am, ^{243}Am, ^{244}Cm, and lanthanides (Ln), are attractive in terms of reprocessing for further use. Methods such as the PUREX (plutonium–uranium redox extraction), UREX (uranium extraction), TRUEX (transuranic extraction), DIAMEX (diamide extraction), and UNEX (universal extraction) processes, among others, have been successfully utilized for decades to affect the industrial-scale separation of certain actinides and radioactive elements, including uranium, plutonium, strontium, and cesium, and to some extent, americium and curium.[43] Most of these processes rely on a manipulation of the metal cation oxidation state followed by liquid–liquid extraction mediated by extractants, such as tri-*n*-butyl phosphate (TBP), octyl(phenyl)-*N,N*-dibutyl carbamoylmethyl phosphine oxide (CMPO), bis(2-ethylhexyl)phosphoric acid (HDEHP), diethylene-triaminepentaacetic acid (DTPA), and ethylenediamineteraacetic acid (EDTA). The reader interested in learning more about solvent extraction and other separation methods commonly used for radioactive waste reprocessing is referred to Volume 19 of this series.

2.3.1 Liquid–Liquid Extraction of Minor Actinides, Lanthanides, and Other Fission Products

Despite the advances in SNF reprocessing achieved over the last 50 years, researchers still continue to search for improved separations. Supramolecular receptors are attractive as possible extractants in the context of this over-arching goal. Traditionally, the focus has been on the synthesis of novel receptors that can be useful in separating FP, such as Sr, Cs, U, Th, and Pu. However, in recent years, efforts have been devoted to the preparation of receptors that can aid in the isolation of highly radiotoxic MA (Np, Am, and Cm) from lanthanides (Ln). This is an important goal since lanthanides can function as neutron poisons on subsequent recycle of the actinides in nuclear reactors. Moreover, removal of MA from SNF followed by transmutation into shorter-lived radionuclides would reduce the potential hazard from MAs and could bring the lifetime of the resulting waste down from 10,000 years to ca. 300–500 years.[45]

Unfortunately, the isolation of MA from their chemically similar Ln congeners is not a trivial task. This is animating efforts to generate new extractants that could be used in liquid–liquid extraction-based scenarios. Some of the extractants produced to date in this context have relied on substituted calixarenes.[46] Most of the systems in question are mono- and di-crown, and even thio-substituted functionalized calixarenes. Several have been used successfully in the isolation of ^{137}Cs and ^{90}Sr.[46] Phosphine oxide calixarenes derivatives were reported to extract An and Ln nitrate salts from aqueous acidic phases into organic phases with quantitative extraction efficiencies.[46] Herein, we present examples of supramolecular systems other than these classic systems that have been reported for the extraction of MA and Ln. The rationale for their use is also outlined briefly.

Strontium-90 is the product of radioactive fission and is commonly present in SNFs. It is of concern due to its high energy of β-decay and long half-life (28.8 years), which can lead to severe and undesirable physiological effects. In 2012, Shimojo and coworkers reported the synthesis of receptor **20** consisting of a diaza-18-crown-6 bearing two 4-acyl-5-pyrazolone-type β-diketones that exhibited an ability to extract radioactive strontium cations (^{90}Sr^{2+}) from aqueous solutions into imidazolium bis-imide **21** (an ionic liquid, IL) or chloroform phases (Figure 2.14).[47] To test receptor **20** as an extractant, it was dissolved in **21** or in chloroform to produce 1 mM solutions that were then contacted with an aqueous solution of Sr(NO$_3$)$_2$ (0.01 mM; buffered with N-(2-hydroxyethyl)-piperazine-N'-2-ethanesulfonic acid, HEPES).[47] The extracted Sr^{2+} was recovered via back-extraction with 0.01 M aqueous HNO$_3$. When IL **21** was used as the non-aqueous phase, receptor **20** proved more effective for the extraction of Sr^{2+} than a mixture of the components that make it up, namely a specific crown ether referred to as DBzDA18C6 and HPMBP (see Figure 2.14 for structures). In contrast, receptor **20** provided little benefit compared to this control system when chloroform was used as the non-aqueous phase. In both the IL **21** and in chloroform, the amount of Sr^{2+} extracted by receptor **20** plateaued at around 80%.[47] The superior extraction performance by receptor **20** relative to its constituents seen in the case of the IL receiving phase (i.e., **21**) was attributed to an intramolecular cooperative effect that involves an IL-induced pre-organization of the binding sites in **20**. In chloroform, no such pre-organization is believed to occur and, as a consequence, little augmentations in the strontium transfer is observed.

The need to handle large amounts of high-level nuclear waste continues to drive searches for new extractants that can separate actinides, MA, and Ln. In 2013, Feng, Yuan, and coworkers reported a series of phosphine oxide- and diglycolamide-substituted pillarenes that demonstrated good selectivities and extraction efficiencies toward certain actinides and lanthanides, with the specifics depending on the groups attached to the macrocycle. The phosphine oxide-based pillar[5]ene **22**, shown in Figure 2.15, was found to extract uranyl and thorium cations into dichloromethane

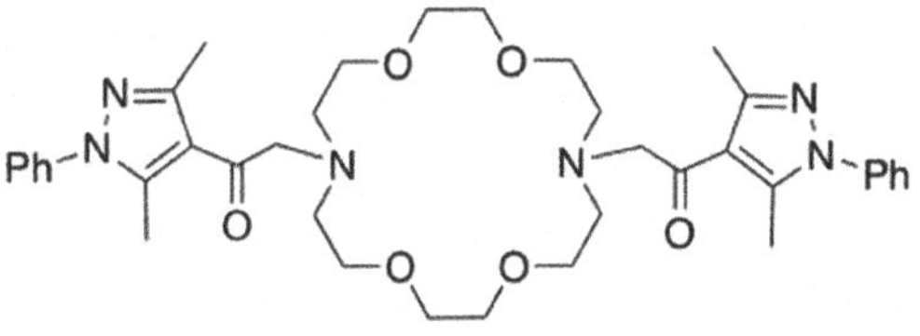

20

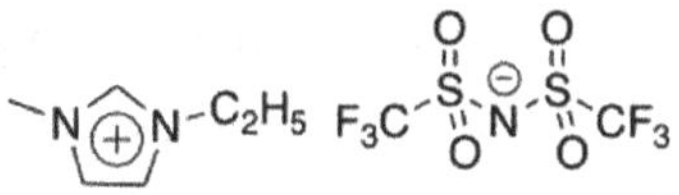

21

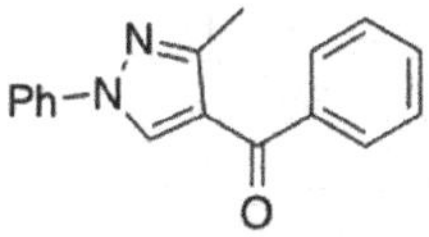

HPMBP

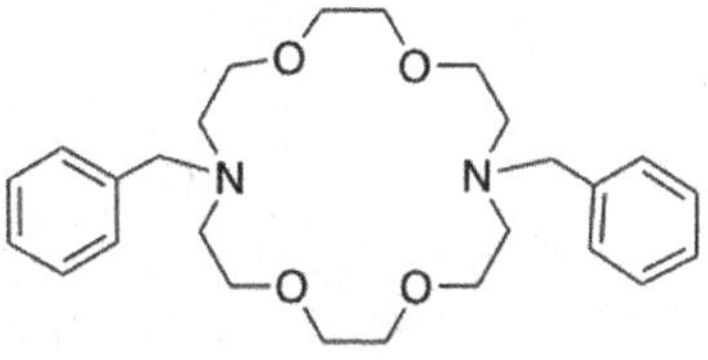

DBzDA18C6

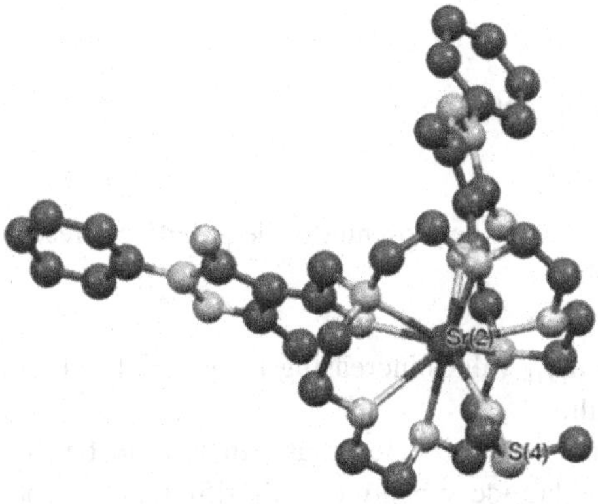

[(20·Sr)(DMSO)]

FIGURE 2.14 Chemical structures of receptor **20**, the ionic liquid (IL) **21**, the components HMPBP and CBzDA18C6 that make up receptor **20**, and the crystal structure of the complex [(20·Sr)(DMSO)]. (Data from the Cambridge Crystallographic Data Centre [CCDC # 917773].[47])

from 1.0 M nitric acid aqueous solutions and do so with selectivity over several lanthanide cations (e.g., La^{3+}, Ce^{3+}, Pr^{3+}, Nd^{3+}, Sm^{3+}, Eu^{3+}, Gd^{3+}, Yb^{3+}, and Lu^{3+}).[48] Relatively good extraction percentages (46% and 92% for Th^{4+} and UO_2^{2+}, respectively) were seen when **22** was used at 1.0×10^{-3} M. In contrast, under similar conditions the extraction percentages for the lanthanide cations varied from 10% to 18%. With trioctylphosphine oxide (TOPO), studied at a 10 times greater concentration than **22** (i.e., 1.0×10^{-2} M), extraction percentages of only 12% and 10% were seen for Th^{4+} and UO_2^{2+}, respectively.[48] The pre-organization of the phosphine oxide groups in **22** provided by the macrocycle was thought to underlie the extraction selectivity for the Th^{4+} and UO_2^{2+} cations. These authors also reported that the extraction of Th^{4+} and UO_2^{2+} was enhanced in the presence of $NaNO_3$ in acidic nitric

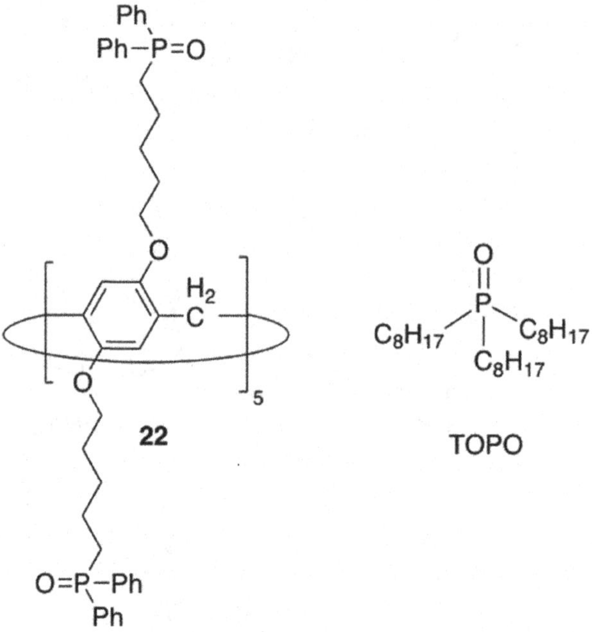

FIGURE 2.15 Structures of the phosphine oxide-pillar[5]ene receptor **22** and the phosphine oxide known as TOPO.

acid solutions with $A_{U/Th}$ values increasing from 4.12 to 11.0 in the presence of 4 M of this particular salt.[48]

In 2014, the same team reported that Am^{3+} could be differentiated from Eu^{3+} by using the diglycolamide-substituted pillar[5]arene receptors **23–25** as extractants. Tests were carried out using the receptors (at 1.0 mM) and N,N,N',N'-tetra-i-propyldiglycolamide (TiPrDGA) (at 10 mM) in 1-octanol as the non-aqueous phase, Figure 2.16a. In these studies, receptors **23–25** were contacted with an equal volume of aqueous solutions containing Am^{3+} and Eu^{3+} at multiple HNO_3 concentrations (1.0–3.0 M).[49] The results obtained led to the conclusion that in a highly acidic environment (>3.0 M in HNO_3), high K_D values for Eu^{3+} were observed. In fact, for both, Am^{3+} and Eu^{3+}, the pillarenes **23–25** exhibited K_D values 10–1000 times larger than the ones obtained using TiPrDGA alone under comparable conditions.[49] This differentiation was ascribed to the probable co-extraction of nitrate together with Am^{3+} and Eu^{3+}. Highly efficient separations of Eu^{3+} from Am^{3+} were achieved when 1.0 M HNO_3 was used; here, separation factors ($SF_{Eu/Am} = K_{DEu}/K_{DAm}$) of 9.71 and 12.16 were seen for **24** and **25**, respectively.[49] These values exceed those reported when N,N,N',N'-tetraoctyl diglycolamide, TODGA, was used as the extractant and the non-aqueous phase was 1-octanol ($A_{Eu/Am}$ ~6.17).[50] The best-reported separation efficiency for Eu^{3+} over Am^{3+} ($SF_{Eu/Am} = 15.2$) was seen when 25 was tested using a source phase containing 3.0 M HNO_3.

A later report from the same team that relied on the use of room-temperature ionic liquids (RTILs) in conjunction with receptors **23**, **24**, and **25** provided evidence

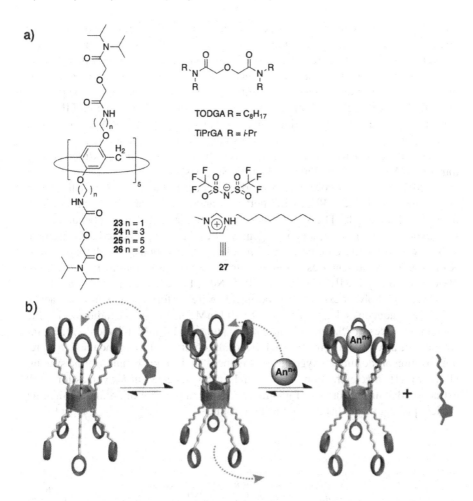

FIGURE 2.16 (a) Chemical structures of diglycolamide-substituted pillar[5]arenes **23–26**, the diglycolamides TODGA and T*i*PrGA, and the room-temperature ionic liquid (RTIL) **27**; (b) schematic representation of proposed competitive host–guest interaction mechanism underlying the extraction of An ions by **23–26**. (This scheme was reproduced from Wang and Horrocks,[50] with permission from the Royal Society of Chemistry.)

for efficient actinide extraction. The suggested extraction mechanism involves competitive host–guest interactions and the displacement of the cation of the RTIL (**27**; which functions as both solvent and guest) by a metal ion, Figure 2.16b. The ability to extract actinide ions, such as UO_2^{2+}, PuO_2^{2+}, Pu^{4+}, and Am^{3+}, along with fission products, such as Eu^{3+}, Sr^{2+}, and Cs^+, was investigated at different aqueous source phase acidities.[51] Receptors **23**, **24**, and **25** were found to be very efficient at transferring Am^{3+} from an aqueous phase (containing 1.0 M HNO_3) into the RTIL phase over other actinides (e.g., K_D=4098 vs. 0.19 and 0.5 for Am^{3+}, PuO_2^{2+}, and UO_2^{2+}, respectively).[51] For instance, when the RTIL **27** was used as the non-aqueous phase, a 485 times higher distribution values for Am^{3+} was seen with receptors **25–27** (using a 1.0 M HNO$_3$ source phase) than observed when 1-octanol was used as the receiving phase under otherwise similar conditions (i.e., K_{DAm} = 4089 vs. 15 in 27 and 1-octanol, respectively). With 27, more than 92% of Am^{3+} was extracted.[51] Receptor 25 also proved effective for the extraction of PuO$_2$$^{2+}$, with a 60% greater efficiency being observed than when diglycolamide-substituted calixarenes were used as the putative extractants.[52] Although receptors 23, 24, and 26 in conjunction with an RTIL phase (e.g., 27) displayed good extraction efficiencies and selectivities for Am^{3+} over other the actinides and some fission products, a slightly higher distribution ratio was still seen for Eu^{3+} relative to Am^{3+}. The overall selectivity order for these extractants was thus Eu^{3+} ~ Am^{3+} > Pu^{4+} >> PuO$_2$$^{2+}$ ~ UO$_2$$^{2+}$ > Sr^{2+} ~ Cs$^+$.

In 2017, Jansone-Popova and coworkers reported the bis-lactam-1,10-phenanthroline receptors 28–30 bearing both soft and hard donor atoms within a pre-organized framework. The use of lactam moieties attached to a phenanthroline framework was used to create convergent cavities that were expected to be particularly effective for the coordination of trivalent metal cations.[53] In the solid state, X-ray single-crystal analyses confirmed that receptor 28 is able to form 1:1 complexes with many Ln(III) cations (i.e., Pr^{3+}, Nd^{3+}, Eu^{3+}, Gd^{3+}, and Tb^{3+}), as shown in Figure 2.17. Solvent extraction experiments were performed by contacting acidic (3 M HNO$_3$) aqueous solutions of Eu(NO$_3$)$_3$ (0.1 mM spiked with the radiotracer ^{152}Eu) containing traces of ^{241}Am with an equal volume of 1.0 mM 1,2-dichloromethane solutions of 28 or 39.[53] On the basis of the results obtained, it was concluded that pre-organization of the amide oxygen groups in these bis-lactam phenanthroline systems allows for effective extraction of Am^{3+} and Eu^{3+} as reflected in K_D values for Am^{3+} extraction that are six orders of magnitude higher in the case of 28 and 29 than for the less pre-organized bis-amide phenanthroline system 30.[53,54] In fact, receptor 29,

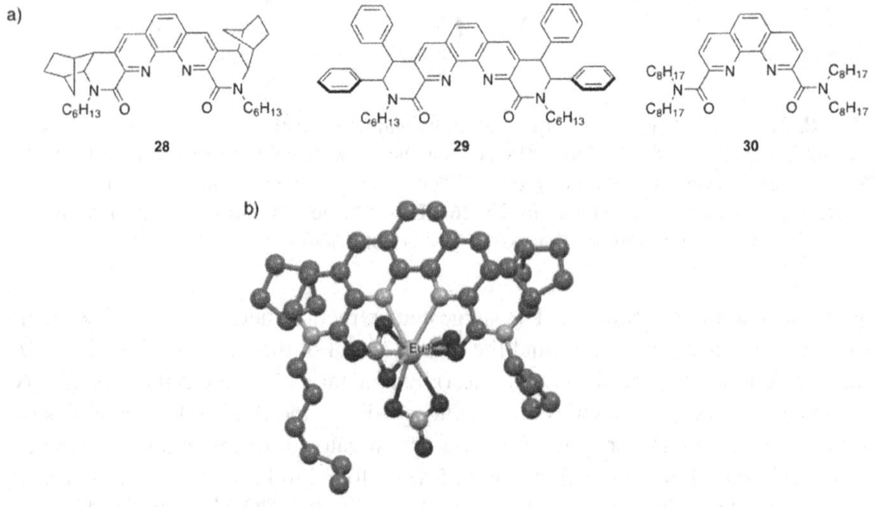

FIGURE 2.17 (a) Chemical structures of receptors 28–30, and (b) single-crystal structure of the Eu(III) complex of receptor 28. (Data from the Cambridge Crystallographic Data Centre [CCDC # 1545804].[53])

which contains unsaturated lactam rings, was found to display an unprecedented selectivity for Am^{3+} over Eu^{3+} as reflected in a separation factor of $SF_{Am/Eu} \approx 211$.[53] This receptor **29** was found to function as an extractant over a range of nitric acid concentrations between 0.03 and 3 M. Density-functional theory (DFT) calculations provided support for the notion that pre-organization of the ligand plays a key role in enabling the selectivity for americium, as well as the good extraction efficacy.

2.3.2 LIQUID–LIQUID EXTRACTION OF TRANSITION METALS

Metal ion recognition is highly dependent on the stability of the complex that forms with a given ligand. Usually defined by the corresponding formation constant, the stability of a complex is governed by factors such as the geometry, the number of binding sites provided by the ligand, and the stereoelectronic preferences of the metal ion. In the case of macrocyclic ligands, size-complementarity between the metal ion and the cavity of the ring can also affect the stability of the metal complex. This, in turn, can influence the extent of complexation and the selectivity of the system. For applications related to extraction, the hydration energy of the metal cation, reflected in the Irving–Williams series, also comes in to play.[55] In fact, most of the ligands developed for the extraction of transition metals have targeted copper, lead, mercury, etc. whose cations are relatively poorly hydrated

As mentioned above, calixarenes derivatives have been widely used as metal cation extractants. That they are available in different sizes, amenable to modification, and endowed with flexible conformations that have made them attractive in the context of ligand design. Early on, Yang and coworkers reported that a thiacalix[4]arene bearing two aza crowns (i.e., compound **31**; Figure 2.18) was effective at extracting "soft" metal cations from water into chloroform. Extraction efficiencies of 81.7 and 72.2% were recorded for Ag^+ and Hg^{2+}, respectively.[56] In contrast, when receptor **31** was exposed to alkali metal cations, such as Na^+, K^+, and Cs^+, poor extraction percentages ($\leq 5\%$) were observed.[56]

Lippolis and coworkers reported that the aza-dithia-phenanthroline macrocycles **32–37** were effective at extracting the soft cations Ag^+, Cu^{2+}, Zn^{2+}, and Hg^{2+} (with nitrate or perchlorate as counterinons) out of an aqueous phase and into chloroform, Figure 2.19.[57] Under the experimental conditions employed (i.e., [ligand] = 1 mM in chloroform, [cation] = 0.1 mM in water at pH 3.1), the extraction efficiency of **32–37** follows the order $Ag^+ > Cu^{2+} > Zn^{2+}$. All ligands exhibited nearly quantitative extraction yields toward Ag^+ (97–99%), an efficiency that was only affected by the presence of Cu^{2+} or Hg^{2+}, as inferred from competition studies. Effective separation of Hg^{2+} (extraction yield >60%) was obtained with **34**.[57] In the case of Cu^{2+}, ligand **37** bearing a pyridine subunit in the macrocyclic framework gave the highest extraction yield (98%) within the series **32–37**. In competitive extraction experiments using Cd^{2+} and Pb^{2+} ([ligand] = 1 mM in chloroform, [cation] = 0.1 mM in water at pH 6.1) with receptor **37**, good Cu^{2+} extraction yields of $\geq 95\%$ were still seen. The Cd^{2+} and Pb^{2+} cations could be also be extracted by **34**, albeit in modest yields of 24% and 25%, respectively.[57]

Wen and coworkers prepared a series of aromatic oligothioamides, **38–40**, containing varying numbers of sulfur atoms, Figure 2.20. Extraction experiments, involving metal cations, including Zn^{2+}, Cd^{2+}, Co^{2+}, Ni^{2+}, Cu^{2+}, and Pb^{2+} (at 2.0×10^{-5} M

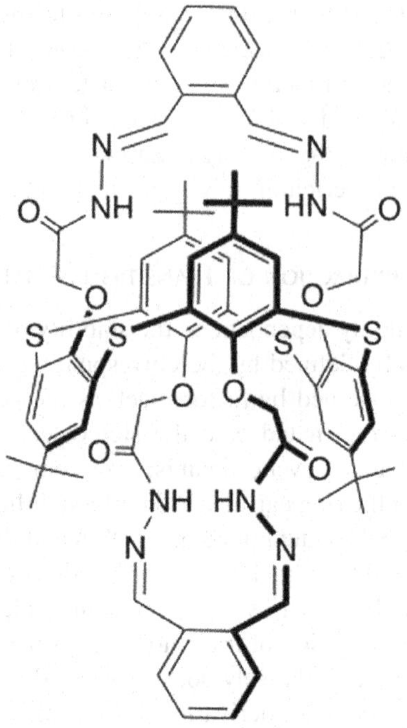

31

FIGURE 2.18 Chemical structure of the thiacalix[4]arene bis-aza crown receptor **31**.

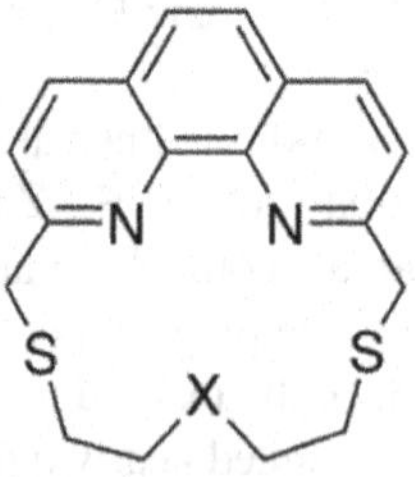

32 X = S
33 X = O
34 X = CH
35 X = NH
36 X = N–CH$_3$

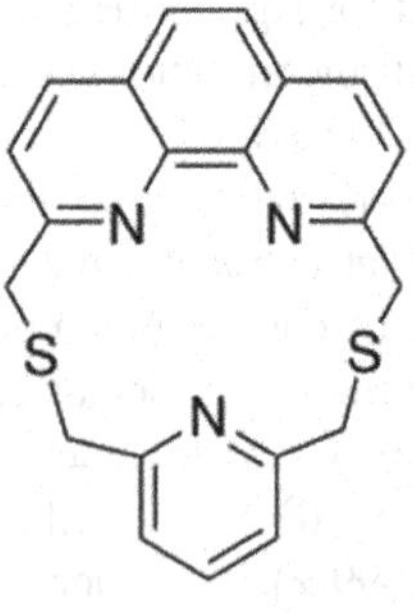

37

FIGURE 2.19 Chemical structures of the aza-dithia-phenanthroline macrocycles **32–37**.

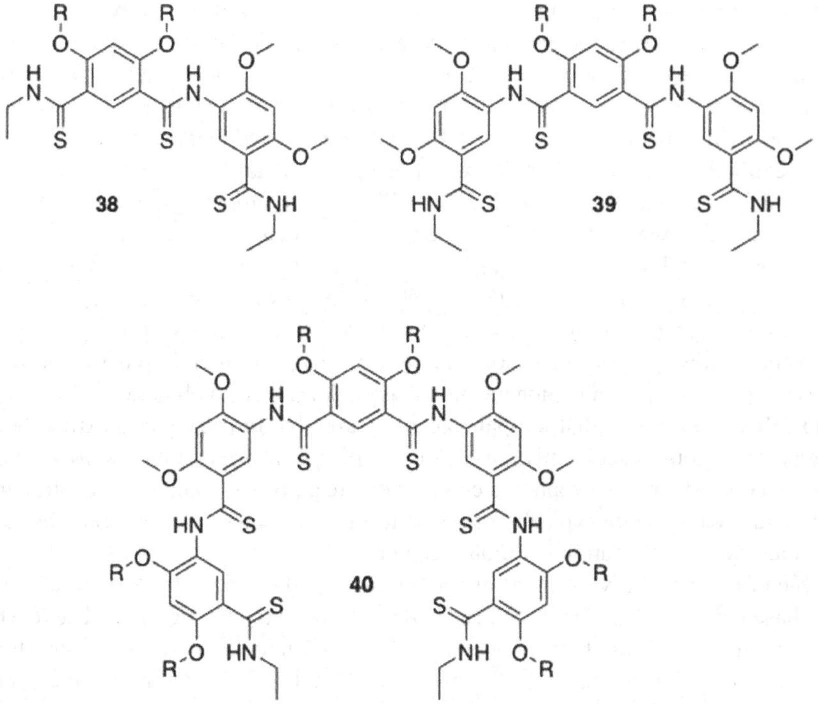

FIGURE 2.20 Chemical structures of the aromatic oligothioamides 38–40.

in water) into a non-aqueous dichloromethane phase containing the ligand in question (at 2.0×10^{-4} M). Based on the results obtained, it was concluded that ligands 38–40 were effective at extracting Cu^{2+} with the extraction percentages being in the 84 to 96% range. In contrast, the other test cations, including the Zn^{2+}, Cd^{2+}, Co^{2+}, Ni^{2+}, and Pb^{2+} ions, were not extracted well.[58] The authors attributed the higher relative extraction efficiencies seen for Cu^{2+} relative to the other test metal cations to the inherent affinity of the Cu^{2+} cations for the sulfur donors provided by the thioamide groups, as well as a good pre-organized spatial orientation for these atoms. This match was reflected in the particularly high Cu^{2+} cation extraction displayed by ligand 40 (96.4%; log $K_D = 11.02$), which has six thioamide groups relative to ligand 38 which provides only three thioamide groups (83.5%; log $K_D = 12.08$).[58] However, it is also important to note that the stability of the metal complexes and the higher extraction efficiencies seen in the case of the copper complexes simply follows the Irving–Williams order.[55]

2.4 ION-PAIR RECOGNITION AND LIQUID–LIQUID EXTRACTION BY SUPRAMOLECULAR SYSTEMS

Since Reetz's pioneer work on ion-pair receptors roughly two decades ago, tremendous effort has been devoted to the preparation of new recognition systems capable of recognizing and extracting such guests.[59] Several critical review papers on the generalized topic have appeared during the past 10 years.[60–68] In addition, three limiting interaction modes, namely contact ion pair, solvent-separated ion pair, and host–guest separated ion pair, were codified during this period. In this section, we will focus primarily on ion-pair receptors that have been studied in the context of liquid–liquid extraction. The liquid–liquid extraction of base metal cations (e.g., copper and nickel) using salicylaldimine- or salicylaldoxime-based ditopic ligands was reviewed by Tasker and coworkers in 2007.[69] Therefore, the focus will be on the extraction of non-base metal salts. Since liquid extraction and liquid membrane transport are closely related in terms of the underlying chemistry, examples where receptors have been used for ion-pair transport are also discussed briefly in this section.

In the context of extraction, ion-pair receptors are often endowed with advantages relative to cation and anion receptors, used alone or in combination. This stems primarily from the fact that ion-pair receptors are able in principle to extract both anions and cations concurrently (i.e., as salts) without the need for charge compensation. In contrast, free hydrophobic counter ions (e.g., organic anions or quaternary ammonium cations) are typically required to make effective an extraction process mediated by a "pure" cation or anion receptor.

One of the very early examples involving the liquid–liquid extraction of targeted salts based on the use of an ion-pair receptor came from the Beer group.[70] The receptor in question was a multitopic tripodal tris(amido benzo-15-crown-5) (cf. structure **41** in Figure 2.21). It was prepared by condensing tris(2-aminoethyl)amine (tren) with three equivalents of 4-chlorocarbonylbenzo-15-crown-5 in the presence of

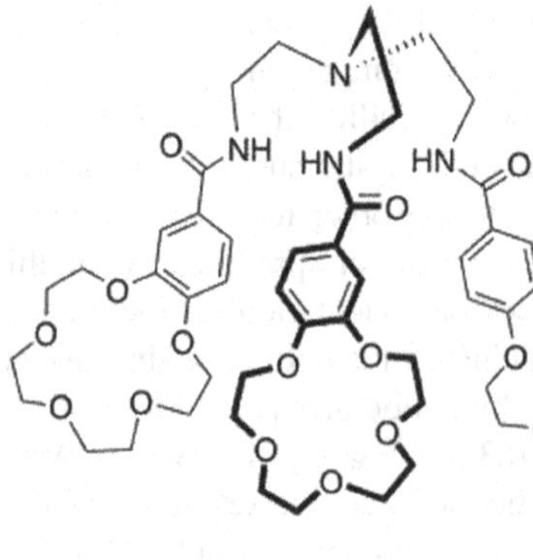

41

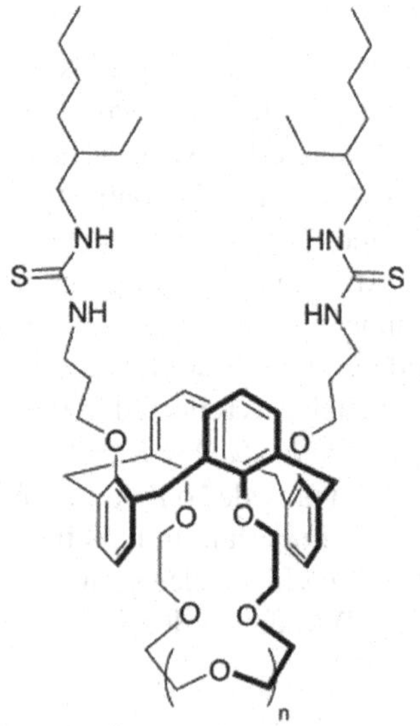

42 n = 2
43 n = 3

FIGURE 2.21 Structures of tripodal tris(amido benzo-15-crown-5) **41** and calixarene-based ditopic receptors **42** and **43**.

Et$_3$N in CH$_2$Cl$_2$ to give **41** in 55% yield. Receptor **41** was found to bind halide (e.g., Cl$^-$ and I$^-$) and the perrhenate (ReO$_4$$^-$) anions as their sodium salts, with the latter cations being complexed within the crown ether moieties. Pertechnetate anion (TcO$_4$$^-$) extraction experiments led to the conclusion that receptor **41** is capable of extracting NaTcO$_4$ from an aqueous source phase into dichloromethane as an ion pair. The extraction efficiency was estimated to be ca. 70%, as inferred from inductively coupled plasma mass spectrometry (ICP-MS) and ^{99}Tc NMR spectroscopic analyses. Based on preliminary U-tube membrane transport investigations, receptor **41** can also act as a carrier capable of transporting the pertechnetate anion across a membrane-like organic phase at a rate of 6.3×10^{-8} mol h^{-1}.

In separate work, Barboiu and coworkers reported a neutral heteroditopic (ureido) crown ether receptor, namely 4-phenylurea-benzo-15-crown-5, which could self-assemble into dimeric or polymeric superstructures as the result of cooperative macrocyclic cation complexation, anion–hydrogen-bonding, and π–π donor–acceptor interactions.[71] Membrane transport experiments led to the conclusion that this crown ether-based receptor could transport NaX salts across liquid chloroform membranes.

In a quest to explore the relationship between the simple ion receptors and ion-pair receptors, Reinhoudt and coworkers studied the facilitated transport of hydrophilic salts by mixtures of anion and cation carriers as well as by ditopic carriers.[72] This was done by comparing the membrane transport properties of calixarene-based ditopic receptors with those of cation + anion receptor mixtures. In the case of simple cation receptors, the extraction efficiency and efficacy of salt transport proved to be affected by the nature of the co-transported anion. It was found that mixtures, wherein both cation receptors and anion receptors are present in the non-aqueous membrane phase could be used to promote the transport of salts (e.g., KCl and KH$_2$PO$_4$) under conditions where the corresponding cation carriers alone were ineffective. The transport rate was found to be limited by diffusion, with the mean diffusion coefficient being found to be highly dependent on the number of carriers involved in the transport process.

Ion-pair transport was not always improved by simply adding an anion receptor to a hydrophobic layer containing a cation receptor. Salt concentrations were also found to affect the transport efficiency. In the studies in question, mixtures of a cation receptor and an anion receptor, at low salt concentration, proved effective for transport. In contrast, at high salt concentrations, cation receptors alone were found to promote effective transport. The ditopic receptors **42** and **43** (Figure 2.21) were found to transport CsCl or KCl much faster than either the constituent monotopic anion receptor or the corresponding cation receptor at relatively low salt concentrations. Conversely, at high salt concentrations, the transport of salts (e.g., KCl) by single-ion receptors can be competitive, even superior to that observed for ditopic anion-cation (ion-pair) receptors.

In 2001, Smith and coworkers reported the synthesis of the ditopic receptor **44** (Figure 2.22),[73] which provided the first unambiguous example of a system capable of capturing alkali halide salts in the form of contact ion pairs. In the solid state, various salts (e.g., NaBr, KBr, NaI, and KI) are bound as contact ion pairs as inferred from X-ray diffraction analyses. Binding studies revealed that the counter-cations Na$^+$ and K$^+$ produce a moderate (and similar) enhancement on the Cl$^-$ affinity. Similar contact

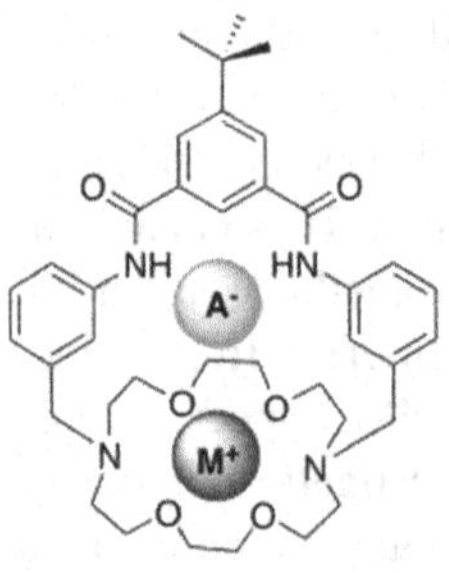

44 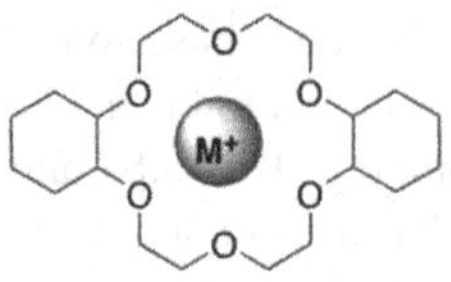   45 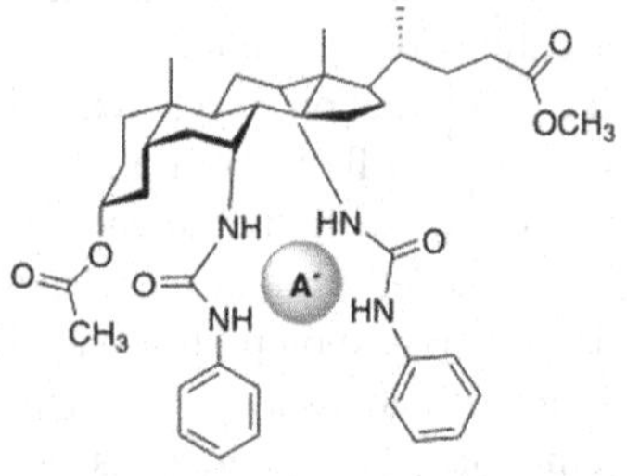   46

FIGURE 2.22 Structures of the ditopic receptor **44**, the monotopic cation receptor **45**, and the monotopic anion receptor **46**.

ion-pair complexes were also observed in the case of a 2,5-diamidopyrrole-capped macrobicyclic receptor reported by the same group.[74]

Smith's ion-pair receptor **44** was found capable of effectively transporting cation-anion salts across vesicle membranes.[75] In contrast, almost no transport activity was seen when a binary mixture of a crown derivative and an isophthalamide (the two recognition components making up **44**) were tested in a control experiment. This led the authors to suggest that ion-pair transporters, such as **44**, would be likely to induce interesting biological effects.

Subsequently, Smith and coworkers established that receptor **44** acts as an extractant for ion pairs.[75] Specifically, they found that this system could recognize and extract various alkali halide salts from an aqueous phase and facilitate their transfer into CDCl$_3$. In the course of liquid–liquid extraction studies, improved transport of hydrophilic salts through a liquid organic membrane could be achieved by using either a binary mixture of the monotopic cation receptor **45** and the anion receptor **46** or by using the ditopic receptor **44** as carriers. Interestingly, **44** was able to transport alkali halide salts across the membrane up to 10 times faster than either **45** or **46** alone and two times faster than a binary mixture of **45** and **46**. Receptor **44** was also found to bind selectively and extract solid NaCl, KCl, NaBr, and KBr into CDCl$_3$ as their contact ion pairs, as inferred from NMR spectroscopic and X-ray crystallographic analyses.[76] The ability of receptor **44** to function as an extractant under liquid–liquid and solid–liquid conditions and as a carrier in transport experiments likely reflects the reduced polarity that accrues from concurrent anion + cation complexation.

In order to build up extraction systems capable of extracting "hard" ion pairs, such as KF or KCl, from aqueous environments, Sessler and coworkers developed a mixed methyl methacrylate (MMA) copolymer system containing both anion and cation recognition subunits.[31] The lead system (**47**) bears both pendant calix[4]pyrrole subunits and benzo-[15]crown-5 moieties and was designed to permit the concurrent binding of both halide and potassium ions. Two control systems, **49** and **50**, containing only one type of ion-recognition moiety were also prepared. The extraction properties of **48–50** were examined in parallel, Figure 2.23. As inferred

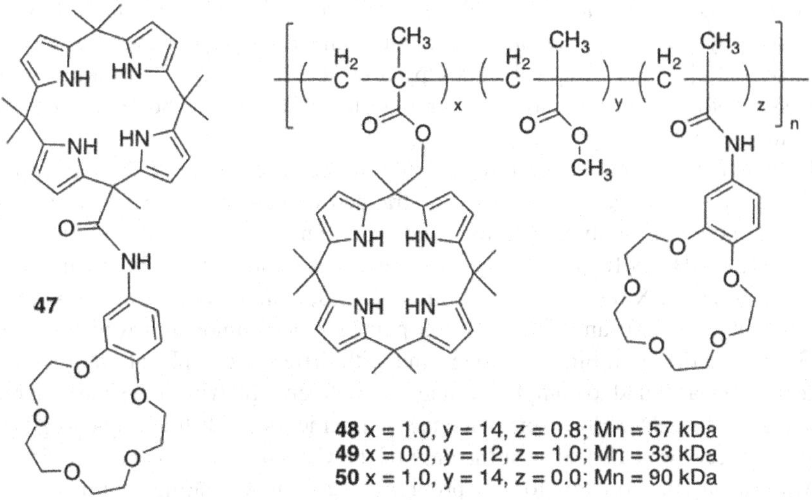

47

48 x = 1.0, y = 14, z = 0.8; Mn = 57 kDa
49 x = 0.0, y = 12, z = 1.0; Mn = 33 kDa
50 x = 1.0, y = 14, z = 0.0; Mn = 90 kDa

FIGURE 2.23 Structures of polymers **48–50** and control **47**.

from ^{19}F NMR spectroscopy and flame emission spectroscopy (FES), copolymer **48** was found more effective in extracting both KCl and KF from an aqueous source phase into CD_2Cl_2 than either polymer **49** or **50**. A polymer-free control (**47**) was also prepared and tested; it proved ineffective as an extractant for either KF or KCl. Presumably, the incorporation of both anion and cation binding sites into the same copolymer allows the system to work as a whole to overcome the relatively high hydration energies of KF and KCl and extract these ion pairs efficiently and with a degree of selectivity. The presence of multiple recognition sites was thought to contribute to the effect through multivalency.

As noted earlier in this chapter, calix[4]pyrrole (*meso*-octamethylcalix[4]pyrrole) (**9**, Figure 2.8) was first reported by Baeyer in 1886[28] and has been the subject of extensive recent study as an easy-to-prepare platform for anion[29,77,78] and ion-pair recognition.[30] In 2003, Sessler and coworkers reported several polyfluorinated calix[n]pyrroles with high affinities for anions.[79] Subsequently, it was found by Sessler and Moyer and coworkers that two of these systems, namely β-octafluoro-*meso*-octamethylcalix[4]pyrrole (**52**) and β-decafluoro-*meso*-decamethylcalix[5]pyrrole (**53**),[80] were able to extract cesium salts (e.g., CsCl, CsBr, and $CsNO_3$) into nitrobenzene (NB) from aqueous source phases. This finding was of importance at the time, as it represented a clear demonstration of overcoming the Hofmeister bias normally observed for normal liquid–liquid processes. As discussed previously, the Hofmeister bias favors the extraction of less-hydrophilic anions.

The same team later reported the ion-pair recognition-based liquid–liquid extraction of CsCl and CsBr using **9**.[81] Solvent extraction studies involving the transfer of salts from water to nitrobenzene provided evidence for the formation of a 1:1:1 (cesium:calix[4]pyrrole:halide) ternary ion-pair complex (**51**), a formulation consistent with what was seen in the solid state.[30] By contrast, **9** was found ineffective as an extractant for $CsNO_3$ under conditions identical to where it functions well for

CsCl and CsBr. Presumably, this reflects the relatively low nitrate anion affinity of the calix[4]pyrrole. CsNO$_3$ extraction into nitrobenzene could, however, be achieved using the calix[4]arene-strapped calix[4]pyrrole 54 (Figure 2.24). This latter system possesses both a strong Cs$^+$ cation recognition site and an anion binding motif within the same cavity.[82]

In an effort to create a multitopic ion-pair receptor, receptor 54 was further modified to produce a calix[4]crown-5 strapped calix[4]pyrrole 55 that contains at several potential binding sites for cations and one binding site for anions.[83] As inferred from ^{1}H NMR spectroscopic studies carried out in 10% CD$_3$OD in CDCl$_3$ and single-crystal X-ray crystal structure analyses, compound 55 was found to act as a receptor for CsF and KF. Both ion pairs are accommodated well by receptor 55 but *via* different binding modes and with different complexation dynamics. Receptor 55 was found to complex K$^+$ and Cs$^+$ preferentially within the calix[4]arene crown-5 binding site. However, evidence of interaction with the ethylene glycol-derived spacers was also seen in the case of the Cs$^+$ salt. Ion-pair receptor 55 was shown to act as an extractant for KF, proving effective in mediating its transfer from an aqueous phase into nitrobenzene, thus overcoming the high hydration energies of the K$^+$ and F$^-$ ions.

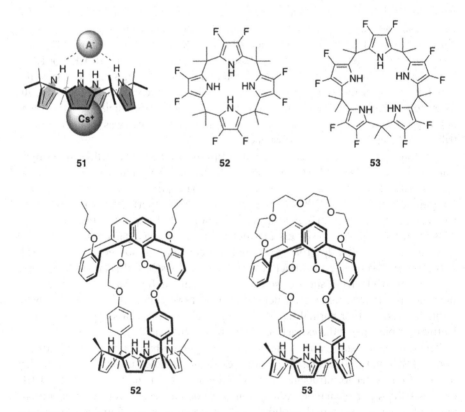

FIGURE 2.24 Structures of the CsA ion-pair complex 51 and the chemical structures of receptors 52–55 (A = anion).

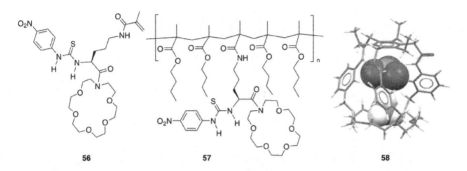

FIGURE 2.25 Structures of receptors **56**, polymer **57**, and the DFT optimized structure of the NH₄NO₃ complex of **58**. The DFT optimized structure was generated using the coordinates given in Romanski and Piątek.[84] (Copyright © 2013, with permission of the American Chemical Society.)

In 2012, Piatek and coworkers described a polymerizable ditopic ion-pair receptor **56**, shown in Figure 2.25,[84] which proved able to associate with sodium salts (e.g., NaCl, NaOAc, and NaNO₃), as inferred from the ¹H NMR spectroscopic studies carried out in CD₃CN. It is noteworthy that the selectivity of **56** toward anions in the presence of sodium cations follows the order $Cl^- > AcO^- > NO_3^-$, while in the presence of TBA⁺ cations, it is in the order $AcO^- > Cl^- > NO_3^-$. When tested in the context of solid–liquid extraction experiments, receptor **56** displays a pronounced preference for NaNO₃ with a near 100% extraction efficiency being recorded. In contrast, the extraction efficiencies for NaCl (30%) and NaAcO (80%) were notably lower. However, receptor **56** failed to act as an extractant for any of these salts under liquid–liquid extraction conditions.

Receptor **56** was then copolymerized with *n*-butyl methacrylate in the presence of 2% AIBN to produce **57**, Figure 2.25. This polymer proved effective at extracting NaNO₃, NaCl, and KNO₃ from an aqueous source phase into CHCl₃. Extraction efficiencies of 44%, 19%, and 27% were noted for these three salts, respectively, as determined via atomic emission spectroscopy. Later the same group reported a heteroditopic macrotricyclic ion-pair receptor **58** that is based on a 4,10,16-triaza-18-crown-6 subunit.[85] Receptor **58** allowed for the selective recognition of ammonium nitrate in organic media. It also proved capable of extracting NH₄NO₃ from the solid state into an organic phase, as inferred from ¹H NMR spectroscopic studies and nitrite/nitrate colorimetric analyses. This proposed recognition was supported by DFT calculations (cf. Figure 2.26).

Jabin and coworkers reported the D_{3h}-symmetric tail-to-tail bis-calix[6]arene **59** featuring two divergent cavities triply connected by ureido spacers, Figure 2.26.[86] This calix[6]arene dimer was found to act as a multitopic ion-pair receptor capable of recognizing concurrently two cations within the calix[6]arene subunits and one anion via interactions with the ureido moieties in the presence of a protic solvent. A high selectivity for linear ammonium ions (e.g., RNH₃⁺) associated with doubly charged anions (e.g., SO₄²⁻) was observed. Replacement of the urea units with thiourea groups gave the ostensibly related multitopic ion-pair receptor **60**.[87] This system was expected to stabilize a complex wherein a sulfate anion is sandwiched between

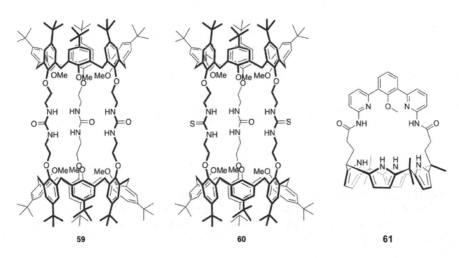

                    59   60   61

FIGURE 2.26 Structures of the bis-calix[6]urea **59**, its bis-calix[6]thiourea analog **60**, and a hemispherand-strapped calix[4]pyrrole **61**.

two ammonium ions. In fact, high selectivity for ammonium sulfate salts was seen in $CD_3OD/CDCl_3$. As revealed by a single-crystal X-ray structure of complex [**60**•(EtNH$_3$)$_2$SO$_4$], the ammonium cations are stabilized by the calix[6]arene units and the sulfate anions are bound by the thiourea linkers, leading to a host–guest cascade complex. In spite of the preference for sulfate anion salts seen in organic media, neither of these receptors proved effective as an extractant for this highly hydrated anion under liquid–liquid experiment conditions. Consistent with the lower hydration energy of nitrate compared to sulfate, receptor **60** proved capable of extracting ammonium nitrate salts from an aqueous source phase. However, its bis-calix[6]urea congener **59** proved ineffective in this regard.

Recently, Sessler and coworkers prepared a hemispherand-strapped calix[4]pyrrole **61** (Figure 2.27)[88] that was inspired by both the elegant recognition features of hemispherands (a classic set of receptors for Li$^+$)[89] and the recognized anion binding properties of the calix[4]pyrroles discussed previously in this chapter. Receptor **61** was found to act as an ion-pair receptor and proved effective in recognizing a number of lithium salts (e.g., LiCl, LiBr, LiI, LiNO$_3$, and LiNO$_2$), as inferred from ^{1}H NMR spectroscopy, single X-ray crystal diffraction analyses, and DFT calculations. When CDCl$_3$ was used as the non-aqueous phase, receptor **61** proved effective as an extractant for LiNO$_2$ under both model solid–liquid and liquid–liquid extraction conditions. High selectivity over the corresponding sodium and potassium salts was seen. The chemistry in question is shown schematically in Figure 2.27.[88]

Receptor **61** was also found capable of stabilizing ion-pair complexes involving MX-type monoanionic salts (i.e., CsF, CsCl, CsBr, and CsOH) and dianionic M$_2$X-type (viz. Cs$_2$CO$_3$) salts on the basis of ^{1}H NMR spectroscopic titrations, single-crystal structural analyses, and theoretical calculations.[88] Under conditions of solid–liquid extraction, **61** was able to extract CsCl selectively from the solid state into an organic CD$_2$Cl$_2$ phase. Proton NMR spectroscopic studies in conjunction with inductively coupled plasma mass spectrometric (ICP-MS) analyses provided

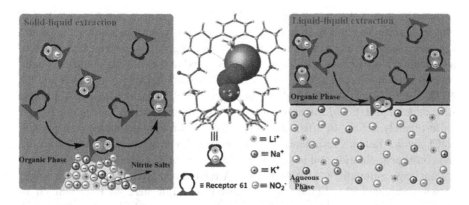

FIGURE 2.27   Cartoon illustration of the solid–liquid extraction and liquid–liquid extraction of LiNO$_2$ effected by receptor **61**. (Reprinted with permission from He et al.[88] Copyright © 2017, Wiley–VCH Verlag GmbH & Co. KGaA, Weinheim.)

support for the conclusion that receptor **61** could be used to extract effectively CsOH and Cs$_2$CO$_3$ under conditions of liquid–liquid extraction where chloroform was used as the organic phase. Receptor **61** was also tested as a carrier. In U-tube experiments where a chloroform layer was used to separate the source and to receive aqueous phases, direct transport of CsOH was seen. This led to a sharp increase in the pH and the cesium concentrations in the receiving phase; see Figure 2.28.[88] The pH of the source phase was also found to decrease.

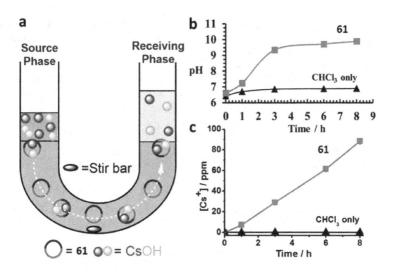

FIGURE 2.28   (a) Illustration of the general setup for the U-tube experiments used to test the ability of **61** to act as a carrier for CsOH. (b) The pH curve of the receiving phase observed as a function of time. (c) Plot of the Cs$^+$ concentration in the receiving phase as a function of time. (Reprinted with permission from He et al.[88] Copyright © 2017, Wiley–VCH Verlag GmbH & Co. KGaA, Weinheim.)

In 2016, Romański and coworkers published an ion-pair receptor **62** that is based on an L-ornithine scaffold (Figure 2.29).[90] It contains two squaramide anion binding domains and one crown ether-based cation binding site. It was expected to act as an effective ion-pair receptor. Indeed, **62** was found to form strong complexes with several sodium salts (e.g., NaCl, NaBr, and NaNO$_3$), as inferred from UV-Vis and ^{1}H NMR spectroscopic studies. It also proved capable of extracting the dye Toluidine Blue under liquid–liquid conditions when chloroform was used as the organic phase. Unfortunately, efforts to affect the extraction and through-membrane transport of NaCl always failed. Presumably, this inability reflects the very "hard" nature of the constituent chloride and sodium ions. Extraction was seen when softer counter-cations (e.g., ammonium salts) were used. A related receptor bearing only a single crown, compound **63** in Figure 2.29, was also tested. It forms complexes with chloride anions that are roughly two orders of magnitude less stable than those produced by receptor **62**. Not surprisingly, it proved less effective as an extractant in test studies involving the extraction of chloride salts from an aqueous source phase into a chloroform layer.

In an effort to extract relatively hard salts (e.g., KCl) from aqueous media, Piatek and coworkers developed the heteroditopic ion-pair receptor **64**.[91] This system

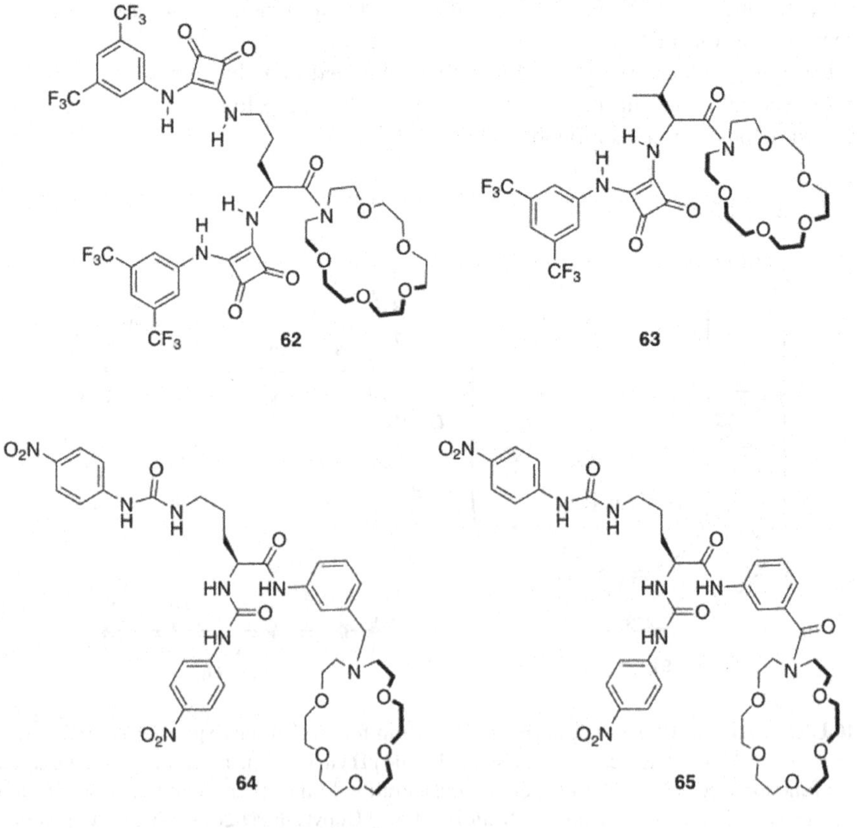

FIGURE 2.29 Structures of receptors **62–65**.

contains both urea units for anion recognition and an *N*-benzyl-aza-18-crown-6 linked together *via* an L-ornithine scaffold. Receptor **64** differs from **65** only in the cation binding domains (3° amine vs. the corresponding amide). Nevertheless, it was found to bind cations much more strongly in acetonitrile. Presumably, this enhancement reflects the benefits of synergistic ion-pair complexation that the presence of a more effective cation receptor subunit confers. Receptor **64** proved to be an effective extractant for both KCl and NH$_4$Cl under liquid–liquid conditions where chloroform was used as the non-aqueous phase. In contrast, receptor **65** was completely inactive as an extractant. On this basis, it was concluded that liquid–liquid extraction efficiencies could be improved by fine-tuning the structures of ostensibly similar extractants.

2.5 CONCLUSIONS

The examples provided in this chapter have helped underscore recent progress in the area of supramolecular-based extractant design. Ultimately, the separations such functioning receptor systems could allow ions and ion pairs of societal interest to be targeted with high specificity and efficacy.

Typically, the systems described in this chapter function via relatively weak interactions and thus rely on carefully engineered receptor frameworks. While early receptor-based extractants contained charged groups or subunits whose charge could be modulated as a result of guest binding, increasing attention is now being devoted to systems that are capable of accommodating an ion pair directly without the need for consideration of counter-ion effects. Where the design expectations are met, the result is powerful systems that allow for the efficient recognition and extraction of ions that would otherwise remain in an uncomplexed state as either solids or solutions in aqueous environments. Obviously, many challenges remain in terms of achieving optimum extraction efficacies and selectivities. However, this chapter serves to underscore just how much progress has been made in recent years.

ACKNOWLEDGMENTS

This work was supported by the U.S. Department of Energy, Office of Basic Energy Sciences (DE-FG02-01ER15186 to J.L.S.) and the Robert A. Welch Foundation (F-0018 to J.L.S.).

REFERENCES

1. Evans, N. H.; Beer, P. D. Advances in Anion Supramolecular Chemistry: From Recognition to Chemical Applications. *Angew. Chem. Int. Ed.* 2014, *53*, 11716–11754.
2. Marcus, Y. Ion Properties. In *Ionic Interactions in Natural and Synthetic Macromolecules*, Ciferri, A., Perico, A., Eds.; Wiley: New York, 2012; pp 1–33.
3. Moss, B. Water Pollution by Agriculture. *Philos. Trans. R. Soc. Lond., B, Biol. Sci.* 2008, *363*, 659–666.
4. Kumar, N.; Seminario, J. N. Solution of Actinide Salts in Water Using Polarizable Continuum Model. *J. Phys. Chem. A* 2015, *119*, 689–703.

5. Marcus, Y. Thermodynamics of Solvation of Ions. Part 5. – Gibbs Free Energy of Hydration at 298.15 K. *J. Chem. Soc. Faraday Trans.* 1991, *87*, 2995–2999.

6. Schneider, H. Ionic Interactions in Supramolecular Complexes. In *Ionic Interactions in Natural and Synthetic Macromolecules*, Ciferri, A., Perico, A., Eds.; Wiley: New York, 2012; pp 35–47.

7. Ciferri, A. Ionic Mixed Interactions and Hofmeister Effects. In *Ionic Interactions in Natural and Synthetic Macromolecules*, Ciferri, A., Perico, A., Eds.; Wiley: New York, 2012; pp 169–209.

8. Vargas-Zúñiga, G. I.; Sessler, J. L. Anion and Ion Pair Recognition Under Interfacial Aqueous Conditions. *Compr. Supramol. Chem. II* 2017, *4*, 161–189.

9. Stephan, H.; Gloe, K.; Schiessl, P.; Schmidtchen, F. P. Lipophilic Ditopic Guanidinium Receptors: Selective Extractants for Tetrahedral Oxoanions. *Supramol. Chem.* 1995, *5*, 273–280.

10. Gloe, K.; Stephan, H.; Grotjahn, M. Where Is the Anion Extraction Going? *Chem. Eng. Technol.* 2003, *26*, 1107–1117.

11. Menon, S. V. G.; Kelkar, V. K.; Manohar, C. Application of Baxter's Model to the Theory of Cloud of Solutions. *Phys. Rev. A* 1991, *43*, 1130–1133.

12. Chiarizia, R.; Stepinski, D.; Antonio, M. R. Study of HCl Extraction by Selected Neutral Organophosphorous Compounds in N-Octane. *Sep. Sci. Technol.* 2010, *45*, 1668–1678.

13. Eller, L. R.; Stępień, M.; Fowler, C. J.; Lee, J. T.; Sessler, J. L.; Moyer, B. A. Octamethyl-Octaundecylcyclo[8]Pyrrole: A Promising Sulfate Anion Extractant. *J. Am. Chem. Soc.* 2007, *129*, 11020–11021.

14. Seidel, D.; Lynch, V.; Sessler, J. L. Cyclo[8]Pyrrole: A Simple-to-Make Expanded Porphyrin with No Meso Bridges. *Angew. Chem. Int. Ed. Engl.* 2002, *41* (8), 1422–1425.

15. Seipp, C. A.; Williams, N. J.; Bryantsev, V. S.; Moyer, B. A. A Simple Guanidinium Motif for the Selective Binding and Extraction of Sulfate. *Sep. Sci. Technol.* 2017, *6395*, 1–10.

16. Smedley, P. L.; Kinniburgh, D. G. A Review of the Source, Behaviour and Distribution of Arsenic in Natural Waters. *Appl. Geochem.* 2002, *17*, 517–568.

17. Ranjan, D.; Talat, M.; Hasan, S. H. Biosorption of Arsenic from Aqueous Solution Using Agricultural Residue "Rice Polish." *J. Hazard. Mater.* 2009, *166*, 1050–1059.

18. Guo, H.; Li, Y.; Zhao, K. Arsenate Removal from Aqueous Solution Using Synthetic Siderite. *J. Hazard. Mater.* 2010, *176*, 174–180.

19. Wei, Y. H.; Zhou, Y. G.; Tomita, B. Study of Hydration Behavior of Wood Cement-Based Composite II: Effect of Chemical Additives on the Hydration Characteristics and Strength of Wood-Cement Composites. *Wood Sci. Technol.* 1999, *33*, 501–517.

20. De Mattia, G.; Bravi, M. C.; Laurenti, O.; De Luca, O.; Palmeri, A.; Sabatucci, A.; Mendico, G.; Ghiselli, A. Impairment of Cell and Plasma Redox State in Subjects Professionally Exposed to Chromium. *Am. J. Ind. Med.* 2004, *46*, 120–125.

21. Cohen, M. D.; Kargacin, B.; Klein, C. B.; Costa, M. Mechanisms of Chromium Carcinogenicity and Toxicity. *Crit. Rev. Toxicol.* 1993, *23*, 255–281.

22. Bayrakcı, M.; Yiğiter, Ş. Synthesis of Tetra-Substituted Calix[4]Arene Ionophores and Their Recognition Studies Toward Toxic Arsenate Anions. *Tetrahedron* 2013, *69*, 3218–3224.

23. Sayin, S.; Eymur, S.; Yilmaz, M. Anion Extraction Properties of a New "Proton-Switchable" Terpyridin-Conjugated Calix[4]Arene. *Ind. Eng. Chem. Res.* 2014, *53*, 2396–2402.

24. Pinkse, M. W. H.; Merkx, M.; Averill, B. A. Fluoride Inhibition of Bovine Spleen Purple Acid Phosphatase: Characterization of a Ternary Enzyme-Phosphate-Fluoride Complex as a Model for the Active Enzyme-Substrate-Hydroxide Complex. *Biochemistry* 1999, *38*, 9926–9936.

25. Moyer, B. A.; Bonnesen, P. V. Physical Factors in Anion Separations in Supramolecular Chemistry of Anions. In *Supramolecular Chemistry of Anions*, Bianchi, A., Bowman-James, K., García-España, E., Eds.; Wiley–VCH: New York, 1997; pp 4–9.

26. Chiu, C. W.; Gabbaï, F. P. Fluoride Ion Capture from Water with a Cationic Borane. *J. Am. Chem. Soc.* 2006, *128*, 14248–14249.

27. Das, P.; Mandal, A. K.; Kesharwani, M. K.; Suresh, E.; Ganguly, B.; Das, A. Receptor Design and Extraction of Inorganic Fluoride Ion from Aqueous Medium. *Chem. Commun. (Camb.)* 2011, *47*, 7398–7400.

28. Baeyer, A.; Rein, U. Condensation product von Pyrrol Mit Aceton. *Ber. Dtsch. Chem. Ges.* 1886, *19*, 2184–2185.

29. Gale, P. A.; Sessler, J. L.; Král, V.; Lynch, V. Calix[4]Pyrroles: Old yet New Anion-Binding Agents. *J. Am. Chem. Soc.* 1996, *118*, 5140–5141.

30. Custelcean, R.; Delmau, L. H.; Moyer, B. A.; Sessler, J. L.; Cho, W. S.; Gross, D. E.; Bates, G. W.; Brooks, S. J.; Light, M. E.; Gale, P. A. Calix[4]Pyrrole: An Old yet New Ion-Pair Receptor. *Angew. Chem. Int. Ed. Engl.* 2005, *44*, 2537–2542.

31. Aydogan, A.; Coady, D. J.; Kim, S. K.; Akar, A.; Bielawski, C. W.; Marquez, M.; Sessler, J. L. Poly(Methyl Methacrylate)s with Pendant Calixpyrroles and Crown Ethers: Polymeric Extractants for Potassium Halides. *Angew. Chem. Int. Ed.* 2008, *120*, 9794–9798.

32. Aydogan, A.; Lee, G.; Lee, C. H.; Sessler, J. L. Reversible Assembly and Disassembly of Receptor-Decorated Gold Nanoparticles Controlled by Ion Recognition. *Chemistry* 2015, *21* (6), 2368–2376.

33. Moyer, B. A.; Sloop, F. V.; Fowler, C. J.; Haverlock, T. J.; Kang, H. A.; Delmau, L. H.; Bau, D. M.; Hossain, M. A.; Bowman-James, K.; Shriver, J. A.; et al. Enhanced Liquid–Liquid Anion Exchange Using Macrocyclic Anion Receptors: Effect of Receptor Structure on Sulphate–Nitrate Exchange Selectivity. *Supramol. Chem.* 2010, *22*, 653–671.

34. Kim, S. K.; Lee, J.; Williams, N. J.; Lynch, V. M.; Hay, B. P.; Moyer, B. A.; Sessler, J. L. Bipyrrole-Strapped Calix[4]Pyrroles: Strong Anion Receptors That Extract the Sulfate Anion. *J. Am. Chem. Soc.* 2014, *136* (42), 15079–15085.

35. Custelcean, R.; Moyer, B. A. Anion Separation with Metal–Organic Frameworks. *Eur. J. Inorg. Chem.* 2007, *2007* (10), 1321–1340.

36. Ravikumar, I.; Ghosh, P. Recognition and Separation of Sulfate Anions. *Chem. Soc. Rev.* 2012, *41* (8), 3077–3098.

37. Jia, C.; Wu, B.; Li, S.; Huang, X.; Zhao, Q.; Li, Q.-S.; Yang, X.-J. Highly Efficient Extraction of Sulfate Ions with a Tripodal Hexaurea Receptor. *Angew. Chem. Int. Ed.* 2011, *50*, 486–490.

38. Akhuli, B.; Ravikumar, I.; Ghosh, P. Acid/Base Controlled Size Modulation of Capsular Phosphates, Hydroxide Encapsulation, Quantitative and Clean Extraction of Sulfate with Carbonate Capsules of a Tripodal Urea Receptor. *Chem. Sci.* 2012, *3* (5), 1522–1530.

39. Clare, J. P.; Ayling, A. J.; Joos, J. B.; Sisson, A. L.; Magro, G.; Pérez-Payán, M. N.; Lambert, T. N.; Shukla, R.; Smith, B. D.; Davis, A. P. Substrate Discrimination by Cholapod Anion Receptors: Geometric Effects and the "Affinity-Selectivity Principle." *J. Am. Chem. Soc.* 2005, *127*, 10739–10746.

40. Forsberg, C. W. Rethinking High-Level Waste Disposal: Separate Disposal of High Heat Radionuclides (90Sr and 137Cs). *Nucl. Technol.* 2000, *131*, 252–268.

41. Coleman, C. F.; Leuze, R. E. Some Milestone Solvent Extraction Processes at the Oak Ridge National Laboratory. *J. Tenn. Acad. Sci.* 1978, *53* (3), 102–107.

42. Madic, C.; Hudson, M. J.; Baron, P.; Ouvrier, N.; Hill, C.; Arnaud, F.; Espartero, A. G.; Desreux, J.-F.; Modolo, G.; Malmbeck, R.; et al. EUROPART: European Research Programme for Partitioning of Minor Actinides within High Active Wastes Issuing the Reprocessing Spent Nuclear Fuels. In *6th OECD/NEA Information Exchange Meeting*

(IEM) on Actinide (AN) and Fission Product (FP) Partitioning and Transmutation (P&T); Luxemburg, 2006; pp 1–38.

43. Moyer, B. A.; Lumetta, G. L.; Mincher, B. J. Minor Actinide Separation in the Reprocessing of Spent Nuclear Fuels: Recent Advances in the United States. In *Reprocessing and Recycling of Spent Nuclear Fuels*; Woodhead Publishing: Waltham, MA, 2015; Chap. 11, pp 289–312.

44. Tachimori, S.; Morita, Y. Overview of Solvent Extraction Chemistry for Reprocessing. In *Ion Exchange and Solvent Extraction. A Series of Advances*, Moyer, B. A., Ed.; CRC Press: Boca Raton, 2009; pp 1–63.

45. Hill, C. Overview of Recent Advances in An(III)/Ln(III) Separation by Solvent Extraction. In *Ion Exchange and Solvent Extraction. A Series of Advances*, Moyer, B. A., Ed.; CRC Press: Boca Raton, FL, 2009; pp 119–193.

46. Dozol, J. F.; Ludwig, R. Extraction of Radioactive Elements by Calixarenes. In *Ion Exchange and Solvent Extraction. A Series of Advances*, Moyer, B. A., Ed.; CRC Press: Boca Raton, FL, 2009; pp 195–318.

47. Okamura, H.; Ikeda-Ohno, A.; Saito, T.; Aoyagi, N.; Naganawa, H.; Hirayama, N.; Umetani, S.; Imura, H.; Shimojo, K. Specific Cooperative Effect of a Macrocyclic Receptor for Metal Ion Transfer into an Ionic Liquid. *Anal. Chem.* 2012, *84* (21), 9332–9339.

48. Fang, Y.; Wu, L.; Liao, J.; Chen, L.; Yang, Y.; Liu, N.; He, L.; Zou, S.; Feng, W.; Yuan, L. Pillar[5]Arene-Based Phosphine Oxides: Novel Ionophores for Solvent Extraction Separation of f-Block Elements from Acidic Media. *RSC Adv.* 2013, *3* (30), 12376.

49. Wu, L.; Fang, Y.; Jia, Y.; Yang, Y.; Liao, J.; Liu, N.; Yang, X.; Feng, W.; Ming, J.; Yuan, L. Pillar[5]Arene-Based Diglycolamides for Highly Efficient Separation of Americium(III) and Europium(III). *Dalton Trans.* 2014, *43*, 3835–3838.

50. Wang, Y.; Horrocks, W. D. Characterization of Lanthanide Ion Binding to Macrocyclic Tricarboxylate Ligands Containing 12-, 15- and 18-Membered Rings by Europium(III) Luminescence Spectroscopy. *Inorg. Chim. Acta* 1997, *263*, 309–314.

51. Li, C.; Wu, L.; Chen, L.; Yuan, X.; Cai, Y.; Feng, W.; Liu, N.; Ren, Y.; Sengupta, A.; Murali, M. S.; et al. Highly Efficient Extraction of Actinides with Pillar[5] Arene-Derived Diglycolamides in Ionic Liquids via a Unique Mechanism Involving Competitive Host–Guest Interactions. *Dalton Trans.* 2016, *45*, 19299–19310.

52. Mohapatra, P. K.; Sengupta, A.; Iqbal, M.; Huskens, J.; Godbole, S. V.; Verboom, W. Remarkable Acidity Independent Actinide Extraction with a Both-Side Diglycolamide-Functionalized Calix[4]Arene. *Dalton Trans.* 2013, *42*, 8558.

53. Jansone-Popova, S.; Ivanov, A. S.; Bryantsev, V. S.; Sloop, F. V.; Custelcean, R.; Popovs, I.; Dekarske, M. M.; Moyer, B. A. Bis-Lactam-1,10-Phenanthroline (BLPhen), a New Type of Preorganized Mixed N,O-Donor Ligand That Separates Am(III) over Eu(III) with Exceptionally High Efficiency. *Inorg. Chem.* 2017, *56*, 5911–5917.

54. Galletta, M.; Scaravaggi, S.; Macerata, E.; Famulari, A.; Mele, A.; Panzeri, W.; Sansone, F.; Casnati, A.; Mariani, M. 2,9-Dicarbonyl-1,10-Phenanthroline Derivatives with an Unprecedented Am(III)/Eu(III) Selectivity under Highly Acidic Conditions. *Dalton Trans.* 2013, *42*, 16930.

55. Irving, H.; Williams, R. J. P. 637The Stability of Transition-Metal Complexes. *J. Chem. Soc.* 1953, 3192–3210.

56. Yang, F.; Huang, C.; Guo, H.; Lin, J.; Peng, Q. Synthesis and Extraction Property of Novel Thiacalix[4]Biscrown: Thiacalix[4]-1,3-2,4-Aza-Biscrown. *J. Incl. Phenom. Macrocycl. Chem.* 2007, *58*, 169–172.

57. Aragoni, M. C.; Arca, M.; Bencini, A.; Biagini, S.; Blake, A. J.; Caltagirone, C.; Demartin, F.; De Filippo, G.; Devillanova, F. A.; Garau, A.; et al. Interaction of Mixed-Donor Macrocycles Containing the 1,10-Phenanthroline Subunit with Selected

Transition and Post-Transition Metal Ions: Metal Ion Recognition in Competitive Liquid–Liquid Solvent Extraction of CuII, ZnII, PbII, CdII, AgI, and Hg II. *Inorg. Chem.* 2008, *47*, 8391–8404.

58. He, L.; Jiang, Q.; Gao, R.; Yang, X.; Feng, W.; Luo, S.; Yang, Y.; Yang, L.; Yuan, L. Crescent Aromatic Oligothioamides as Highly Selective Receptors for Copper(II) Ion. *Sci. China Chem.* 2014, *57*, 1246–1256.

59. Reetz, M. T.; Niemeyer, C. M.; Harms, K. Crown Ethers with a Lewis Acidic Center – A New Class of Heterotopic Host Molecules. *Angew. Chem. Int. Ed. Engl.* 1991, *30*, 1472–1474.

60. Kim, S. K.; Sessler, J. L. Ion Pair Receptors. *Chem. Soc. Rev.* 2010, *39*, 3784–3809.

61. Kim, S. K.; Sessler, J. L. Calix[4]Pyrrole-Based Ion Pair Receptors. *Acc. Chem. Res.* 2014, *47* (8), 2525–2536.

62. Kirkovits, G. J.; Shriver, J. A.; Gale, P. A.; Sessler, J. L. Synthetic Ditopic Receptors. *J. Incl. Phenom. Macrocycl. Chem.* 2001, *41*, 69–75.

63. McConnell, A. J.; Beer, P. D. Heteroditopic Receptors for Ion-Pair Recognition. *Angew. Chem. Int. Ed. Engl.* 2012, *51*, 5052–5061.

64. Molina, P.; Tárraga, A.; Alfonso, M. Ferrocene-Based Multichannel Ion-Pair Recognition Receptors. *Dalton Trans.* 2014, *43*, 18–29.

65. Naseer, M. M.; Jurkschat, K. Organotin-Based Receptors for Anions and Ion Pairs. *Chem. Commun. (Camb.)* 2017, *53*, 8122–8135.

66. Severin, K. Self-Assembled Organometallic Receptors for Small Ions. *Coord. Chem. Rev.* 2003, *245*, 3–10.

67. Smith, B. Macrocyclic Chemistry: Current Trends and Future Prospectives. In *Ion-Pair Recognition by Ditopic Receptors*; Springer Netherlands: London, U.K., 2005; pp 137–152.

68. Itsikson, N. A.; Geide, I. V.; Morzherin, Y. Y.; Matern, A. I.; Chupakhin, O. N. Heteroditopic Receptors. *Heterocycles* 2007, *72*, 53–77.

69. Tasker, P. A.; Tong, C. C.; Westra, A. N. Co-Extraction of Cations and Anions in Base Metal Recovery. *Coord. Chem. Rev.* 2007, *251* (13–14), 1868–1877.

70. Beer, P. D.; Hopkins, P. K.; McKinney, J. D. Cooperative Halide, Perrhenate Anion-Sodium Cation Binding and Pertechnetate Extraction and Transport by a Novel Tripodal Tris(Amido Benzo-15-Crown-5) Ligand. *Chem. Commun.* 1999, 1253–1254.

71. Barboiu, M.; Vaughan, G.; van der Lee, A. Self-Organized Heteroditopic Macrocyclic Superstructures. *Org. Lett.* 2003, *5*, 3073–3076.

72. Chrisstoffels, L. A. J.; de Jong, F.; Reinhoudt, D. N.; Sivelli, S.; Gazzola, L.; Casnati, A.; Ungaro, R. Facilitated Transport of Hydrophilic Salts by Mixtures of Anion and Cation Carriers and by Ditopic Carriers. *J. Am. Chem. Soc.* 1999, *121*, 10142–10151.

73. Mahoney, J. M.; Beatty, A. M.; Smith, B. D. Selective Recognition of an Alkali Halide Contact Ion-Pair. *J. Am. Chem. Soc.* 2001, *123*, 5847–5848.

74. Mahoney, J. M.; Marshall, R. A.; Beatty, A. M.; Smith, B. D.; Camiolo, S.; Gale, P. A. Complexation of Alkali Chloride Contact Ion-Pairs Using A 2,5-Diamidopyrrole Crown Macrobicycle. *J. Supramol. Chem.* 2001, *1*, 289–292.

75. Koulov, A. V.; Mahoney, J. M.; Smith, B. D. Facilitated Transport of Sodium or Potassium Chloride across Vesicle Membranes Using a Ditopic Salt-Binding Macrobicycle. *Org. Biomol. Chem.* 2003, *1*, 27–29.

76. Mahoney, J. M.; Beatty, A. M.; Smith, B. D. Selective Solid-Liquid Extraction of Lithium Halide Salts Using a Ditopic Macrobicyclic Receptor. *Inorg. Chem.* 2004, *43*, 7617–7621.

77. Wu, Y. D.; Wang, D. F.; Sessler, J. L. Conformational Features and Anion-Binding Properties of Calix[4]Pyrrole: A Theoretical Study. *J. Org. Chem.* 2001, *66*, 3739–3746.

78. Gale, P. A.; Sessler, J. L.; Král, V. Calixpyrroles. *Chem. Commun.* 1998, 1–8.

79. Anzenbacher, P.; Try, A. C.; Miyaji, H.; Jursíková, K.; Lynch, V. M.; Marquez, M.; Sessler, J. L. Fluorinated Calix[4]Pyrrole and Dipyrrolylquinoxaline: Neutral Anion Receptors with Augmented Affinities and Enhanced Selectivities. *J. Am. Chem. Soc.* 2000, *122*, 10268–10272.

80. Levitskaia, T. G.; Marquez, M.; Sessler, J. L.; Shriver, J. A.; Vercouter, T.; Moyer, B. A. Fluorinated Calixpyrroles: Anion-Binding Extractants That Reduce the Hofmeister Bias. *Chem. Commun. (Camb.)* 2003, 2248–2249.

81. Wintergerst, M. P.; Levitskaia, T. G.; Moyer, B. A.; Sessler, J. L.; Delmau, L. H. Calix[4] Pyrrole: A New Ion-Pair Receptor as Demonstrated by Liquid–Liquid Extraction. *J. Am. Chem. Soc.* 2008, *130*, 4129–4139.

82. Kim, S. K.; Sessler, J. L.; Gross, D. E.; Lee, C. H.; Kim, J. S.; Lynch, V. M.; Delmau, L. H.; Hay, B. P. A Calix[4]Arene Strapped Calix[4]Pyrrole: An Ion-Pair Receptor Displaying Three Different Cesium Cation Recognition Modes. *J. Am. Chem. Soc.* 2010, *132*, 5827–5836.

83. Kim, S. K.; Lynch, V. M.; Young, N. J.; Hay, B. P.; Lee, C. H.; Kim, J. S.; Moyer, B. A.; Sessler, J. L. KF and CsF Recognition and Extraction by a Calix[4]Crown-5 Strapped Calix[4]Pyrrole Multitopic Receptor. *J. Am. Chem. Soc.* 2012, *134*, 20837–20843.

84. Romanski, J.; Piątek, P. Tuning the Binding Properties of a New Heteroditopic Salt Receptor through Embedding in a Polymeric System. *Chem. Commun.* 2012, *48*, 11346–11348.

85. Romanski, J.; Piątek, P. Selective Ammonium Nitrate Recognition by a Heteroditopic Macrotricyclic Ion-Pair Receptor. *J. Org. Chem.* 2013, *78*, 4341–4347.

86. Moerkerke, S.; Ménand, M.; Jabin, I. Calix[6]Arene-Based Cascade Complexes of Organic Ion Triplets Stable in a Protic Solvent. *Chemistry* 2010, *16*, 11712–11719.

87. Moerkerke, S.; Le Gac, S.; Topić, F.; Rissanen, K.; Jabin, I. Selective Extraction and Efficient Binding in a Protic Solvent of Contact Ion Triplets by Using a Thiourea-Based Bis-Calix[6]Arene Receptor. *Eur. J. Org. Chem.* 2013, *2013*, 5315–5322.

88. He, Q.; Peters, G. M.; Lynch, V. M.; Sessler, J. L. Recognition and Extraction of Cesium Hydroxide and Carbonate Using a Neutral Multitopic Ion-Pair Receptor. *Angew. Chem. Int. Ed.* 2017, *56*, 1–6.

89. Cram, D. J.; Lein, G. M. Host–Guest Complexation.36. Spherand and Lithium and Sodium-Ion Complexation Rates and Equilibria. *J. Am. Chem. Soc.* 1985, *107*, 3657–3668.

90. Zdanowski, S.; Piątek, P.; Romanski, J. An Ion Pair Receptor Facilitating the Extraction of Chloride Salt from the Aqueous to the Organic Phase. *New J. Chem.* 2016, *40*, 7190–7196.

91. Zakrzewski, M.; Załubiniak, D.; Piątek, P. An Ion-Pair Receptor Comprising Urea Groups and N-Benzyl-Aza-18-Crown-6: Effective Recognition and Liquid–Liquid Extraction of KCl Salt. *Dalton Trans.* 2018, *47* (2), 323–330.

3 Task-Specific Ionic Liquids for Metal Ion Extraction
Progress, Challenges, and Prospects

Mark L. Dietz and Cory A. Hawkins

CONTENTS

3.1 INTRODUCTION

Although examples of the low-melting organic salts now known as ionic liquids (ILs) were first reported more than a century ago (1), it is only in the last two decades that the unique properties and immense practical utility of these unusual solvents have been widely recognized (2–13). During this time, countless ionic liquids have been described, among them "conventional" ILs comprising a bulky, asymmetric organic cation paired with any of a wide range of organic or inorganic anions (14), solvate (i.e., complex cation) ionic liquids (15–20), protic ILs (21–23), and "inorganic liquids" (24–27). "Task-specific" ionic liquids (TSILs) represent an increasingly important subclass of ionic liquids consisting of a conventional IL to which is appended a functional group chosen to impart the solvent with specific physicochemical, catalytic, or solute-binding properties (28). Like essentially all areas of investigation involving ionic liquids (29), the study of the synthesis, characterization, and application of TSILs has experienced explosive recent growth. What began with a literal handful of publications nearly two decades ago (30–32) has come to include a vast and rapidly expanding body of literature encompassing hundreds of papers, patents, and reviews. While many of these have focused on the preparation of TSILs and their applications in synthesis or catalysis (33–35), a significant portion of published work has concerned their potential utility in separations, most commonly of metal ions. In this report, we examine the progress in this area, describe the limitations of TSILs that may impede further progress, and outline promising areas for future work. As will be shown, while much has been accomplished, significant additional effort will be required if TSILs are to achieve their full potential as media for metal ion extraction.

3.2 BACKGROUND

In 1998, Davis et al. prepared a series of novel ionic liquids in which the cationic component of the IL consisted of an imidazolium core functionalized to yield a species closely resembling miconazole, an antifungal medication (30). The size and (for that time) unusual complexity of the cation led the investigators to consider the possibility that ILs might be prepared from ions not only capable of functioning as the constituents of a solvent, but also of exhibiting specific, tailored interactions with dissolved solutes (28). The following year, Davis (31) showed that a thiazolium IL could serve as both solvent and catalyst for the benzoin condensation, thereby demonstrating that an IL could indeed be designed to interact with a solute in a pre-determined manner. Further demonstration of the capabilities of such TSILs soon followed (36–38), and since this time numerous studies describing their applications have appeared. In general terms, these applications can be broadly classified into three categories: altering the solvent properties of the IL relative to a more "conventional" analog bearing a simple hydrocarbon substituent (39); increasing the rate of environmental degradation of the IL (40, 41), and imparting the IL with the ability to catalytically activate or covalently bind a solute (36). It is the last of these areas that is most relevant to the application of TSILs in metal ion separations.

The separation of metal ions using TSILs commonly involves liquid–liquid extraction (also known as solvent extraction, abbreviated as SX), in which an aqueous solution containing the ion of interest and one or more other metal ions is contacted with an immiscible (i.e., hydrophobic) liquid in which is contained a metal ion-selective organic extractant. By proper choice of conditions (e.g., extractant, aqueous phase pH), the partitioning of the metal ions present can be adjusted to achieve the preferential extraction of the ion of interest. In conventional SX, the water-immiscible diluent comprises a molecular solvent (e.g., a paraffinic, aromatic, or chlorinated hydrocarbon). More recently, in an effort to avoid the adverse environmental impact associated with the use of many of these organic solvents, ionic liquids have been proposed as alternative diluents (42–48). Despite decreased fugitive emissions for IL-based SX systems (due to the near absence of vapor pressure of typical ionic liquids (49)), however, these systems remain susceptible to loss of the extractant to the aqueous phase. Many are also plagued by the loss of the IL constituents through the dissolution or ion-exchange processes involving aqueous phase components (50, 51). In addition, the solubility of a number of extractants in ionic liquids has proven to be limited (52–54). In principle, the use of TSILs can provide a means of reducing or eliminating these problems. That is, as shown in Figure 3.1A, a TSIL typically

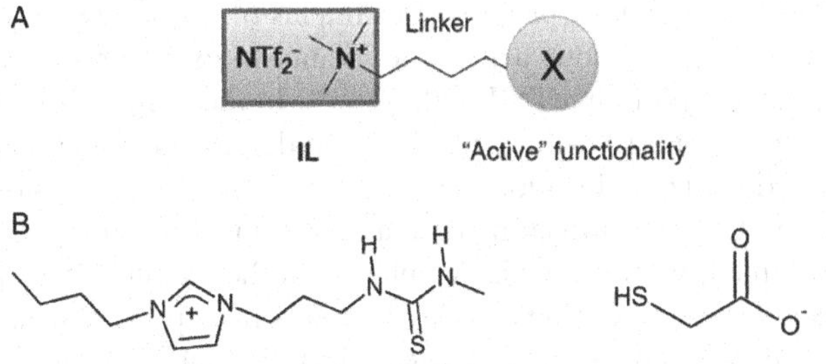

FIGURE 3.1 A: "Generic" structure of a TSIL; B: Thiourea-derivatized *N,N'*-dialkylimidazolium cation (left) and a thiol-functionalized carboxylate anion (right).

consists of three components: the cationic and anionic constituents of the IL "backbone," the active functional group responsible for carrying out the "task" of interest (in this case, metal ion extraction), and a linker joining this group to the IL backbone. Here then, a TSIL comprises an extractant *bound to* rather than *dissolved in* the ionic liquid, thereby eliminating the problem of extractant insolubility in the IL and (ideally) reducing the loss of extractant or IL constituents to the aqueous phase.

As is illustrated by the thiourea-derivatized *N,N'*-dialkylimidazolium cation (55) and the thiol/thioether-derivatized carboxylate anions (56) shown in Figure 3.1B, either the IL cation or anion may incorporate the functional group. To date, however, the overwhelming majority of TSILs incorporate functionalized cations. The popularity of functionalized IL cations is undoubtedly a consequence of the comparative ease of incorporating active groups into the cation. Figure 3.2, for example, depicts a representative synthesis, in this case for the preparation of a TSIL incorporating a phosphoryl group suitable for the extraction of uranium from acidic media (57).

As can be seen, only two basic steps are involved: quaternization of an alkylimidazole, followed by anion metathesis to yield a water-insoluble analog of the original halide salt. Of course, if the appropriate alkyl halide is not available for the quaternization reaction, it obviously must be prepared prior to that step, adding to the cost and overall synthetic difficulty. While synthetic challenges also characterize the preparation of functionalized IL anions for use in TSILs (58), a more

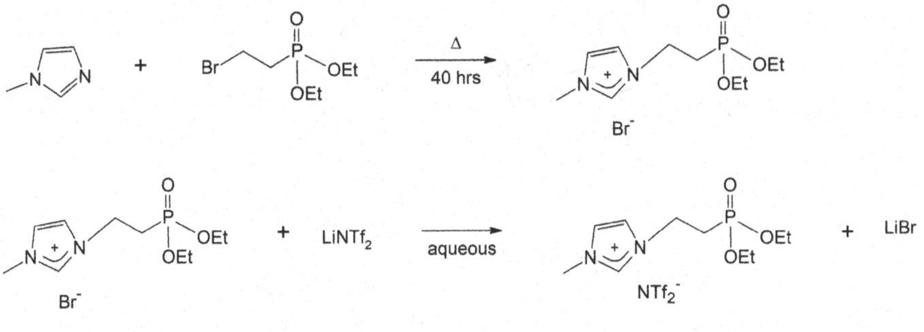

FIGURE 3.2 Representative synthetic scheme for the preparation of a cation-functionalized task-specific ionic liquid.

important factor in the relative paucity of examples of derivatized anions is likely the ready availability of various acidic extractants whose deprotonated form can serve as a functional anion in a TSIL (54, 59–61). In the sections that follow, each of these three types of TSILs—those in which the IL cation is functionalized, those based on a functionalized IL anion, and those in which the IL anion comprises the anionic form of a conventional metal ion extractant (e.g., an organophosphorus acid)—are examined, with particular emphasis on the extent to which regardless of type, the use of these "working salts" (62) results in improved metal ion separation performance (e.g., efficiency and selectivity) vis-à-vis conventional IL-based extraction systems. Our objective is not the encyclopedic coverage of the application of TSILs in metal ion separations, but rather to provide an overview of the most significant results reported to date and an indication of their implications for future efforts in this area.

3.3 TSILs INCORPORATING FUNCTIONALIZED CATIONS

Although TSILs involving the attachment of a functional group to all of the most widely known families of IL cations, including pyridinium (63), piperidinium (64), pyrrolidinium (64), and quaternary ammonium ions (65), have been described, TSILs based on imidazolium cations are by far the most common. Indeed, the first report of a TSIL incorporating a metal ion complexing moiety, a 2001 study by Visser et al. (36), involved an imidazolium cation to which was appended a urea, thiourea, or thioether functional group. A combination of the resultant cations (examples of which are shown in Figure 3.3) and hexafluorophosphate anion (PF_6^-) yielded liquids capable of serving as both a hydrophobic solvent and a metal ion extractant. In particular, Hg^{2+} and Cd^{2+} were found to be efficiently extracted by certain of the compounds prepared, with distribution ratios (D_M, a measure of extraction efficiency, defined as $[M]_{org}/[M]_{aq}$) as high as 710 for Hg^{2+} for one of the thioether derivatives (as a 1:1 w/w solution in $C_4mim^+PF_6^-$). Not surprisingly, extraction was far less efficient ($D < 1$) for analogous unfunctionalized ILs, $C_nmim^+PF_6^-$ ($n = 4$–8).

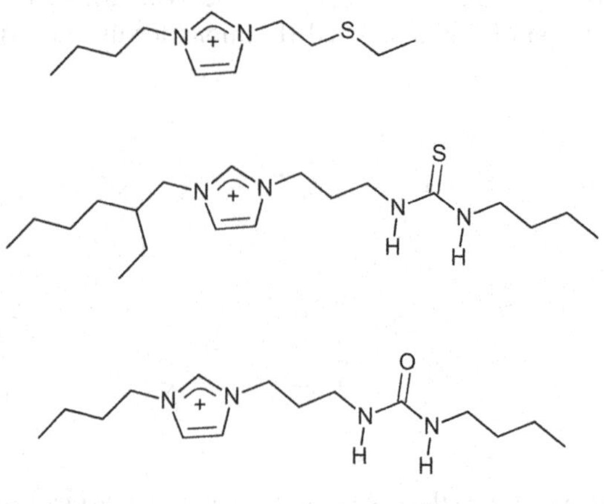

FIGURE 3.3 Thioether-, urea-, and thiourea-functionalized imidazolium cations.

Curiously, however, no results were provided for any urea-, thiourea-, or thioether-based extractant dissolved in the latter ILs. It is, therefore, unclear if, in terms of extraction efficiency, any of the TSILs actually provide an improvement over the results achievable with a more "conventional" extraction system. Also notably absent are data demonstrating the recovery of extracted metal ions from the TSIL phase. Despite these significant shortcomings, this work nonetheless did establish the feasibility of appending a metal complexing moiety to an IL cation and employing the resultant material as a metal ion extractant.

The following year, Visser (55) described the results of a more detailed characterization of these same TSILs and a further examination of their utility in the extraction of Hg^{2+} and Cd^{2+}. All of the TSILs were found to be markedly more viscous than $C_n mim^+ PF_6^-$ ($n = 4$–8), the apparent result of the presence of heteroatoms and long, bulky alkyl chains in the TSILs. This made it necessary to dilute them with a conventional ionic liquid prior to their use in extraction studies. More problematic, however, was the near absence of pH dependency of metal ion extraction. That is, unexpectedly, decreasing solution pH did not lead to back-extraction of the extracted metal ions from the IL phase. Thus, the issue of how best to effect metal ion recovery remained unaddressed.

Since these early reports, a number of studies have appeared describing TSILs based on imidazolium cations to which have been appended any of a wide variety of functional groups capable of interacting with metal ions (66–84). In certain instances, these compounds have been applied not in solvent extraction *per se*, but rather in solid-phase extraction (85, 86) or extraction chromatography (87–89), a specialized form of liquid chromatography in which the stationary phase comprises an extractant, or its solution in an appropriate organic solvent, dispersed on a porous solid support. Because the application of TSILs in the preparation of such materials has been treated elsewhere (90), however, it will not be considered in any detail here. While a handful of these studies have described the extraction of transition elements (66, 69, 72, 79, 84) or heavy metals (70, 72, 81), most have concerned the design of systems suitable for the removal of actinides and/or fission products from aqueous solution. In 2006, for example, Luo et al. (67) reported the preparation of a pair of TSILs incorporating an aza-crown ether fragment (Figure 3.4), and evaluated these compounds in the biphasic extraction of two important fission products, Cs-137 and Sr-90 (as represented by their nonradioactive analogs, CsCl and $SrCl_2$, respectively).

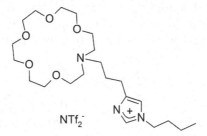

FIGURE 3.4 *N*-(3-butylimidazoliumpropyl)aza-18-crown-6 ether *bis*[(trifluoromethyl)sulfonyl]imide.

Unexpectedly, the values of D_M for both Cs^+ and Sr^{2+} were found to be *lower* with the TSILs than with either dicyclohexano-18-crown-6 (DCH18C6) or *N*-octylaza-18-crown-6, dissolved in a conventional IL ($C_4mim^+Tf_2N^-$). This observation was attributed to the proximity of the cationic imidazolium group to the complexing moiety of the TSILs and the resultant repulsion of the target metal cations. Also unexpected (given the ostensible ease with which the aza-crown moiety can be protonated) were the lower recoveries of both metal ions from the TSILs, rendering recycle of these TSILs inferior to that of a conventional IL containing an extractant.

Somewhat more promising results were obtained that same year by Ouadi et al. (68) in the extraction of americium using an imidazolium substructure onto which a hydroxybenzyl-amine moiety had been grafted (Figure 3.5). Here, the Am(III) extraction efficiency of the TSIL was found to be superior (by up to an order of magnitude) to that of the analogous extractant, 2-propylaminomethylphenol, dissolved in $C_4mim^+Tf_2N^-$. In addition, the recovery of extracted Am(III) (and thus, the recycle of the TSIL) was found to be facile, with stripping exceeding 99% in each of several trials upon contact of the Am-loaded TSIL phase with 1 M $HClO_4$. Lastly, in contrast to the results reported previously by Visser et al. (55), which showed TSILs to be too viscous to be employed in undiluted form, the viscosity of the *bis*[(trifluoromethyl)-sulfonyl]imide salt of the 2-hydroxybenzylamine-functionalized butylimidazolium cation was considered to be sufficiently "low" ($\eta = 2070$ cP) as to allow for the direct use of the compound as the extracting phase. These benefits, however, are partly outweighed by modeling data suggesting that Am partitioning occurs via an anion exchange process involving an anionic Am complex and the Tf_2N^- anion of the ionic liquid, an observation having significant negative implications for the "greenness" of the TSIL. Along these same lines, solubilization losses of the TSILs at the high pH values required for efficient extraction, conditions under which the TSIL is deprotonated, may be substantial, but were not determined.

As was the case for Am stripping from the hydroxybenzylamine-functionalized TSILs of Ouadi et al. (68), facile recovery of extracted metal ion was also achieved by Wang et al. (74) in the extraction of yttrium from aqueous solution by a carboxylic acid-functionalized imidazolium-based TSIL (Figure 3.6). In this system, however, the viscosity of each of the compounds prepared—all derivatives of C_nmim^+ (with $n = 4$–8)—was so high as to preclude their use without dilution by the corresponding nonfunctionalized IL. Further restricting their use was the observation of emulsion formation for the *n*-octyl-derivative, thus limiting extraction studies to derivatives bearing a short alkyl chain. Most importantly, despite the expected reduction in the prevalence of ion-exchange processes arising from incorporation of the extracting

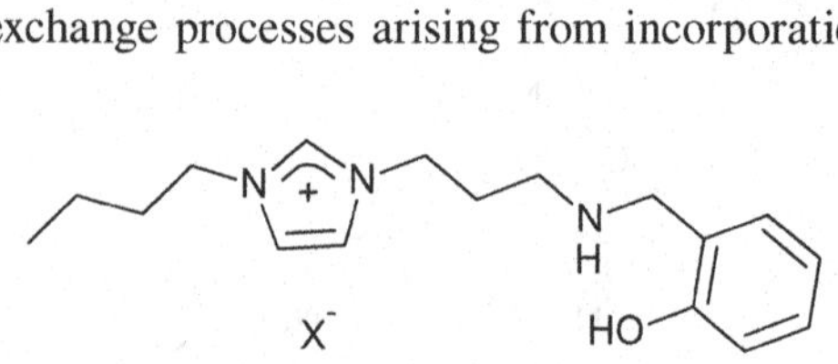

FIGURE 3.5   1-Butyl-3-[3-(2-hydroxybenzoamino)propyl]-3H-imidazol-1-ium TSILs (X= hexafluorophosphate (PF_6^-) or *bis*[(trifluoromethylsulfonyl)]imide (Tf_2N^-).

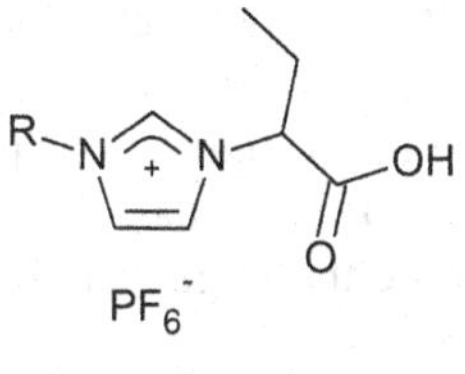

FIGURE 3.6   1-Alkyl-3(1-carboxylpropyl)imidazolium hexafluorophosphate.

moiety into the IL cation, exchange of aqueous Y^{3+} for the $C_n mim^+$ component of the IL diluent was reported to be the predominant mode of yttrium extraction. The problems associated with such processes have been noted previously (50, 91–93). These results thus appear to call into question the desirability of incorporating ionizable extractants into the cationic constituent of a TSIL. This is particularly true given that other work, in which the acidic organophosphorus extractant bis-2-ethylhexylphosphoric acid (HDEHP) was simply dissolved in $C_n mim^+$-based ILs for application in the extraction of several trivalent lanthanide and actinide ions, has established that an extraction mechanism not involving such ion exchange governs extraction into both the ILs and dodecane for this extractant (94).

A variety of other extractant fragments have also been incorporated into TSILs. The well-established chemistry and widespread application of neutral organophosphorus reagents in SX systems employing molecular solvents, for example, has led to considerable interest in their application as the active functional group of a TSIL. Among such reagents, none is better known than tri-*n*-butylphosphate (TBP), and an imidazolium cation has been functionalized with a TBP-like fragment to yield a TSIL. In particular, Li et al. (80) described the preparation and characterization of 1-dibutylphosphorylpropyl-3-methyl-imidazolium *bis*[(trifluoromethyl)sulfonyl] imide (Figure 3.7), along with its possible application in the extraction of U(VI) from nitric acid solution. The U(VI) distribution ratios with this TSIL were found to rise as the aqueous acidity increased, reaching 10 at the upper end of the range of acidities considered (0.1–5 M).

No data were provided for a solution of TBP in analogous, unfunctionalized ILs, but these D_U values are consistent with those reported previously by Dietz et al. (95) for a 1.2 M solution of TBP in $C_n mim^+ Tf_2 N^-$ ($n = 5$–10) over the same acid range. The D_U value at high acidity, while modest, along with the decline in D_U with decreasing acidity would seem to suggest that it could be possible to devise a TSIL-based analog to the conventional PUREX process for the recovery of uranium (96). According to

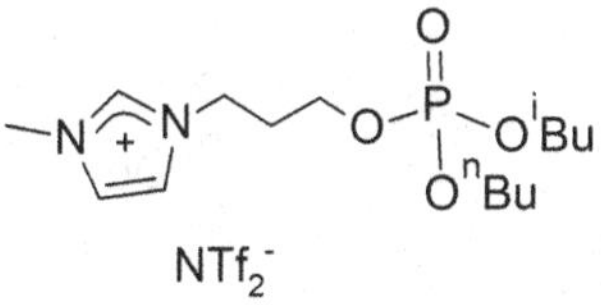

FIGURE 3.7   1-Dibutylphosphorylpropyl-3-methylimidazolium *bis*[(trifluoromethyl)sulfonyl]imide.

the authors, however, the overall extraction behavior of the TSIL may represent a composite of contributions from the partitioning of a neutral uranium complex and (in contrast to PUREX) of cationic or anionic complexes, processes whose undesirability is well known (50, 51). Because data demonstrating the involvement of ion exchange are sparse, a far more thorough examination of this system is needed before its utility can be reliably assessed.

Rout et al. (57) examined the behavior of a closely related TSIL, diethyl-2-(3-me thylimidazolium)ethylphosphonate bis[(trifluoromethyl)sulfonyl]imide (Figure 3.8), analogous to the well-known uranium extractant diamyl amylphosphonate (DAAP), which serves as the basis of a commercial extraction chromatographic resin for the separation and preconcentration of uranium (97). In particular, the efficacy of this TSIL in the extraction of Am(III), Pu(IV) and U(VI) from nitric acid solution was determined. As a first step in this evaluation, the extraction of the three ions into analogous, unfunctionalized ILs containing no extractant was measured. Interestingly, the extraction of Pu(IV) was found to be significant (with D_{Pu} as high as 37.5) for $C_8mim^+Tf_2N^-$ alone, the apparent result of the exchange of one or more anionic nitrato complexes of Pu(IV) for the anionic constituent of the ionic liquid. Such behavior was not observed for either Am(III) or U(VI) (which are less prone to nitrato complex formation) under the same conditions, resulting in a system exhibiting excellent extraction selectivity for Pu(IV) over both Am(III) and U(VI). Introduction of the TSIL to the system (0.2–0.5 M in $C_4mim^+Tf_2N^-$) had two effects. First, distribution ratios at intermediate acidities (i.e., 1 M HNO_3) increased substantially (by ca. 3–4 orders of magnitude) vs. those obtained in the absence of TSIL. At the same time, the direction of the acid dependency of D_{Pu} reversed from that seen in $C_8mim^+Tf_2N^-$ alone, so that D_{Pu} values fell steeply over the 1–5 M HNO_3 range. No explanation for these observations was provided, aside from a suggestion that it may be the result of a combination of extraction mechanisms, including both neutral complex formation and ion exchange. Whatever its origin, the effect of TSIL addition was to reduce the Pu(IV)/U(VI) separation factors under many conditions. In addition, the recovery of extracted plutonium was rendered more difficult. In fact, Pu(IV) recovery from a 0.4 M solution of the TSIL in $C_4mim^+Tf_2N^-$ was incomplete even after 10 stripping steps with 0.03 M diethylenetriamine pentacetic acid (DTPA). Given this and the lack of reported data for solutions of DAAP in conventional ILs, it remains unclear what advantages, if any, are afforded by using a TSIL in this situation.

Mohapatra et al. prepared and characterized a pair of carbamoylmethylphosphine oxide (CMPO)-functionalized TSILs (Figure 3.9), and have evaluated them in the extraction of actinides and lanthanides from acidic aqueous phases (75). Both TSILs

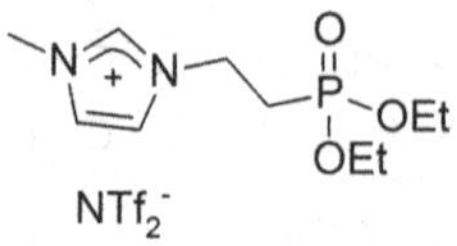

FIGURE 3.8 Diethyl-2-(3-methylimidazolium)ethylphosphonate bis[(trifluoromethyl) sulfonyl]-imide.

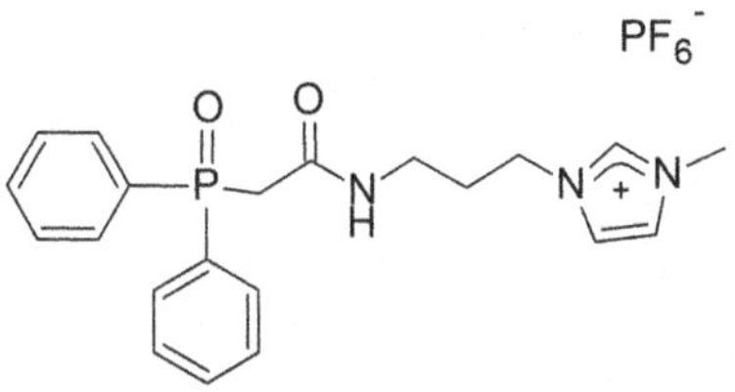

FIGURE 3.9 CMPO-imidazole propanamide.

examined, which differed only in anion (PF_6^- vs. Tf_2N^-), were found to be "pasty semi-solids," rendering their use without a diluent impractical and again illustrating the problem of high viscosity associated with many TSILs. Unfortunately, the addition of a diluent, while solving the problem of viscosity, led to the emergence of ion exchange as a significant pathway for metal ion extraction. Finally, and unexpectedly, the metal ion distribution ratios observed with the TSILs were generally lower, in some cases substantially, than those obtained with a solution of CMPO in a conventional IL under identical conditions.

Trialkylphosphine oxides (e.g., trioctylphosphine oxide, TOPO) have long been of interest in conventional SX (98), and in 2016 Paramanik et al. (82) reported the first example of a trialkylphosphine oxide-functionalized TSIL (designated PO–TSIL; Figure 3.10). Studies of the extraction of a range of metal ions, including UO_2^{2+}, Pu^{4+}, Am^{3+}, and representative fission products (Eu^{3+}, Sr^{2+}, and Cs^+) by the PO–TSIL showed that only uranium and plutonium are appreciably extracted. Again, the high viscosity of the TSIL made it necessary to dilute it with a conventional IL prior to use, but despite the dilution, the rates of extraction and back-extraction of these ions remained slow relative to those observed with molecular diluents. Curiously, while the nitric acid dependency of the extraction of uranium into $C_8mim^+Tf_2N^-$ by TOPO resembled that observed for a dodecane solution of the same extractant, thus suggesting that both systems involve the extraction of a neutral species, analogous studies using the PO–TSIL were found to yield a completely different acid dependency, one consistent with a significant contribution of cation exchange to the overall extraction process.

Like organophosphorus-based extractants, diglycolamides (DGAs) have also attracted significant interest as the basis for TSILs. In 2013, Mohapatra et al. (99) reported the first preparation of DGA-functionalized ionic liquids (Figure 3.11; DGA–TSILs), and compared their performance in the extraction of actinides (i.e.,

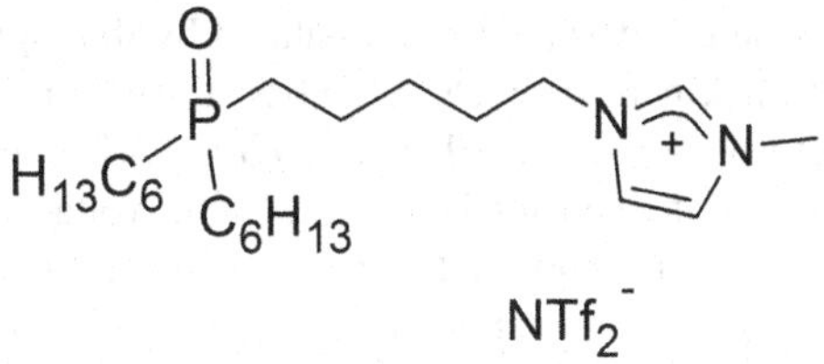

FIGURE 3.10 Phosphine oxide-functionalized imidazolium-based ionic liquid (PO–TSIL).

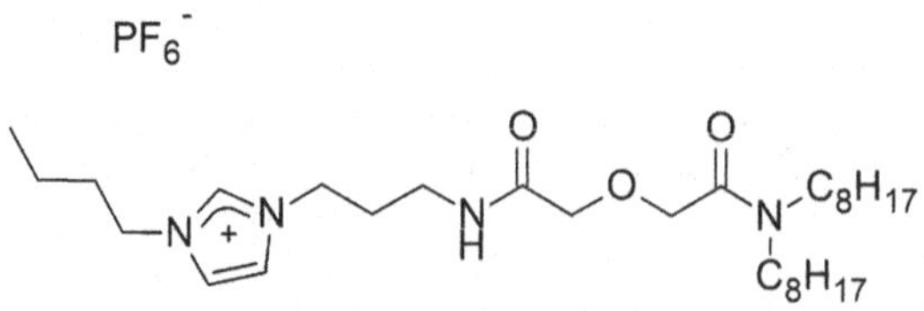

FIGURE 3.11 Diglycolamide-functionalized imidazolium-based ionic liquid (DGA–TSIL).

Am^{3+}, Pu^{4+}, Np^{4+}, and UO_2^{2+}) and selected fission products (i.e., Eu^{3+}, Sr^{2+}, Cs^+) from acidic media to that of N,N,N',N'-tetra-n-octyl-diglycolamide (TODGA), dissolved in either an analogous conventional IL (i.e., $C_4mim^+Tf_2N^-$) or dodecane. For each of the tested ions, higher distribution ratios were observed with TODGA dissolved in the IL than with the corresponding solution in dodecane. In turn, even higher distribution ratios were obtained for the DGA–TSILs. Consistent with observations made for other families of TSILs, however, the viscosity of the DGA–TSILs was high, and as a result the rate of extraction of Am^{3+} was found to be ca. 5 times slower than into a solution of TODGA in an IL. Although dilution of the TSIL with an IL was effective in reducing this viscosity, it was accompanied by a significant (up to an order of magnitude) reduction in Am^{3+} extraction efficiency. In addition, the presence of the IL diluent led to an undesirable change in the mechanism of extraction, from a process involving the transfer of a neutral species to one involving the exchange of the metal ion for the cationic constituent of the IL. Also, because the DGA–TSILs exhibited high D_{Am} values over a wide range of acidities, the recovery of extracted Am^{3+} could not be accomplished simply by changing the acidity of the aqueous phase. Instead, a solution of a complexing agent (such as EDTA or DTPA in 1 M guanidine carbonate) was required. When a 1:4 dilution of the DGA–TSIL in $C_4mim^+Tf_2N^-$ was employed as the extracting phase, eight strip stages were required to effect complete Am recovery. Finally, on a more positive note, the radiolytic stability of the two DGA–TSILs was reported to be superior to that exhibited by other IL-based solvent extraction systems proposed for nuclear reprocessing applications.

In a follow-up paper (76), the authors expanded their investigation of DGA–TSILs to include a more detailed examination of the extraction behavior of actinides other than Am^{3+}, in particular, UO_2^{2+}, NpO_2^{2+}, PuO_2^{2+}, Np^{4+}, and Pu^{4+}. Studies of the nitric acid dependency of D_M for the various ions showed that in all cases, the extraction proceeds by an ion-exchange mechanism in which transfer of the metal ion into the IL phase is accompanied by loss of the cationic component of the diluent, $C_nmim^+Tf_2N^-$, whose use was again necessitated by the high viscosity (and the resultant slow extraction kinetics) of DGA–TSILs. Further investigation revealed the extracted species to be a cationic 1:1 metal:DGA–TSIL complex, regardless of metal ion, a result later confirmed for UO_2^{2+} via cyclic voltammetry (100). In this respect then, these systems are similar to those involving the extraction of alkali and alkaline earth cations into conventional ILs by macrocyclic polyethers, where the formation and transfer of cationic 1:1 metal:extractant complexes underlies the propensity of the systems toward ion exchange (101). Although the recovery of extracted metal ions from ionic liquid-based systems is normally (in the words of the authors) "a challenging task," aqueous solutions of various complexing agents—1 M sodium carbonate for hexavalent actinides and 0.5 M oxalic acid for tetravalent actinides—were found to yield satisfactory stripping.

In a subsequent publication (102), the authors evaluated the utility of one of their DGA–TSILs, as a solution in $C_4mim^+Tf_2N^-$, in the remediation of nuclear wastes. Of the nearly 20 components of the synthetic high-level waste (SHLW) solution examined, several (e.g., Y^{3+}, Zr^{4+}, Pd^{2+}) exhibited a D_M value comparable to or exceeding that seen previously (76) for Am^{3+} (ca. 34) from 3 M HNO_3. A more detailed examination of the extraction of Am^{3+} and Eu^{3+} as a function of SHLW acidity showed that extraction of Eu exceeded that of Am by up to an order of magnitude, depending on the nitric acid concentration. In addition, high levels of Eu extraction were found to depress D_{Eu} values somewhat. Several other lanthanides (e.g., Nd^{3+}) were also found to be well extracted. As a result, the extraction of tri- and tetravalent actinide ions is certain to be significantly (and adversely) affected in the presence of HLW constituents.

That same year, Yun et al. (78) prepared a series of related short-chain DGA–TSILs (Figure 3.12) and found them and/or their lanthanide complexes to be sufficiently water soluble as to permit the use of the TSILs as stripping agents. For example, while Eu^{3+} was efficiently extracted ($D_{Eu} = 145$) by 10 mM TODGA in $C_6mim^+Tf_2N^-$ from dilute (0.01 M) nitric acid, addition of as little as 45 mM of one of the TSILs to the aqueous phase reduced the value of D_{Eu} to <0.02, a decrease of nearly 4 orders of magnitude. Unfortunately, the stripping process was reported to lack selectivity.

Amide- and phosphoramide-based TSILs have also attracted interest, although far less than that garnered by DGA–TSILs. In 2012, for example, Rout et al. (73) synthesized the new amide-functionalized ionic liquid N,N-dioctyl-2-(3-methylimidazolium)acetamide *bis*[(trifluoro-methyl)sulfonyl]imide (Figure 3.13) and examined its behavior in the extraction of Am(III), Pu(IV), and U(VI) from nitric acid. Of the three ions, only Pu(IV) was appreciably extracted when the TSIL was

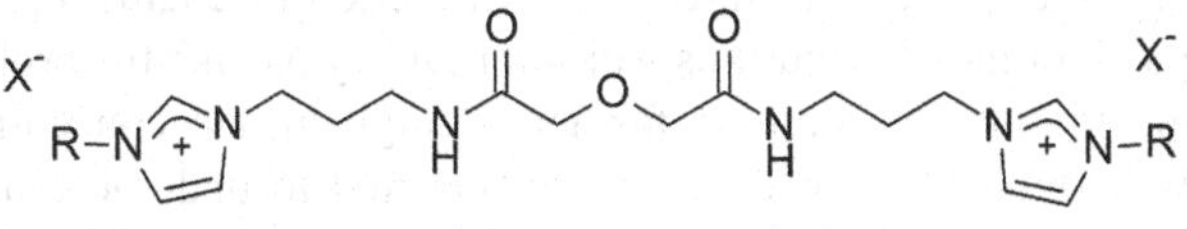

FIGURE 3.12 Dicationic diglycolamide-functionalized ionic liquids (X = Br-, Tf_2N^-).

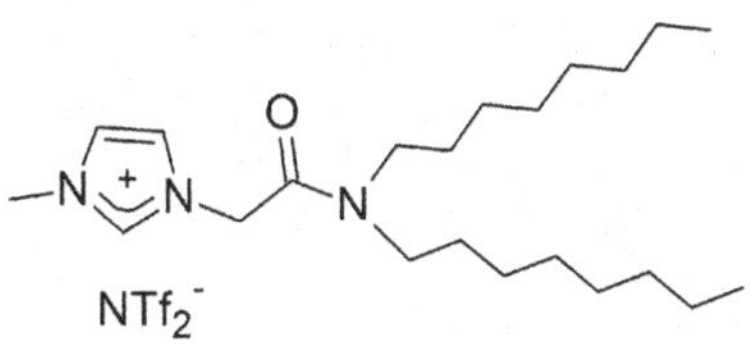

FIGURE 3.13 N,N-dioctyl-2-(3-methylimidazolium)acetamide *bis*[(trifluoromethyl)sulfonyl]imide.

employed (as a solution in $C_4mim^+Tf_2N^-$). The result was separation factors for the extraction of Pu(IV) over both Am(III) and U(VI) exceeding 3000 at certain acidities. Comparison of these results to those obtained for an amide-based molecular extractant, dioctylbutyramide (DOBA), in either a conventional IL ($C_4mim^+Tf_2N^-$) or a molecular solvent (dodecane), clearly showed the superior extraction selectivity of the amide-TSIL. Unfortunately, this improvement was found to come at the cost of substantially greater difficulty in recovering (i.e., stripping) extracted Pu(IV). That is, D_{Pu} exhibits a complex dependency upon nitric acid concentration, initially declining with increasing acidity up to ca. 4 M HNO_3, and then rising as the acidity is increased further. Thus, as is the case for various other TSILs, an acid swing is ineffective for recovering extracted metal ions, making necessary the use of various complexing agents. In this instance, a solution of either citric or acetohydroxamic acid (0.5 M in 0.3 M HNO_3) eventually (after 10 strip stages) provided complete recovery of extracted Pu(IV). It is important to note that in an earlier investigation (57), Pu(IV) was found to be well extracted into imidazolium-based ionic liquids *even in the absence of extractant* from aqueous phases containing high concentrations (>4 M) of nitric acid, while U(VI) and Am(III) were not. It is therefore difficult to characterize the results obtained with the amide-TSIL as constituting a substantial improvement over this simpler alternative.

More recently, Rama et al. (103) described the preparation and characterization of a novel phosphoramide-based TSIL, diethyl-3-(3-butylimidazolium)propylphosphoramide *bis*[(trifluoromethyl)sulfonyl]imide (Figure 3.14), along with its use in the extraction of Pu(IV), U(VI), and Am(III) from acidic nitrate media. As was the case for the amide-functionalized TSIL described by Rout et al. (73), in this system, Pu(IV) was found to be well extracted ($D \approx 35$ at 5 M HNO_3), while U(VI) and Am(III) were not, when a solution of the TSIL in $C_4mim^+Tf_2N^-$ (0.3 M) was employed. Here, however, Pu(IV) extraction increased monotonically with acidity, eventually reaching a plateau at high acidity ($[HNO_3] \geq 5$ M). In addition, comparison of Pu(IV) extraction to that observed for the unfunctionalized IL alone showed that the presence of the TSIL increased the distribution ratios (D_{Pu}) as much as 1–2 orders of magnitude. The extraction behavior, in particular, the shape of the nitric acid dependency of D_{Pu}, was attributed to an extraction mechanism predominantly involving the partitioning of a neutral species formed by pairing the protonated TSIL with an anionic Pu–nitrato complex formed at high nitrate concentrations. This same complex, it should be noted, has been reported to undergo anion exchange with $C_4mim^+Tf_2N^-$, the diluent employed here. Accordingly, the overall extraction is apparently a hybrid of at least two extraction processes.

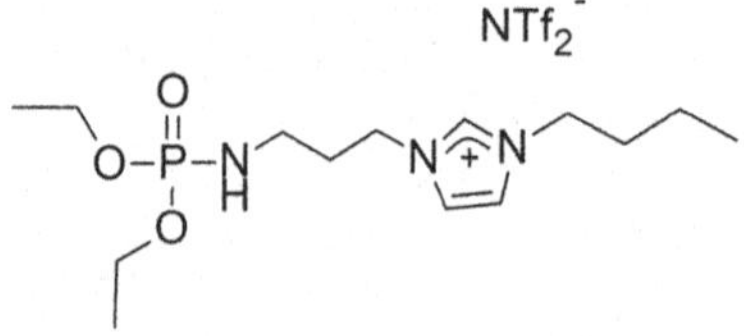

FIGURE 3.14   Diethyl-3-(-3-butylimidazolium)propylphosphoramide *bis*[(trifluoromethyl)-sulfonyl]imide.

As noted earlier, while extractant-functionalized *N,N'*-dialkylimidazolium salts comprise the majority of reported TSILs, other types of ionic liquid cations have also served as the basis of TSILs. Several reports have appeared, for example, in which an extractant fragment has been appended to an *N*-alkylpyridinium cation. In 2008, for example, Papaiconomou et al. (64) reported the preparation of a series of TSILs (Figure 3.15) comprising either a disulfide or nitrile moiety attached to a 4-methyl-pyridinium, 1-methyl-pyrrolidinium, or 1-methyl-piperidinium cation, paired with a tetrafluoroborate (BF_4^-), trifluoromethylsulfonate ($CF_3SO_3^-$), or nonafluoro- butyl-sulfonate (NfO^-) anion, and described their application in the extraction of a several toxic (e.g., Hg^{2+}) and precious (e.g., Ag^+) metal ions. As might be expected, most of the unfunctionalized ILs were found to be incapable of extracting the metal ions tested, Hg^{2+} extraction by ILs based on the pyridinium cation being a notable exception. Incorporation of the nitrile group led to improved extraction of Ag^+ and Pd^{2+} ions, while TSILs with a disulfide exhibited improved extraction of Hg^{2+} and Cu^{2+}. Interestingly, the authors observed a significant influence of the IL cation type (e.g., pyridinium vs. imidazolium) on the extraction behavior of a series of TSILs bearing a given functional group (e.g., nitrile), with pyridinium and pyrrolidinium cations yielding distribution ratios higher than those observed for analogous piperidinium or imidazolium ILs. It is important to note here, however, that Hg^{2+} is well extracted ($D > 1000$) by conventional pyridinium-based ionic liquids, even in the absence of added extractant, thus complicating interpretation of this observation. Clearly, careful examination of a variety of appended extractant fragments will be required to determine if, in fact, the performance of cation-functionalized TSILs is routinely impacted by the choice of cation.

More recent work by Li et al. (63) has also employed *N*-alkylpyridinium cations as the basis of a TSIL (designated as Phos–C_3–pyr$^+$Tf$_2$N$^-$), in this instance, one incorporating a tributyl phosphate (TBP)-like functional group designed to extract uranyl ion (Figure 3.16). As is the case for most other TSILs (see above), the viscosity of

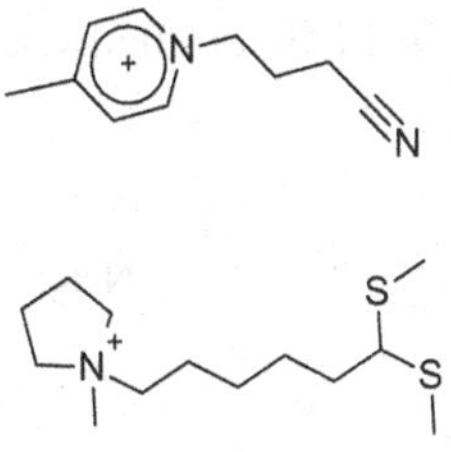

FIGURE 3.15   A: 1-butyronitrile-4-methylpyridinium cation; B: 1-methyl-1-[4,5]-*bis*(methyl sulfide)pentylpyrrolidinium cation.

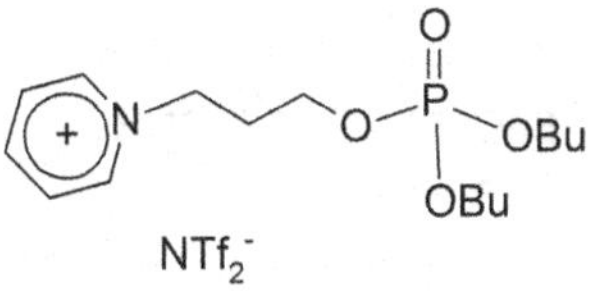

FIGURE 3.16   1-dibutylphosphorylpropyl-3-pyridine *bis*[(trifluoromethyl)sulfonyl]imide.

Phos–C$_3$–pyr$^+$Tf$_2$N$^-$ was found to be high (>1000 cP at 20°C), thereby requiring that it be diluted (1:2 v/v) with C$_4$mim$^+$Tf$_2$N$^-$ prior to use in extraction studies. Measurements of the nitric acid dependency of D_U for the TSIL (as a solution in C$_4$mim$^+$Tf$_2$N$^-$) and a solution of TBP in C$_4$mim$^+$PF$_6^-$ indicated that there is little difference in the extraction behavior of the two systems. Thus, as was the case in the work of Papaiconomou et al. (described above), in which highly efficient extraction of Hg^{2+} by an unfunctionalized IL was observed (64), these results raise a question as to what, if any, advantage is offered by a TSILs vis-à-vis a system employing a conventional IL.

That same year, Chayama et al. (104) described the preparation and character-ization of a TSIL incorporating a monothioether-functionalized pyridinium cation, specifically 3-thiapentyl-pyridinium *bis*[(trifluoromethyl)sulfonyl]imide, and exam-ined its suitability as an extractant/solvent for Ag(I), Cu(I), Pd(II), and Pt(II), so-called "class b" metal ions. Highly efficient extraction and excellent selectivity over a number of other cations (e.g., transition metals, alkali, and alkaline earth cations), which was attributed to the significant affinity of soft bases (here, a thioether) for soft acids, was observed. Unfortunately, no data were presented to demonstrate the superiority of the TSIL over a system simply comprising a thioether dissolved in a conventional IL.

Like pyridinium-based TSILs, those based on quaternary ammonium cations ("quats") have received comparatively little attention as media for metal ion extrac-tion. In 2007, Ouadi et al. (65) prepared a series of quaternary ammonium-based TSILs incorporating a phosphoryl group, yielding several novel, hydrophobic ILs designed to extract U(VI) from acidic nitrate media. Of the three phosphoryl-containing ILs prepared (Figure 3.17), the phosphoramidate (as a 1.1 M solution in trimethylbutylammonium *bis*[(trifluoromethyl)sulfonyl]imide) provided the most efficient extraction (D_U=170 from 3 M HNO$_3$), while analogous phosphate- and phosphonate-functionalized TSILs yielded extraction poorer than that observed for a solution of TBP in a conventional quaternary ammonium IL.

More recently, Fagnant et al. (105) have carried out a preliminary investigation of the phase behavior of betaine *bis*[(trifluoromethyl)sulfonyl] imide–water mixtures

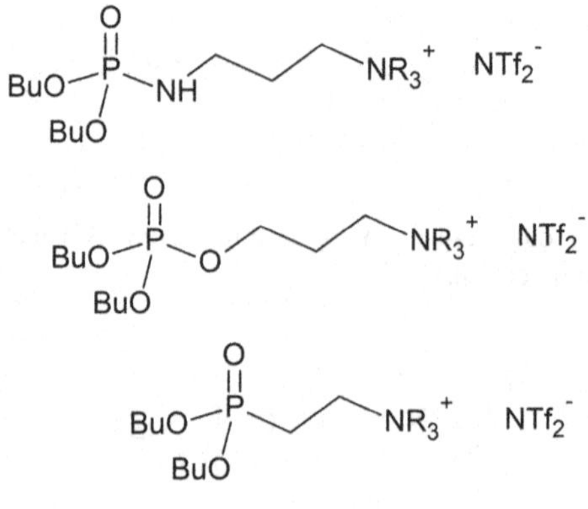

FIGURE 3.17 Phosphoramidate- (top), phosphate- (middle), and phosphonate- (bottom) functionalized quaternary ammonium-based ionic liquids.

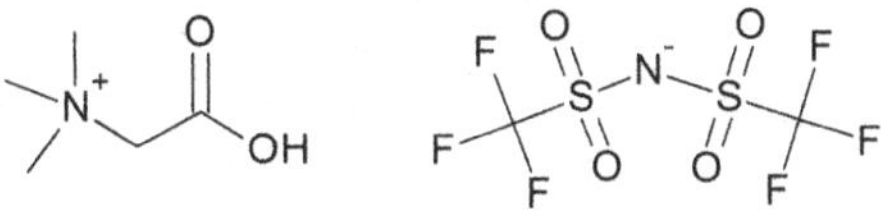

FIGURE 3.18   Betaine *bis*[(trifluoromethyl)sulfonyl]imide.

(where betaine = Hbet⁺ = *N,N,N*-trimethylglycine, a carboxylate-functionalized qua-ternary ammonium cation; Figure 3.18), a system of possible utility in metal ion separations. This study emphasized the fundamental aspects of the phase equilib-rium, however, and although temperature-dependent partitioning of neodymium was observed, no demonstration of the applicability of the TSIL to metal ion separations was demonstrated.

The following year, Sasaki et al. (106) compared the extraction of U(VI) from aqueous nitric acid by Hbet⁺Tf₂N⁻ to that obtained using *N,N,N*-trimethyl-*N*-propylammonium *bis*[(tri-fluoromethyl)sulfonyl]imide (TMPA), a structural ana-log lacking the carboxylate functionality. Extraction into the TSIL was found to be more efficient than that seen for TMPA at low to moderate acidities (≤0.3 M HNO₃). In addition, unlike U(VI) extraction by TBP in dodecane (i.e., the conventional PUREX process), which occurs via an ion-pair mechanism, uranium extraction into Hbet⁺Tf₂N⁻ was found to take place by formation of a complex between UO₂²⁺ and the COO⁻ group produced upon deprotonation of betaine. In contrast to most prior studies of metal ion extraction by TSILs, no diluent was employed to reduce the TSIL viscosity in these extraction studies, making the attainment of equilibrium slow (ca. 7 minutes) by the standards of conventional SX. Curiously, no attempt was made to exploit the temperature-dependent phase behavior of Hbet⁺Tf₂N⁻–water mixtures as the basis for separation, despite the rapid mass transfer kinetics that such a system would exhibit. Systems such as this then would seem to offer significant opportuni-ties for the development of novel approaches to metal ion separations, opportunities that have yet to be adequately explored.

3.4   TSILS INCORPORATING FUNCTIONALIZED ANIONS

Although in principle, TSILs can be prepared by functionalization of either the IL cation or anion, as has already been noted, examples of the latter are few. The reason for this is straightforward. That is, numerous anions capable of interacting strongly and/or selectively with a variety of metal ions already exist, and it is a simple mat-ter to incorporate any of a number of these directly into an IL via ion-exchange reactions, thus eliminating the need for complicated syntheses. Nonetheless, a few TSILs have been reported in which functionalization of an otherwise "conventional" anion was carried out. In 2016, for example, Iampolska et al. (58) described the preparation of a pair of novel TSILs in which 1-methyl-3-alkylimidazolium cations (alkyl = C₁₀H₂₁ or C₁₂H₂₅) were paired with a thiacalix[4]arene to whose upper rim four SO₃⁻ groups had been attached. Although the resultant TSILs were described as being suitable for the extraction of transition metals and *f*-elements, no actual extrac-tion data were provided to demonstrate this.

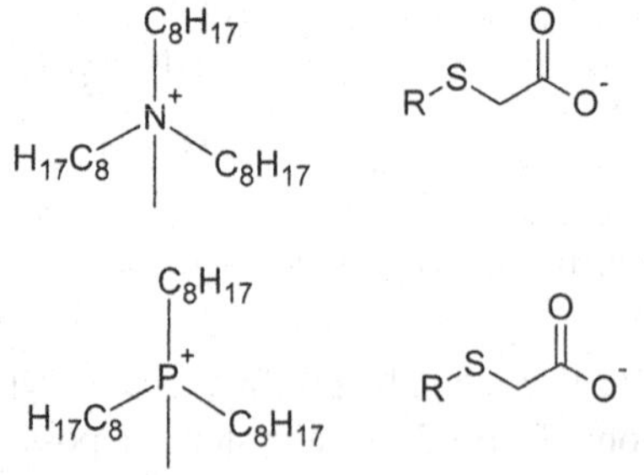

FIGURE 3.19 Upper: Methyltrioctylammonium alkylsulfanyl acetate; Lower: Methyltrioctylphosphonium alkylsulfanyl acetate.

More recently, Platzer et al. (56) outlined the synthesis and characterization of a series of thioglycolate-based TSILs (Figure 3.19) designed for the extraction and recovery of cadmium and copper ions. By pairing any of various benzyl- or alkylsulfanyl acetate anions with either quaternary ammonium or phosphonium cations, the investigators obtained eight ionic liquids, all of which were found to be viscous oils or solids at ambient temperature. Saturation of the TSILs with water, however, rendered them all liquid at room temperature and thus, suitable for use in liquid–liquid extraction. Both Cd^{2+} and Cu^{2+} were found to be well extracted by all of the tested TSILs, as would be expected from the known affinity of thioglycolate anions for these ions. Complete (>90%) extraction, however, required 30 minutes, consistent with reports of slow extraction of other metal ions into other viscous TSILs (see above). Surprisingly, no effort was made to examine the use of a diluent to reduce the TSIL viscosity and therefore, to increase the rates of extraction. Also, no comparison of the performance of the TSILs to that of an analogous extractant simply dissolved in an ordinary quaternary or phosphonium IL was carried out, leaving the question open of the actual utility of the TSIL relative to a simpler, conventional system.

3.5 TSILS INCORPORATING CONVENTIONAL EXTRACTANT ANIONS

That TSILs incorporating a functional anion can be prepared simply by deprotonation of an acidic extractant and anion metathesis (59) or by an acid–base reaction (107) was first recognized nearly a decade ago. (Interestingly, examples of this same type of compound were first prepared in the 1960s (108). Although their application to the extraction of metal ions into various organic diluents was described soon after (109, 110), they were not recognized as ionic liquids for nearly four decades.) In 2008, Kogelnig et al. (59), building on the work of Mikkola et al. (111), reported the synthesis of a series of new TSILs derived from Aliquat 336™ (tricaprylmethylammonium chloride) that incorporated a benzoate, hexanoate, or thiosalicylate anion, and described their evaluation in the extraction of Cd^{2+} from natural waters. Of the three, only the thiosalicylate TSIL (Figure 3.20) provided appreciable extraction, a result of the presence of a soft donor group (thiol) in the anion. Although the authors described the metal extraction process as "fast," no kinetics data in support of this

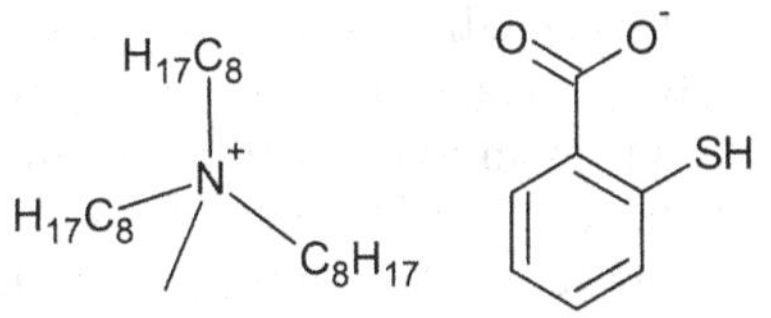

FIGURE 3.20 Tricaprylmethylammonium thiosalicylate.

statement were provided, and the high viscosity reported for Aliquat 336™ (112) suggests that there is reason for skepticism. Nonetheless, this study does represent the first demonstration that a deprotonation–metathesis route constitutes a simple approach to the preparation of TSILs incorporating functional anions.

The following year, Sun et al. (87) prepared a TSIL (dubbed [A336][CA-100]) via an acid–base reaction between the hydroxide form of Aliquat 336™ and *sec-*nonylphenoxyacetic acid, and examined its suitability as an extractant for Sc(III). Comparison of the results to those obtained for the analogous nitrate IL showed that presence of the functional anion (here, a carboxylate) led to much more efficient Sc(III) extraction. As has proven to be the case for nearly all TSILs described to date, however, [A336][CA-100] was found to be highly viscous. As a result, the authors proposed its use as a coating on a porous solid support rather than in conventional liquid–liquid extraction.

Since these initial studies, a number of investigators have described the preparation and characterization of a variety of analogous TSILs. Much of this work has involved salicylate/thiosalicylate (89, 113–121) or the anions of various acidic organophosphorus extractants (e.g., HDEHP) (61, 87, 122), but TSILs incorporating β-diketonates (123), phthalate (124), and dicyanamide (125) have also been described. Of those studies employing salicylate or thiosalicylate anions, several (89, 116, 117) have examined the use of these TSILs as the basis for solid sorbents for various metal ions. Having been discussed elsewhere (90), however, they will not be considered further here.

In 2010, Egorov et al. (113) prepared trioctylmethylammonium salicylate (TOMAS) and examined its utility in the extraction of several transition metal ions (Fe^{3+}, Cu^{2+}, Ni^{2+}, Mn^{2+}) from aqueous solution. Although the TSIL was found to be viscous, it was nonetheless employed in undiluted form in the extraction studies. Extraction of both Fe^{3+} and Cu^{2+}, which are known to form stable salicylate complexes, was observed to be efficient, while that of Ni^{2+} and Mn^{2+}, which form weaker salicylate complexes, was much less so. In contrast to the majority of extraction studies involving TSILs, a detailed treatment of a possible mechanism for extraction, which for iron was proposed to involve the stepwise formation of an iron bisalicylate complex to eventually yield $TOMA^{+}FeSal_2^{-}$ in the IL phase, was provided.

That same year, Stojanovic et al. (114) prepared a series of hydrophobic quaternary ammonium and phosphonium ionic liquids incorporating a variety of functional aromatic anions, among them thiosalicylate, benzoate, vanillate, and (phenylthio) acetate, by a metathesis route employing Aliquat 336™ and Cyphos IL101 (i.e., trihexyl(tetradecyl)phosphonium chloride, $P_{666,14}^{+}Cl^{-}$) as precursors. While most of the report concerned the physicochemical characterization of the new compounds, selected thio- and thioether-functionalized TSILs (i.e., the thiosalicylate (TS$^-$) and 2-methylthiobenzoate (MTBA$^-$) salts of Aliquat 336™ and Cyphos IL101) were evaluated as extractants for platinum removal from an aqueous phase comprising 10 mM CaCl$_2$ at pH 7.5. Of the tested TSILs, best results were obtained with P$_{666,14}^+$ MTBA$^-$, for which nearly complete extraction was achieved in 30 minutes. Curiously, the extraction efficiency was found to decrease with decreasing TSIL viscosity, an observation for which no explanation was offered. No correlation was observed between extraction efficiency and IL-phase water content, as had been anticipated.

More recently, Rajendran et al. (115) evaluated one of these same TSILs, the thiosalicylate form of Aliquat 336™, along with eight other closely related quaternary ammonium-based TSILs incorporating an aromatic anion, as potential extracting agents for the removal of various metal ions (Pb^{2+}, Zn^{2+}, Ni^{2+}, Fe^{3+}, Cu^{2+}) from contaminated water. With the exception of tetrabutylammonium benzoate, which was found to be a solid, all of the TSILs were viscous liquids at ambient temperature. A number of them were found to provide efficient extraction of Pb^{2+} and Fe^{3+} from neutral solution, but none yielded satisfactory Cu^{2+} extraction.

At least one report has appeared concerning the preparation of a TSIL incorporating a phthalate anion. Like several other investigators, Biswas et al. (124) employed Aliquat 336™ as a precursor, obtaining a TSIL that despite its viscosity, was employed in undiluted form in subsequent extraction experiments. Measurements of the extraction of actinides (e.g., U(VI) and Th(IV)) and lanthanides (e.g., Nd(III)) by the TSIL indicate that it is selective for the extraction of uranium from acidic (pH 1–2) nitrate-, sulfate-, and phosphate- containing aqueous phases, although nonnegligible extraction of thorium is also observed under certain conditions.

The anion of any of several organophosphorus acids has also served as the basis for a TSIL. Among the TSILs described by Sun et al. (107), for example, were those combining the tricaprylmethylammonium ion (i.e., the cationic constituent of Aliquat 336™) with bis(2-ethylhexyl)phosphinate, bis(2,4,4-trimethylpentyl)phosphinate, and bis(2,4,4-trimethyl-pentyl)monothiophosphinate (the latter two corresponding to the anions of Cyanex 272 and Cyanex 302, respectively). A follow-up study by the same authors (61) focused on the utility of several such "bifunctional ionic liquids" (so named because both the anionic and cationic constituents of the TSIL are derived from extractants), here Aliquat 336™ (as IL$^+$) and an organophosphorus reagent (as IL$^-$) in the extraction of Eu(III) from nitric and hydrochloric acids. Due apparently to the high viscosity of the TSILs, for the extraction studies, all were dissolved in either cyclohexane or chloroform, the latter a choice clearly inconsistent with the purported "greenness" of IL-based extraction systems. In all instances, the extraction of Eu(III) was found to be more efficient for the TSILs than for a mixture of the individual extractants in the same solvent, an observation attributed to the breaking of the intramolecular hydrogen bonds characteristic of both organophosphorus and carboxylic acids, and an increase in the hydrophobicity of the extracted ion-association complex. This increased extraction efficiency vis-à-vis the individual reagents was dubbed an "internal synergistic effect," a name chosen to indicate that the effect arises from two components of a single species (i.e., the TSIL) rather than from two chemically distinct entities in the ionic liquid, as was the case for the first report of synergism in an ionic liquid (126). Subsequent studies (127) to extend this work revealed that synergistic effects are also observed between pairs of bifunctional ILs in a given solvent, and that these effects can be greater than those observed for combinations of the acidic extractants employed to provide the IL anions. The mechanism underlying these effects remains unclear, and given the complexity of such systems, is likely to require substantial additional effort to discern.

The application of TSILs incorporating organophosphorus acid anions to the solvent extraction of other rare-earth elements (REE) has also been described. By pairing the anion of bis(2-ethylhexyl)phosphoric acid (i.e., DEHP$^-$) with a quaternary ammonium or tetraalkylphosphonium cation, Sun (54) obtained a series of DEHP$^-$-based extractants that were evaluated for possible use in an IL-based TALSPEAK process. All of the tested compounds were found to provide significantly higher D_M values when dissolved in $C_6mim^+Tf_2N^-$ than in diisopropylbenzene (DIPB). Perhaps most noteworthy, however, is the observation that the tetraoctylammonium salt of DEHP$^-$ is completely miscible with $C_6mim^+Tf_2N^-$, while the solubility of HDEHP itself in the same IL is limited to only ca. 0.04 M. This suggests that one important use for TSILs may be in increasing the solubility of extractants whose phase compatibility with ILs (or even molecular diluents) is limited.

REE extraction by TSILs derived from the HDEHP anion was also treated in some detail by Rout et al. (128). Measurements of the extraction of Nd(III) as a function of various system parameters (e.g., pH, nature of the diluent) showed that the extraction by $C_6mim^+DEHP^-$ and $C_6mpyr^+DEHP^-$ (each dissolved in $C_6mim^+Tf_2N^-$) differ markedly, especially regarding the pH dependency of D_{Nd}, from extraction by an analogous quaternary ammonium-based TSIL, $N_{4444}^+DEHP^-$, in the same diluent. Thus, claim the authors, extraction processes employing TSILs for REE extraction can be "tuned" simply by proper choice of the TSIL cation. Extension of these studies to other rare earths led to the observation that heavier REE are extracted much more efficiently than are light rare earths and yttrium, a clearly desirable result given the relative scarcity of the heavy REE (129). As might be anticipated from these results, Dutta et al. (122) observed efficient extraction of Gd(III) from a pH 4 aqueous phase using a quaternary ammonium-based TSIL derived from bis(2-ethylhexyl) phosphonic acid, dissolved in dodecane or toluene. Interestingly, far poorer extraction was observed in isodecanol, a solvent that could be expected to form hydrogen bonds with the TSIL anion. This observation is consistent with the notion that the high extraction efficiency observed for "bifunctional ILs" (see above) is the result of disruption of the H-bonding observed for organophosphorus reagents in the absence of the TSIL cation.

The extraction of lanthanides by TSILs incorporating the anion of various β-diketones has also been reported. Following work by Himmler et al. (130), which led to the preparation of a water-soluble hexafluoroacetylacetonate TSIL ($C_2mim^+hfac^-$), Mehdi et al. (131) described the synthesis and characterization of a series of hydrophobic 1-alkyl-3-methylimidazolium hexafluoroacetylacetonates ($C_nmim^+hfac^-$, with $n=4$–18), along with a pair of analogous compounds employing the 1-butyl-2,3-dimethylimidazolium or N-butyl-N-methylpyrrolidinium cations. Attempts to prepare TSILs incorporating other β-diketonate anions yielded either water-miscible compounds or materials exhibiting limited stability. That these ILs could serve as metal ion extractants was demonstrated by contacting an aqueous solution of $Nd(Tf_2N)_3$ or $Co(Tf_2N)_2$ with neat $C_4mim^+hfac^-$. Transfer of the colored complex formed from the aqueous to the IL phase was noted. Unfortunately, no effort was made to quantify the results obtained, nor was any comparison made of the extraction efficiency achieved to that observed when a β-diketone is simply dissolved in a conventional IL.

A more thorough study of the utility of β-diketonate-based ILs was subsequently carried out by Rout et al. (132), who examined the extraction of Pu(IV), U(VI), and Am(III) from nitric acid solution by several TSILs combining the thenoyltrifluoroacetonate (TTA) anion with a quaternary ammonium (e.g., trioctylmethylammonium, abbreviated "TOMAN") or dialkylimidazolium (e.g., 1-octyl-3-methylimidazolium) cation. Preliminary investigations employing the nonfunctionalized IL TOMAN$^+$Tf$_2$N$^-$ alone (i.e., in the absence of TSIL) showed that significant extraction of Pu(IV) occurs at sufficiently high (>1 M) aqueous phase nitric acid concentrations. Such an observation is consistent with results reported by Rout et al. (57) for $C_8mim^+Tf_2N^-$ alone, which were attributed to the exchange of one or more anionic nitrato complexes of Pu(IV) for the anionic constituent of the ionic liquid. Although the addition of the TTA-based TSIL to the system (0.2 M) increased the extraction of Pu(IV) over the entire range of acidities examined (0.5–5 M), the increase was far less pronounced at high acidities than at low. That enhancement of Pu(IV) extraction is observed at all is somewhat unexpected, given that decreasing extraction with rising acidity is characteristic of β-diketone extractants. Examination of the behavior of the TSIL in a molecular solvent (e.g., xylene) suggests that in the presence of sufficient concentrations of acid, the TSIL may be partially converted to the quaternary ammonium nitrate and thenoyltrifluoroacetone, thus providing an explanation for these observations. More importantly, the results suggest that in these systems, the extraction behavior of a TSIL can be predicted from the behavior of the extractants comprising the TSIL, measured individually.

More recently, Pandey et al. (123) prepared and characterized several of TSILs comprising either choline or tetrabutylphosphonium cation in combination with hexafluoroacetylacetone, 2-thenoyltrifluoroacetone 4,4,4-trifluoro-1-phenyl-butanedione, or 6,6,7,7,8,8-heptafluoro-2,2-dimethyl-3,5-octanedione. Although the metal ion extraction ability of the new TSILs was not evaluated, the results are significant in that they include a convenient, high-yield, one-pot route to the preparation of a number of β-diketonate-based TSILs.

Finally, in 2014, Rout (133) described a TSIL pairing a quaternary ammonium cation (e.g., Aliquat 336™) with an anion (i.e., dioctyl diglycolamate) derived from DGA (*N,N,N',N'*-tetra-*n*-octyldiglycol-amide) for use in the solvent extraction of neodymium ion. Upon contact with a nitric acid-containing aqueous phase at a pH < 2, an undetermined fraction of the TSIL was converted to the nitrate form. When the ionic liquid A336$^+$ NO$_3^-$ was employed as a diluent, the dependency of D_{Nd} on TSIL concentration (at an aqueous pH of 2–5, which encompasses the pK_a of DGA) was found to have a slope of 1, suggesting the following extraction mechanism:

$$Nd^{3+} + 3NO_3^- + A336^+DGA^- \rightleftharpoons Nd(NO_3)_3 \cdot [A336][DGA]$$

Such an observation is noteworthy in that it indicates that despite the overall complexity of the extraction mechanism, there are conditions under which the extraction of a metal salt can be effected, thus avoiding undesirable ion-exchange processes.

3.6 CONCLUSIONS AND FUTURE DIRECTIONS

It is clear from the foregoing that in the two decades that have passed since the description of the first TSILs, substantial effort has been directed toward their application in the extraction of metal ions. In fact, TSILs derived from essentially every major type of IL cation have by now been described. At the same time, representative examples of most (if not all) extractant families of current interest, particularly for actinide and fission product separations, have been employed as the active functional group of a TSIL. In evaluating what has been accomplished thus far, and in seeking to identify worthwhile directions for future work, it is instructive to consider the ways in which TSILs could potentially improve metal ion separations and then, to assess the extent to which this potential has been realized.

As already noted, TSILs could, in principle, reduce or eliminate the problem of the limited solubility of various extractants in ionic liquids. In addition, TSILs could exhibit different metal ion partitioning mechanisms or, more accurately, a different balance among possible mechanisms, than that observed for systems comprising an extractant dissolved in a conventional ionic liquid. That is, when an extractant molecule is tethered to the cationic constituent of the IL, reduced solubility of the cation in the aqueous phase, and thus a reduced contribution of ion-exchange processes to the overall extraction mechanism, would be anticipated. As a result, TSILs could provide improved selectivity and/or higher extraction efficiency than possible with systems comprising an analogous extractant in an IL or an ordinary molecular solvent. Given all of this, TSILs could be expected to add, perhaps significantly, to the existing range of capabilities of IL-based extraction systems for metal ions. But to what extent has this actually been achieved?

From the limited data available, it appears that TSILs can indeed address the issue of poor extractant solubility in ionic liquids. For example, work by Sun et al. (54) has shown that while the solubility of HDEHP in $C_6mim^+Tf_2N^-$ is only 0.04 M, the solubility of the TSIL comprising the trioctylmethylammonium salt of DEHP$^-$ is at least 20× greater (ca. 0.8 M). Similarly, as reported by Morita et al. (72), the solubility in $C_4mim^+PF_6^-$ of the 8-sulfonamidoquinoline derivative of $C_2mim^+Cl^-$ is >1 M, while that of the analogous extractant in the same IL is only 0.01 M. Although additional investigation is needed to confirm the generality of these observations, these initial reports are clearly very encouraging.

The effectiveness of TSILs in providing higher extraction efficiencies than more "conventional" systems is less clear, however. As already noted, in an evaluation of a series of quaternary ammonium-based TSILs incorporating a phosphoryl group designed to extract U(VI) from acidic nitrate media, Ouadi et al. (65) found that a phosphoramidate-functionalized quaternary ammonium IL could provide extraction far more efficient than that observed for a solution of TBP in a conventional quat. In contrast, as also noted above, Luo et al. (67) have observed that the extraction efficiency for Cs^+ and Sr^{2+} of TSILs incorporating an aza-crown ether fragment is actually *lower* than that observed with either dicyclohexano-18-crown-6 (DCH18C6) or *N*-octylaza-18-crown-6, dissolved in a conventional IL ($C_4mim^+Tf_2N^-$). Other, similarly contradictory results have been reported elsewhere (134). Thus, at best, it can be stated that a *properly designed* TSIL can provide superior performance vis-à-vis an analogous extractant dissolved in a conventional IL. Aside from recent work by Dehaudt et al. (135) suggesting that an important factor in optimizing the performance of TSILs incorporating a functionalized cation may be minimization of the repulsions between the complexed metal ion and the charged moiety of the IL cation, exactly what constitutes "proper design" remains uncertain, however. Also uncertain, in this case a consequence not of conflicting results, but rather of the absence of systematic study, are the relative merits of TSIL-based solvent extraction systems and those employing ordinary molecular diluents. These areas thus represent additional fruitful areas for future investigations.

As has been stated elsewhere (129), progress in the application of TSILs will require a greater understanding of the mechanistic aspects of their behavior than is now available. At present, the need to overcome the high viscosity of these materials means that it has generally proven difficult to study them without dilution with a nonfunctionalized IL. This dilution, however, often has the unfortunate consequence of introducing ion-exchange processes into the overall mechanism for metal ion extraction by the TSIL, processes whose undesirability has long been recognized (50). In an effort to avoid this, a few investigators have examined the use of undiluted TSILs supported on various porous silica, polymer, or carbonaceous substrates (85–89) for metal ion separations. To date, however, the preliminary nature of the results reported and the lack of emphasis on the mechanistic aspects of extraction make it difficult to draw any firm conclusions as to the utility of this approach (90). Further work in this area, along with efforts to design lower viscosity TSILs, is clearly warranted. Along these same lines, an increase in temperature is known to reduce, in some instances substantially (27), the viscosity of ionic liquids. Given this, it is clear that a systematic investigation of the effect of temperature on the viscosity and extraction properties of TSILs is also needed. Finally, heretical though it may seem, extraction systems comprising a TSIL in combination with a conventional organic diluent also warrant further consideration. Despite the obvious violation of the tenets of green chemistry such a system could represent (136), it seems reasonable to conclude that the possibilities afforded by this combination (not just in reducing the extractant viscosity, but also in effecting needed separations) could outweigh environmental considerations. This is particularly true now that the "greenness" of many ionic liquids has itself been called into question (137, 138).

Although an incomplete understanding of structure-property relationships and metal ion extraction mechanisms in TSILs has undoubtedly slowed the development of practical TSIL-based extraction systems, progress has nonetheless been made on several fronts. Among the recent studies most worthy of note is work by Shkrob et al. (139), in which an organic-immiscible TSIL was employed as a replacement for the *aqueous* phase in a SX system designed to mimic the well-known TALSPEAK process (140). In addition to providing a novel route to achieving group separation of minor actinides from lanthanides, this work demonstrates that TSILs should not be thought of solely as functionalized substitutes for organic solvents. Along these same lines, Sadeghi et al. (120) recently described the preparation of an aqueous biphasic system (ABS) based on a TSIL. Although ABS systems based on conventional ionic liquids have long been known (141), the use of TSILs to prepare such systems represents a new opportunity for the design of systems for metal ion separations. In this instance, a series of $C_n mim^+$-based TSILs ($n = 4$–8) incorporating salicylate or thio-salicylate anions were prepared, and their utility in the aqueous biphasic extraction of chromium for subsequent determination by flame atomic absorption spectrometry was evaluated. The high recoveries and low detection limits observed suggest that the use of TSIL-based ABS systems might represent a broadly applicable approach to efficient metal ion extraction. Finally, these same authors (142, 143) have recently demonstrated the utility of TSILs in the "*in-situ*" dispersive liquid–liquid microextraction (DLLME) of metal ions. In conventional DLLME, a metal ion extractant is dissolved in a dense, water-immiscible extraction solvent (e.g., $CHCl_3$), which in turn, is dissolved in a water-miscible "disperser" solvent (e.g., ethanol). Injection of a small volume of this mixture into the (aqueous) sample results in the rapid and efficient dispersal of the extractant through the sample and, after a period of standing or centrifugation, the accumulation of a small droplet of the extraction solvent containing the metal ion of interest (144). For *in-situ* DLLME, the extraction solvent, extractant, and disperser solvent are all replaced with one reagent, the water-soluble form of a TSIL. Addition of the salt of an anion such as hexafluorophosphate results in anion metathesis and the formation of the water-immiscible form of the TSIL. The overall effect is a significant reduction in the consumption of conventional organic solvents. Undoubtedly, similarly creative applications for TSILs will continue to emerge, and taken together with expected progress in defining the fundamental aspects of their behavior, suggest that interest in these unique solvent/extractant hybrids will only continue to grow.

REFERENCES

1. J.S. Wilkes, "A short history of ionic liquids – From molten salts to neoteric solvents," *Green Chem.* 2002, *4*, 73–80.
2. V.I. Pârvulescu, C. Hardacre, "Catalysis in ionic liquids," *Chem. Rev.* 2007, *107*, 2615–2665.
3. X. Han, D.W. Armstrong, "Ionic liquids in separations," *Acc. Chem. Res.* 2007, *40*, 1079–1086.
4. A. Taubert, Z. Li "Inorganic materials from ionic liquids," *Dalton Trans.* 2007, 723–727.
5. N.V. Plechkova, K.R. Seddon, "Applications of ionic liquids in the chemical industry," *Chem. Soc. Rev.* 2008, *37*, 123–150.
6. F. Zhou, Y. Liang, W. Liu, "Ionic liquid lubricants: Designed chemistry for engineering applications," *Chem. Soc. Rev.* 2009, *38*, 2590–2599.
7. M. Armand, F. Endres, D.R. MacFarlane, H. Ohno, B. Scrosati, "Ionic-liquid materials for the electrochemical challenges of the future," *Nat. Mater.* 2009, *8*, 621–629.
8. J. Lu, F. Yan, J. Texter, "Advanced applications of ionic liquids in polymer science," *Prog. Polym. Sci.* 2009, *34*, 431–448.
9. T. Torimoto, T. Tsuda, K. Okazaki, S. Kuwabata, "New frontiers in materials science opened by ionic liquids," *Adv. Mater.* 2010, *22*, 1196–1221.

10. P. Sun, D.W. Armstrong, "Ionic liquids in analytical chemistry," *Anal. Chim. Acta* 2010, *661*, 1–16.
11. J.P. Hallett, T. Welton, "Room-temperature ionic liquids: Solvents for synthesis and catalysis. 2," *Chem. Rev.* 2011, *111*, 3508–3576.
12. X.P. Zhang, X.C. Zhang, H.F. Dong, Z.J. Zhao, S.J. Zhang, Y. Huang, "Carbon capture with ionic liquids: Overview and progress," *Energy Environ. Sci.* 2012, *5*, 6668–6681.
13. Y. Zhou, J. Qu, "Ionic liquids as lubricant additives: A review," *ACS Appl. Mater. Interfaces* 2017, *9*, 3209–3222.
14. M.J. Earle, K.R. Seddon, "Ionic liquids: Green solvents for the future," *Pure Appl. Chem.* 2000, *72*, 1391–1398.
15. J.-F. Huang, H. Luo, S. Dai, "A new strategy for the synthesis of novel classes of room-temperature ionic liquids based on complexation reaction of cations," *J. Electrochem. Soc.* 2006, *153*, J9–J13.
16. J.-F. Huang, H. Luo, C. Liang, D. Jiang, S. Dai, "Advanced liquid membranes based on novel ionic liquids for selective separation of olefin/paraffin via olefin-facilitated transport," *Ind. Eng. Chem. Res.* 2008, *47*, 881–888.
17. T.M. Anderson, D. Ingersoll, A.J. Rose, C.L. Staiger, J.C. Leonard, "Synthesis of an ionic liquid with an iron coordination cation," *Dalton Trans.* 2010, *39*, 8609–8612.
18. Y. Song, H. Jing, B. Li, D. Bai, "Crown ether complex cation ionic liquids: Preparation and applications in organic reactions," *Chemistry* 2011, *17*, 8731–8738.
19. S.D. Jagadale, M.B. Deshmukh, A.G. Mulik, D.R. Chandam, P.P. Patil, D.R. Patil, S.A. Sankpal, "Crown ether complex cation-like ionic liquids: Synthesis and catalytic applications in organic reactions," *Der Pharma Chemica* 2012, *4*, 202–207.
20. Y. Song, C. Cheng, H. Jing, "Aza-crown ether complex cation ionic liquids: Preparation and applications in organic reactions," *Chemistry* 2014, *20*, 12894–12900.
21. C.A. Angell, N. Byrne, J.P. Belieres, "Parallel developments in aprotic and protic ionic liquids: Physical chemistry and applications," *Acc. Chem. Res.* 2007, *40*, 1228–1236.
22. T.L. Greaves, C.J. Drummond, "Protic ionic liquids: Properties and applications," *Chem. Rev.* 2008, *108*, 206–237.
23. T.L. Greaves, C.J. Drummond, "Protic ionic liquids: Evolving structure-property relationships and expanding applications," *Chem. Rev.* 2015, *115*, 11379–11448.
24. L. Dai, S. Yu, Y. Shan, M. He, "Novel room-temperature inorganic ionic liquids," *Eur. J. Inorg. Chem.* 2004, *2004*, 237–241.
25. A.B. Bourlinos, K. Raman, R. Herrera, Q. Zhang, L.A. Archer, E.P. Giannelis, "A liquid derivative of 12-tungstophosphotungstic acid with unusually high conductivity," *J. Am. Chem. Soc.* 2004, *126*, 15358–15359.
26. P.G. Rickert, M.R. Antonio, M.A. Firestone, K.A. Kubatko, T. Szreder, J.F. Wishart, M.L. Dietz, "Tetraalkylphosphonium polyoxometalates: Electroactive, 'task-specific' ionic liquids," *Dalton Trans.* 2007, 529–531.
27. P.G. Rickert, M.R. Antonio, M.A. Firestone, K.A. Kubatko, T. Szreder, J.F. Wishart, M.L. Dietz, "Tetraalkylphosphonium polyoxometalate ionic liquids: Novel organic-inorganic hybrid materials," *J. Phys. Chem. B* 2007, *111*, 4685–4692.
28. J.H. Davis, Jr., "Task-specific ionic liquids," *Chem. Lett.* 2004, *33*, 1072–1077.
29. M. Deetlefs, M. Fanselow, K.R. Seddon, "Ionic liquids: The view from Mount Improbable," *RSC Adv.* 2016, *6*, 4280–4288.
30. J.H. Davis, Jr., K.J. Forrester, T. Merrigan, "Novel organic ionic liquids (OILs) incorporating cations derived from the antifungal drug miconazole," *Tetrahedron Lett.* 1998, *39*, 8955–8958.
31. J.H. Davis, Jr., K.J. Forrester, "Thiazolium-ion based organic ionic liquids (OILs). Novel OILs which promote the benzoin condensation," *Tetrahedron Lett.* 1999, *40*, 1621–1622.

32. T.L. Merrigan, E.D. Bates, S.C. Dorman, J.H. Davis, Jr., "New fluorous ionic liquids function as surfactants in conventional room-temperature ionic liquids," *Chem. Commun.* 2000, 2051–2052.

33. M. Pucheault, M. Vaultier, "Task specific ionic liquids and task specific onium salts," *Top. Curr. Chem.* 2010, *290*, 83–126.

34. A.D. Sawant, D.G. Raut, N.B. Darvatkar, M.M. Salunkhe, "Recent developments of task-specific ionic liquids in organic synthesis," *Green Chem. Lett. Rev.* 2011, *4*, 41–54.

35. R.L. Vekariya, "A review of ionic liquids: Applications towards catalytic organic transformations," *J. Molec. Liq.* 2017, *227*, 44–60.

36. A.E. Visser, R.P. Swatloski, W.M. Reichert, J.H. Davis, Jr., R.D. Rogers, R. Mayton, S. Sheff, A. Wierzbickil, "Task-specific ionic liquids for the extraction of metal ions from aqueous solutions," *Chem. Commun.* 2001, *1*, 135–136.

37. A.E. Visser, J.D. Holbrey, R.D. Rogers, "Hydrophobic ionic liquids incorporating *N*-alkylisoquinolinium cations and their utilization in liquid-liquid separations," *Chem. Commun. (Camb.)* 2001, *23*, 2484–2485.

38. E.D. Bates, R.D. Mayton, I. Ntai, J.H. Davis, Jr., "CO_2 capture by a task-specific ionic liquid," *J. Amer. Chem. Soc.* 2002, *124*, 926–927.

39. S.V. Dzyuba, R.A. Bartsch, "Expanding the polarity range of ionic liquids," *Tetrahedron Lett.* 2002, *43*, 4657–4659.

40. N. Gathergood, P.J. Scammells, "Design and preparation of room-temperature ionic liquids containing biodegradable side chains," *Aust. J. Chem.* 2002, *55*, 557–560.

41. N. Gathergood, M.T. Garcia, P.J. Scammells, "Biodegradable ionic liquids. Part 1. Concept, preliminary targets and evaluation," *Green Chem.* 2004, *6*, 166–175.

42. J.G. Huddleston, H.D. Willauer, R.P. Swatloski, A.E. Visser, R.D. Rogers, "Room-temperature ionic liquids as media for 'clean' liquid-liquid extraction," *Chem. Commun.* 1998, 1765–1766.

43. S. Dai, Y.H. Ju, C.E. Barnes, "Solvent extraction of strontium nitrate by a crown ether using room-temperature ionic liquids," *Dalton Trans.* 1999, 1201–1202.

44. H. Zhao, S.Q. Xia, P.S. Ma, "Use of ionic liquids as 'green' solvents for extractions," *J. Chem. Technol. Biotechnol.* 2005, *80*, 1089–1096.

45. M.L. Dietz, "Ionic liquids as extraction solvents: Where do we stand?," *Sep. Sci. Technol.* 2006, *41*, 2047–2063.

46. D. Han, K.H. Row, "Recent applications of ionic liquids in separation technology," *Molecules* 2010, *15*, 2405–2426.

47. X. Sun, H. Luo, S. Dai, "Ionic liquids-based extraction: A promising strategy for the advanced nuclear fuel cycle," *Chem. Rev.* 2012, *112*, 2100–2128.

48. M. Cvjetko Bubalo, S. Vidović, I. Radojčić Redovniković, S. Jokić, "Green solvents for green technologies," *J. Chem. Technol. Biotechnol.* 2015, *90*, 1631–1639.

49. M.J. Earle, J.M.S.S. Esperança, M.A. Gilea, J.N.C. Lopes, L.P.N. Rebelo, J.W. Magee, K.R. Seddon, J.A. Widegren, "The distillation and volatility of ionic liquids," *Nature* 2006, *439*, 831–834.

50. M.L. Dietz, J.A. Dzielawa, "Ion-exchange as a mode of cation transfer into room-temperature ionic liquids containing crown ethers: Implications for the 'greenness' of ionic liquids as diluents in liquid-liquid extraction," *Chem. Commun. (Camb.)* 2001, 2124–2125.

51. M.P. Jensen, J. Neuefeind, J.V. Beitz, S. Skanthakumar, L. Soderholm, "Mechanisms of metal ion transfer into room-temperature ionic liquids: The role of anion exchange," *J. Am. Chem. Soc.* 2003, *125*, 15466–15473.

52. K. Shimojo, M. Goto, "First application of calixarenes as extractants in room-temperature ionic liquids," *Chem. Lett.* 2004, *33*, 320–321.

53. H.M. Luo, S. Yu, S. Dai, "Solvent extraction of Sr^{2+} and Cs^+ based on hydrophobic protic ionic liquids," *Zeitschrift für Naturforschung A* 2007, *62*, 281–291.

54. X.Q. Sun, H.M. Luo, S. Dai, "Solvent extraction of rare-earth ions based on functionalized ionic liquids," *Talanta* 2012, *90*, 132–137.
55. A.E. Visser, R.P. Swatloski, W.M. Reichert, R. Mayton, S. Sheff, A. Wierzbicki, J.H. Davis, R.D. Rogers, "Task-specific ionic liquids incorporating novel cations for the coordination and extraction of Hg^{2+} and Cd^{2+}: Synthesis, characterization, and extraction studies," *Environ. Sci. Technol.* 2002, *36*, 2523–2529.
56. S. Platzer, M. Kar, R. Leyma, S. Chib, A. Roller, F. Jirsa, R. Krachler, D.R. MacFarlane, W. Kandioller, B.K. Keppler, "Task-specific thioglycolate ionic liquids for heavy metal extraction: Synthesis, extraction efficacies and recycling properties," *J. Hazard. Mater.* 2017, *324*, 241–249.
57. A. Rout, K.A. Venkatesan, T.G. Srinivasan, P.R.V. Vasudeva Rao, "Unusual extraction of plutonium(IV) from uranium(VI) and americium(III) using phosphonate-based task specific ionic liquid," *Radiochim. Acta* 2010, *98*, 459–466.
58. A.D. Iampolska, S.G. Kharchenko, Z.V. Voitenko, S.V. Shishkina, A.B. Ryabitskii, V.I. Kalchenko, "Synthesis of thiacalix[4]arene task-specific ionic liquids," *Phosphorus Sulfur Silicon Relat Elem* 2016, *191*, 174–179.
59. D. Kogelnig, A. Stojanovic, M. Galanski, M. Groessl, F. Jirsa, R. Krachler, B.K. Keppler, "Greener synthesis of new ammonium ionic liquids and their potential as extracting agents," *Tetrahedron Lett.* 2008, *49*, 2782–2785.
60. W.R. Mohamed, S.S. Metwally, H.A. Ibrahim, E.A. El-Sherief, H.S. Mekhamer, I.M.I. Moustafa, E.M. Mabrouk, "Impregnation of task-specific ionic liquid into a solid support for removal of neodymium and gadolinium ions from aqueous solution," *J. Mol. Liq.* 2017, *236*, 9–17.
61. X.Q. Sun, Y. Ji, F.C. Hu, B. He, J. Chen, D.Q. Li, "The inner synergistic effect of bifunctional ionic liquid extractant for solvent extraction," *Talanta* 2010, *81*, 1877–1883.
62. J.H. Davis, "Working salts: Synthesis and uses of ionic liquids containing functionalized ions," in *Ionic Liquids: Industrial Applications for Green Chemistry*, R.D. Rogers and K.R. Seddon (Eds.), American Chemical Society, Washington, DC, 1999, pp. 247–258.
63. H.Y. Li, B. Wang, S. Liu, "Synthesis of pyridine-based task-specific ionic liquid with alkyl phosphate cation and extraction performance for uranyl ion," *Ionics* 2015, *21*, 2551–2556.
64. N. Papaiconomou, J.-M. Lee, J. Salminen, M. von Stosch, J.M. Prausnitz, "Selective extraction of copper, mercury, silver, and palladium ions from water using hydrophobic ionic liquids," *Ind. Eng. Chem. Res.* 2008, *47*, 5080–5086.
65. A. Ouadi, O. Klimchuk, C. Gaillard, I. Billard, "Solvent extraction of U(VI) by task specific ionic liquids bearing phosphoryl groups," *Green Chem.* 2007, *9*, 1160–1162.
66. J.R. Harjani, T. Friščić, L.R. MacGillivray, R.D. Singer, "Metal chelate formation using a task-specific ionic liquid," *Inorg. Chem.* 2006, *45*, 10025–10027.
67. H. Luo, S. Dai, P.V. Bonnesen, A.C. Buchanan III, "Separation of fission products based on ionic liquids: Task-specific ionic liquids containing an aza-crown ether fragment," *J. Alloy Comp.* 2006, *418*, 195–199.
68. A. Ouadi, B. Gadenne, P. Hesemann, J.J.E. Moreau, I. Billard, C. Gaillard, S. Mekki, G. Moutiers, "Task-specific ionic liquids bearing 2-hydroxybenzylamine units: Synthesis and americium-extraction studies," *Chem. Eur. J* 2006, *12*, 3074–3081.
69. J.R. Harjani, T. Friščić, L.R. MacGillivray, R.D. Singer, "Removal of metal ions from aqueous solutions using chelating task-specific ionic liquids," *Dalton Trans.* 2008, 4595–4601.
70. N. Li, G. Fang, B. Liu, J. Zhang, L. Zhao, S. Wang, "A novel hydrophobic task specific ionic liquid for the extraction of Cd(II) from water and food samples as applied to AAS determination," *Anal. Sci.* 2010, *26*, 455–459.

71. A. Rout, K.A. Venkatesan, T.G. Srinivasan, P.R. Vasudeva Rao, "Unusual extraction of plutonium(IV) from uranium(VI) and americium(III) using phosphonate-based task specific ionic liquid," *Radiochim. Acta* 2010, *98*, 459–466.

72. K. Morita, N. Hirayama, K. Morita, H. Imura, "An 8-sulfonamidoquinoline derivative with imidazolium unit as an extraction reagent for use in ionic liquid chelate extraction systems," *Anal. Chim. Acta* 2010, *680*, 21–25.

73. A. Rout, K.A. Venkatesan, T.G. Srinivasan, P.R. Vasudeva Rao, "Extraction behavior of actinides and fission products in amide-functionalized ionic liquid," *Sep. Purif. Technol.* 2012, *97*, 164–171.

74. W. Wang, Y. Liu, A. Xu, H. Yang, H. Cui, J. Chen, "Solvent extraction of yttrium by task-specific ionic liquids bearing carboxylic group," *Chin. J. Chem. Eng.* 2012, *20*, 40–46.

75. P.K. Mohapatra, P. Kandwal, M. Iqbal, J. Huskens, M.S. Murali, W. Verboom, "A novel CMPO-functionalized task specific ionic liquid: Synthesis, extraction and spectroscopic investigations of actinide and lanthanide complexes," *Dalton Trans.* 2013, *42*, 4343–4347.

76. A. Sengupta, P.K. Mohapatra, M. Iqbal, J. Huskens, W. Verboom, "A diglycolamide-functionalized task specific ionic liquid (TSIL) for actinide extraction: Solvent extraction, thermodynamics and radiolytic stability studies," *Sep. Purif. Technol.* 2013, *118*, 264–270.

77. A. Sengupta, P.K. Mohapatra, R.M. Kadam, D. Manna, T.K. Ghanty, M. Iqbal, J. Huskens, W. Verboom, "Diglycolamide-functionalized task specific ionic liquids for nuclear waste remediation: Extraction, luminescence, theoretical and EPR investigations," *RSC Adv.* 2014, *4*, 46613–46623.

78. W. Yun, Z. Youwen, F. Fuyou, L. Huimin, H. Peizhuo, S. Yinglin, "Synthesis of task-specific ionic liquids with grafted diglycolamide moiety. Complexation and stripping of lanthanides," *J. Radioanal. Nucl. Chem.* 2014, *299*, 1213–1218.

79. S. Chatterjee, H. Gohil, E. Suresh, A.R. Paital, "Copper(II)-specific fluorogenic task-specific ionic liquids as selective fluorescence probes and recyclable extractants," *Chem. Eur. J.* 2015, *21*, 13943–13948.

80. H. Li, B. Wang, L. Zhang, L. Shen, "Task-specific ionic liquids incorporating alkyl phosphate cations for extraction of U(VI) from nitric acid medium: Synthesis, characterization, and extraction performance," *J. Radioanal. Nucl. Chem.* 2015, *303*, 433–440.

81. P. Cardiano, C. Foti, P.G. Mineo, M. Galletta, F. Risitano, S.L. Lo Schiavo, "Sequestration ability of task specific ionic liquids towards cations of environmental interest," *J. Molec. Liq.* 2016, *223*, 174–181.

82. M. Paramanik, D.R. Raut, A. Sengupta, S.K. Ghosh, P.K. Mohapatra, "A trialkyl phosphine oxide functionalized task specific ionic liquid for actinide ion complexation: Extraction and spectroscopic studies," *RSC Adv.* 2016, *6*, 19763–19767.

83. R. Rama, A. Rout, K.A. Venkatesan, M.P. Antony, "A novel phosphoramide task specific ionic liquid for the selective separation of plutonium (IV) from other actinides," *Sep. Purif. Technol.* 2017, *172*, 7–15.

84. W. Xu, L. Wang, J. Huang, G. Ren, D. Xu, H. Tong, "Design and synthesis of piperazine-based task-specific ionic liquids for liquid-liquid extraction of Cu(II), Ni(II), and Co(II) from water," *Aust. J. Chem.* 2015, *68*, 825–829.

85. X. Zhou, P.F. Xie, J. Wang, B.B. Zhang, M.M. Liu, H.L. Liu, X.H. Feng, "Preparation and characterization of novel crown ether functionalized ionic liquid-based solid-phase microextraction coatings by sol-gel technology," *J. Chrom. A* 2011, *1218*, 3571–3580.

86. A. Saljooqi, T. Shamspur, M. Mohamadi, A. Mostafavi, "Application of a thiourea-containing task-specific ionic liquid for the solid-phase extraction cleanup of lead ions from red lipstick, pine leaves, and water samples," *J. Sep. Sci.* 2014, *37*, 1856–1861.

87. X. Sun, Y. Ji, J. Chen, J. Ma, "Solvent impregnated resin prepared using task-specific ionic liquids for rare earth separation," *J. Rare Earths* 2009, *27*, 932–936.

88. I.L. Odinets, E.V. Sharova, O.I. Artyshin, K.A. Lyssenko, Y.V. Nelyubina, G.V. Myasoedova, N.P. Molochnikova, E.A. Zakharchenro, "Novel class of functionalized ionic liquids with grafted CMPO-moieties for actinides and rare-earth elements recovery," *Dalton Trans.* 2010, *39*, 4170–4178.

89. A.A. Ismaiel, M.K. Aroua, R. Yusoff, "Palm shell activated carbon impregnated with task-specific ionic liquids as a novel adsorbent for the removal of mercury from contaminated water," *Chem. Eng. J.* 2013, *225*, 306–314.

90. C.A. Hawkins, Md.A. Momen, M.L. Dietz, "Application of ionic liquids in the preparation of extraction chromatographic materials for metal ion separations: Progress and prospects," *Sep. Sci. Technol.* 2018, *53*, 1820–1833.

91. M.L. Dietz, J.A. Dzielawa, I. Laszak, B.A. Young, M.P. Jensen, "Influence of solvent structural variations on the mechanism of facilitated ion transfer into room-temperature ionic liquids," *Green Chem.* 2003, *5*, 682–685.

92. M.L. Dietz, S. Jakab, K. Yamato, R.A. Bartsch, "Stereochemical effects on the mode of facilitated ion transfer into room-temperature ionic liquids," *Green Chem.* 2008, *10*, 174–176.

93. S.L. Garvey, C.A. Hawkins, M.L. Dietz, "Effect of aqueous phase anion on the mode of facilitated ion transfer into room-temperature ionic liquids," *Talanta* 2012, *95*, 25–30.

94. V.A. Cocalia, M.P. Jensen, J.D. Holbrey, S.K. Spear, D.C. Stepinski, R.D. Rogers, "Identical extraction behavior and coordination of trivalent or hexavalent f-element cations using ionic liquid and molecular solvents," *Dalton Trans.* 2005, 1966–1971.

95. M.L. Dietz, D.C. Stepinski, "Anion concentration-dependent partitioning mechanism in the extraction of uranium into room-temperature ionic liquids," *Talanta* 2008, *75*, 598–603.

96. G.R. Choppin, "Solvent extraction processes in the nuclear fuel cycle," *Solvent Extr. Res. Dev. Jpn.* 2005, *12*, 1–10.

97. E.P. Horwitz, M.L. Dietz, R. Chiarizia, H. Diamond, A.M. Essling, D. Graczyk, "Separation and preconcentration of uranium from acidic media by extraction chromatography," *Anal. Chim. Acta* 1992, *266*, 25–37.

98. F.J. Alguacil, C. Caravaca, S. Martínez, "Extraction of gold from cyanide or chloride media by Cyanex 923," *J. Chem. Technol. Biotechnol.* 1998, *72*, 339–346.

99. P.K. Mohapatra, A. Sengupta, M. Iqbal, J. Huskens, W. Verboom, "Highly efficient diglycolamide-based task-specific ionic liquids: Synthesis, unusual extraction behaviour, irradiation, and fluorescence studies," *Chem. Eur. J.* 2013, *19*, 3230–3238.

100. A. Sengupta, M.S. Murali, P.K. Mohapatra, M. Iqbal, J. Huskens, W. Verboom, "An insight into the complexation of UO_2^{2+} with diglycolamide-functionalized task-specific ionic liquids: Kinetic, cyclic voltammetric, extraction, and spectroscopic investigations," *Polyhedron* 2015, *102*, 549–555.

101. C.A. Hawkins, S.L. Garvey, M.L. Dietz, "Structural variations in room-temperature ionic liquids: Influence on metal ion partitioning modes and extraction selectivity," *Sep. Purif. Technol.* 2012, *89*, 31–38.

102. A. Sengupta, P.K. Mohapatra, R.M. Kadam, D. Manna, T.K. Ghanty, M. Iqbal, J. Huskens, W. Verboom, "Diglycolamide-functionalized task-specific ionic liquids for nuclear waste remediation: Extraction, luminescence, theoretical, and EPR investigations," *RSC Adv.* 2014, *4*, 46613–46623.

103. R. Rama, A. Rout, K.A. Venkatesan, M.P. Antony, "A novel phosphoramide task-specific ionic liquid for the selective separation of plutonium (IV) for other actinides," *Sep. Purif. Technol.* 2017, *172*, 7–15.

104. K. Chayama, Y. Sano, S. Iwatsuki, "Pyridinium-based task-specific ionic liquid with a monothioether group for selective extraction of Class *b* metal ions," *Anal. Sci.* 2015, *31*, 1115–1117.

105. D.P. Fagnant, G.S. Goff, B.L. Scott, W. Runde, J.F. Brennecke, "Switchable phase behavior of [HBet][Tf$_2$N]-H$_2$O upon neodymium loading: Implications for lanthanide separations," *Inorg. Chem.* 2013, *52*, 549–551.

106. K. Sasaki, T. Suzuki, T. Mori, T. Arai, K. Takao, Y. Ikeda, "Selective liquid-liquid extraction of uranyl species using task-specific ionic liquid, betainium bis(trifluoromethylsulfonyl)imide," *Chem. Lett.* 2014, *43*, 775–777.

107. X. Sun, Y. Ji, Y. Liu, J. Chen, D. Li, "An engineering purpose preparation strategy for ammonium-type ionic liquid with high purity," *AIChE J.* 2010, *56*, 989–996.

108. R.I. Wakeman, J.F. Coates, "Quaternary ammonium salts of organophosphorus acids," U.S. patent number 3,280,131, issued October 18, 1966.

109. K.D. MacKay, E.R. Rogier, "Solvent extraction of metals from acidic solution with quaternary ammonium salts of hydrogen ion exchange agents," U.S. patent number 4,058,585, issued November 15, 1977.

110. R.R. Grinstead, J.C. Davis, S. Lynn, R.K. Charlesworth, "Extraction by phase separation with mixed ionic solvents," *Ind. Eng. Prod. Res. Devel.* 1969, *8*, 218–227.

111. J.P. Mikkola, P. Virtanen, R. Sjöholm, "Aliquat 336™ – A versatile and affordable cation source for an entirely new family of hydrophobic ionic liquids," *Green Chem.* 2006, *8*, 250–255.

112. Y. Litaiem, M. Dhahbu, "Measurement and correlations of viscosity, conductivity, and density of an hydrophobic ionic liquid (Aliquat 336™) mixtures with a non-associated aprotic solvent (DMC)," *J. Mol. Liq.* 2012, *169*, 54–62.

113. V.M. Egorov, D.I. Djigailo, D.S. Momotenko, D.V. Chernyshov, I.I. Torocheshnikova, S.V. Smirnova, I.V. Pletnev, "Task-specific ionic liquid trioctylmethylammonium salicylate as extraction solvent for transition metal ions," *Talanta* 2010, *80*, 1177–1182.

114. A. Stojanovic, D. Kogelnig, L. Fischer, S. Hann, M. Galanski, M. Groessl, R. Krachler, B.K. Keppler, "Phosphonium and ammonium ionic liquids with aromatic anions: Synthesis, properties and platinum extraction," *Aust. J. Chem.* 2010, *63*, 511–524.

115. A. Rajendran, D. Ragupathy, M. Priyadarshini, A. Magesh, P. Jaishankar, N.S. Madhavan, K. Sajitha, S. Balaji, "Effective extraction of heavy metals from their effluents using some potential ionic liquids as green chemicals," *E-J. Chem.* 2011, *8*, 697–702.

116. E. Stanisz, J. Werner, H. Matusiewicz, "Task specific ionic liquid-coated PTFE tube for solid-phase microextraction prior to chemical and photo-induced mercury cold vapour generation," *Microchem. J.* 2014, *114*, 229–237.

117. A. Mehdinia, S. Shegefti, F. Shemirani, "A novel nanomagnetic task specific ionic liquid as a selective sorbent for the trace determination of cadmium in water and fruit samples," *Talanta* 2015, *144*, 1266–1272.

118. A. Saljooqi, T. Shamspur, M. Mohamadi, D. Afzali, A. Mostafavi, "A microextraction procedure based on a task-specific ionic liquid for the separation and preconcentration of lead ions from red lipstick and pine leaves," *J. Sep. Sci.* 2015, *38*, 1777–1783.

119. H. Shirkhanloo, M. Ghazaghi, H.Z. Mousavi, "Cadmium determination in human biological samples based on trioctylmethyl ammonium thiosalicylate as a task-specific ionic liquid by dispersive liquid-liquid microextraction method," *J. Mol. Liq.* 2016, *218*, 478–483.

120. S. Sadeghi, A.Z. Moghaddam, "Chromium speciation using task-specific ionic liquid/ aqueous phase biphasic system combined with flame atomic absorption spectrometry," *J. Mol. Liq.* 2016, *221*, 798–804.

121. J. Werner, "Determination of metal ions in tea samples using task-specific ionic liquid-based ultrasound assisted dispersive liquid-liquid microextraction coupled to liquid chromatography with ultraviolet detection," *J. Sep. Sci.* 2016, *39*, 1411–1417.

122. B. Dutta, R. Ruhela, M. Yadav, A.K. Singh, K.K. Sahu, N.P.H. Padmanabhan, J.K. Chakravartty, "Liquid-liquid extraction studies of gadolinium with N-methyl-N,N,N-trioctylammonium-*bis*-(2-ethylhexyl)phosphonate – Task-specific ionic liquid," *Sep. Purif. Technol.* 2017, *175*, 158–163.

123. S. Pandey, G.A. Baker, L. Sze, S. Pandey, G. Kamath, H. Zhao, S.N. Baker, "Ionic liquids containing fluorinated β-diketonate anions: Synthesis, characterization, and potential applications," *New J. Chem.* 2013, *37*, 909–919.

124. S. Biswas, V.H. Rupawate, S.B. Roy, M. Sahu, "Task-specific ionic liquid tetraalkyl-ammonium hydrogen phthalate as an extractant for U(VI) extraction from aqueous media," *J. Radioanal. Nucl. Chem.* 2014, *300*, 853–858.

125. Y. Zhou, S. Boudesocque, A. Mohamadou, L. Dupont, "Extraction of metal ions with task-specific ionic liquids: Influence of a coordinating anion," *Sep. Sci. Technol.* 2015, *50*, 38–44.

126. D.C. Stepinski, M.P. Jensen, J.A. Dzielawa, M.L. Dietz, "Synergistic effects in the facilitated transfer of metal ions into room-temperature ionic liquids," *Green Chem.* 2005, *7*, 151–158.

127. X. Sun, K.E. Waters, "The adjustable synergistic effects between acid-base coupling bifunctional ionic liquid extractants for rare earth separations," *AIChE J.* 2014, *60*, 3859–3868.

128. A. Rout, J. Kotlarska, W. Dehaen, K. Binnemans, "Liquid-liquid extraction of neodymium(III) by dialkylphosphate ionic liquids from acidic medium: The importance of ionic liquid cation," *Phys. Chem. Chem. Phys.* 2013, *15*, 16533–16541.

129. K. Wang, H. Adidharma, M. Radosz, P. Wan, X. Xu, C.K. Russell, H. Tian, M. Fan, J. Yu, "Recovery of rare earth elements with ionic liquids," *Green Chem.* 2017, *19*, 4469–4493.

130. S. Himmler, A. König, P. Wasserscheid, "Synthesis of [EMIM]OH *via* bipolar membrane electrodialysis – Precursor production for the combinatorial synthesis of [EMIM]-based ionic liquids," *Green Chem.* 2007, *9*, 935–942.

131. H. Mehdi, K. Binnemans, K. Van Hecke, L. Van Meervelt, P. Nockemann, "Hydrophobic ionic liquids with strongly coordinating ions," *Chem. Comm. (Camb.)* 2010, *46*, 234–236.

132. A. Rout, K.A. Venkatesan, T.G. Srinivasan, P.R. Vasudeva Rao, "Ionic liquid extract-ants in molecular diluents: Extraction behavior of plutonium(IV) in 1,3-diketonate ionic liquids," *Solvent Extr. Ion Exch.* 2011, *29*, 602–618.

133. A. Rout, K. Binnemans, "Solvent extraction of neodymium(III) by functionalized ionic liquid trioctylmethylammonium dioctyl diglycolamate in fluorine-free ionic liquid diluent," *Ind. Eng. Chem. Res.* 2014, *53*, 6500–6508.

134. J. Dehaudt, C.-L. Do-Thanh, H. Luo, S. Dai, "Liquid-liquid extraction of f-block ele-ments using ionic liquids," in *Ionic Liquids: Current State and Future Directions*, M.B. Shiflett, A.M. Scurto (Eds.), American Chemical Society, Washington, DC, 2017, pp. 157–185.

135. J. Dehaudt, N.J. Williams, I.A. Shkrob, H. Luo, S. Dai, "Selective separation of tri-valent f-ions using 1,10-phenanthroline-2,9-dicarboxamide ligands in ionic liquids," *Dalton Trans.* 2016, *45*, 11624–11627.

136. P.T. Anastas, J.C. Warner, *Green Chemistry: Theory and Practice*, Oxford University Press, 1998.

137. S. Zhu, R. Chen, Y. Wu, Q. Chen, X. Zhang, Z. Yu, "A mini-review on greenness of ionic liquids," *Chem. Biochem. Eng. Q.* 2009, *23*, 207–211.

138. M. Cvjetko Bubalo, K. Radošević, I. Radojčić Redovniković, J. Halambek, V. Gaurina Srček, "A brief overview of the potential environmental hazards of ionic liquids," *Ecotoxicol. Environ. Saf.* 2014, *99*, 1–12.

139. I.A. Shkrob, T.W. Marin, M.P. Jensen, "Ionic liquid-based separations of trivalent lanthanide and actinide ions," *Ind. Eng. Chem. Res.* 2014, *53*, 3641–3653.

140. M. Nilsson, K.L. Nash, "A review of the development and operational characteristics of the TALSPEAK process," *Solvent Extr. Ion Exch.* 2007, *25*, 665–701.

141. K.E. Gutowski, G.A. Broker, H.D. Willauer, J.G. Huddleston, R.P. Swatloski, J.D. Holbrey, R.D. Rogers, "Controlling the aqueous miscibility of ionic liquids: Aqueous biphasic systems of water-miscible ionic liquids and water-structuring salts for recycle, metathesis, and separations," *J. Am. Chem. Soc.* 2003, *125*, 6632–6633.

142. S. Sadeghi, A.Z. Moghaddam, "Task-specific ionic liquid-based in-situ dispersive liquid-liquid microextraction for the sequential extraction and determination of chromium species: Optimization by experimental design," *RSC Adv.* 2015, *5*, 60621–60629.

143. S. Sadeghi, V. Ashoori, "Sequential determination of iron species in food samples by new task specific ionic liquid-based *in-situ* dispersive liquid-liquid microextraction prior to flame atomic absorption spectrometry," *Anal. Methods* 2016, *8*, 5031–5038.

144. C. Bosch Ojeda, F. Sánchez Rojas, "Separation and preconcentration by dispersive liquid–liquid microextraction procedure: A review," *Chromatographia* 2009, *69*, 1149–1159.

4 X-Ray Studies of Liquid Interfaces in Model Solvent Extraction Systems

Wei Bu and Mark L. Schlossman

CONTENTS

4.1 INTRODUCTION

The extractant-assisted transfer of metal ions between aqueous and organic phases underlies the industrial process of solvent extraction, which is important for separating, isolating, and thus purifying metal ions.[1–4] The part of this process known as forward extraction involves the contact of an aqueous phase containing a mixture of ionic species with an organic phase to which the targeted metal species is to be selectively transferred by complexation with an amphiphilic extractant molecule, which serves to solubilize the metal cation in the nonpolar phase (Figure 4.1).

FIGURE 4.1 Schematic illustration of the transfer of a hydrated metal ion from an aqueous phase to an organic phase, where the metal ion is encased within a supramolecular complex of organic extractants. Although the structure of hydrated metal ions in water and of ion–extractant complexes (or reverse micelles) in an organic phase are known for some extraction processes,[5–12] little is known about the chemical speciation and composition at the interface. Importantly, the mechanism of interfacial transfer, represented by the blue arrows, is also unknown.

Complexation of metal ions and extractants is believed to take place at or near the organic–aqueous interface,[13,14] though different authors have suggested that the extractant binds metal ions either in the aqueous phase near the interface, or in the organic phase near the interface, or at the interface itself. For instance, the mass transfer with chemical reaction (MTWCR) mechanism postulates that acidic extractants are transferred into the aqueous boundary layer near the organic–aqueous interface, where they are deprotonated and interact with metal ions to form aqueous ion–extractant complexes, which subsequently diffuse into the organic phase.[14] On the other hand, it has been suggested that when the interface is occupied by stronger amphiphiles that exclude extractants, ion–extractant complexes are formed when fingers of water that contain metal ions reach into the organic phase.[15] Somewhat between these two cases, there is evidence that the amphiphilic character of extractants leads to their interaction with ions directly at the interface;[16–19] these studies include the suggestion that reverse micelles of extractants that enclose metal ions can form at the interface.[20] Largely missing from these investigations has been the application of experimental techniques that can locate and identify metal ions, extractants, and ion–extractant complexes in the liquid–liquid interfacial region, though recent X-ray and neutron scattering studies have begun to do just that.[21–25] This chapter reviews these recent scattering studies of liquid interfaces, including studies of both liquid–liquid and liquid–vapor interfaces, that are relevant to understanding the role of the interface in solvent extraction.

X-ray scattering from liquid interfaces has two principal advantages over optical probes, such as ellipsometry or optical reflectivity, for the study of molecular ordering. X-ray wavelengths of the order of 0.5 to 1.5 Å are well matched to the characteristic lengths of interfacial ordering; these include the interfacial widths of the electron-density profile $\rho(z)$ and the length scales for surface-induced atomic or molecular adsorption. Second, the interaction strength between X-rays and matter is sufficiently weak that the measurements to be described below can usually be interpreted unambiguously in terms of a simple kinematic theory.[26]

In the relatively brief period of time during which X-ray scattering methods have been applied to the study of liquid–liquid interfaces, several discoveries about the fundamental nature of this interface have been made that have a bearing upon the extractant-assisted transport of metal ions across the interface. For example, the structure of a liquid–liquid interface between neat liquids has been debated for many years, with discussions of both a broad "interphase" transition region between the liquids,[27] in which the chemical composition is postulated to vary smoothly from one bulk phase to the other, to a sharp interface for which the transition from one bulk phase to the other takes place within a bulk correlation length, which is roughly one to two molecular dimensions as long as the liquids are far from the critical de-mixing point.[28] X-ray reflectivity measurements have demonstrated that the interface between neat water and an alkane liquid, such as the water–dodecane interface commonly used in the solvent extraction of heavy metal ions, is molecularly sharp.[29–33] This result is consistent with computer simulations and subsequent neutron reflectivity measurements.[34,35] Nevertheless, interfaces between more complex solutions can exhibit a structured region that extends from the interface into the bulk liquid over many molecular lengths.[26]

As a result of their amphiphilic character, however weak, extractants have some preference to be located at the aqueous–organic interface. This has the effect of concentrating extractants at the interface, allowing them to interact closely even under conditions when their average separation in the bulk organic phase is large. This concentrating effect may be relevant for solvent extraction since the configuration of extractants at the interface will likely influence their ability to form complexes with aqueous ions. Earlier understanding of the interfacial configuration of amphiphiles was summarized in an influential textbook,[36] which stated that "molecules of oil penetrate between the hydrocarbon chains and remove all interchain attractions," thus leading to the widely held view "that the $-CH_2-$ groups in the adsorbed film are free to move laterally" and the general expectation that amphiphiles (i.e., surfactants or extractants) at the water–oil interface are more disordered than at the water–vapor interface.[37–39] Recent X-ray results have shown the limitations of this point of view by demonstrating that amphiphiles at the alkane-water liquid–liquid interface adopt a variety of configurations, including ordered 2-dimensional (2D) solid-like phases and 2D disordered liquid-like phases, as well as multilayers.[40–47] The ordering exhibited by any particular chemical system depends upon the type of amphiphile and the environmental conditions, such as temperature and pH, as well as the presence of other components in the bulk phases that may interact with amphiphiles at the interface. As discussed later in this review, thermal transitions between different configurations of amphiphiles at interfaces have a practical application in arresting ion–extractant complexes formed at the liquid–liquid interface to allow them to be further characterized by X-ray scattering before they are fully extracted into the organic phase.

Studies of interfacial width, ion adsorption, and amphiphile configurations at liquid–liquid interfaces form the basis for understanding the role of the interface in solvent extraction, and this review will focus on studies of model interfacial systems designed specifically to understand solvent extraction processes. These studies typically use amphiphiles that are known to extract metal ions, with an emphasis on amphiphiles that are either in use industrially or slight variations of these extractants, as well as metal ions of practical interest. In addition, we review studies of water–vapor interfaces of model extraction systems because the technical advantage of shining X-rays directly onto a water surface, without having X-rays penetrate through a bulk phase to access a buried liquid–liquid interface, leads to higher spatial resolution in characterizing interfacial species and enables some X-ray techniques that cannot be used at the liquid–liquid interface. Although studies of the water–vapor interface cannot reveal the mechanism of ion transport across the interface, they can probe interactions of ions with the polar, or head-group, part of extractants. Nevertheless, it must be kept in mind that the configuration of extractants at the water–vapor and oil–water interfaces are similar in some cases, though in other cases they can be very different.[40]

X-ray studies of interfaces between immiscible liquids promise to reveal the mechanism of ion transport across the interface that occurs during solvent extraction. Nevertheless, one cannot merely shine X-rays on a container undergoing solvent extraction and expect to analyze the ongoing interfacial ion transport in real time. In such a hypothetical experiment, the primary X-ray interactions will occur from molecular and ionic species within the bulk of the liquid phases; only a much weaker signal will be produced by internal interfaces between neighboring bulk phases. Instead, the determination of molecular ordering that occurs at a liquid interface, or within a molecular distance of it, requires the application of specialized techniques. Recent progress in the development and use of these specialized interface-sensitive X-ray techniques has been reviewed elsewhere, and we suggest that readers interested in the details of these techniques consult these references.[26,48,49] In this chapter, techniques will be discussed only to the extent required to understand their contribution to characterizing the interfacial structure.

These specialized X-ray techniques place constraints on the configuration of samples that can be studied. For instance, commercial solvent extraction processes often involve vigorous agitation of the two liquid phases, whereas liquid interface X-ray scattering studies described herein consist of measurements from macroscopically flat and quiescent interfaces. It has been demonstrated that X-ray scattering measurements from single, oriented interfaces yield a sub-nanometer to angstrom spatial resolution of molecular-scale processes at the interface, including the arrangement of amphiphiles and metal ions at the interface. In addition, X-ray surface spectroscopy can measure the metal ion coordination and oxidation state at the aqueous surface. To the extent that molecular-scale or even supramolecular-scale processes at the interface are relevant to the practice of solvent extraction, X-ray scattering studies of single, macroscopic interfaces play a role in increasing our fundamental knowledge that may lead to improving the efficiency and kinetics of solvent extraction.

4.2 LIQUID–LIQUID INTERFACE STUDIES

This section reviews recent advances that used interface-sensitive X-ray techniques to probe molecular ordering at the oil–water interface under conditions relevant to solvent extraction, as well as a recent pair of studies that combined X-ray and neutron reflectivity measurements. Specifically, we present measurements designed to probe the extraction of metal ions (including Er(III), Y(III), Nd(III), and Sr(II)) from water into dodecane using either acidic organo-phosphorous or diamide extractants. Dodecane was chosen because it is an industry standard due to its low toxicity and high flash point.[50] The acidic organo-phosphorous extractant used for these studies was di-hexadecyl phosphate (DHDP). Although DHDP is not used industrially, it was chosen for two reasons. First, the phosphate head-group (PO_4H) of DHDP, which complexes with metal ions, is found in industrial acidic organo-phosphorous extractants such as HDEHP (di-(2-ethylhexyl) phosphoric acid), which is used for lanthanide extraction and in the TALSPEAK process for actinide and lanthanide extraction.[1,50] Partitioning measurements confirmed that DHDP extracts Er(III), Y(III), and Sr(II).[22,23,51] Second, the long (C_{16}) hydrocarbon chains of DHDP enabled the formation of a long-lived (hours-long) interfacial intermediate state of the extraction process, whose structure was then measured with X-ray reflectivity and fluorescence. Diamide extractants studied include DMDOHEMA (*N,N'*-dimethyl *N,N'*-dioctyl hexyl ethoxy malonamide) and DMDBTDMA (*N,N'*-dimethyl-*N,N'*-dibutyl-2-tetradecylmalonamide), which are currently used in the nuclear industry to separate f-elements from high-level radioactive liquid waste (the DIAMEX process) and to recycle rare-earth elements.[52,53]

Two interface-sensitive techniques, X-ray reflectivity and X-ray fluorescence near total reflection (XFNTR), were used to study the DHDP dodecane-aqueous system for the solvent extraction of rare earth Er(III) and Y(III) ions, as well as Sr(II) ions.[22,23] Figure 4.2 illustrates the geometry of X-ray reflectivity and fluorescence,

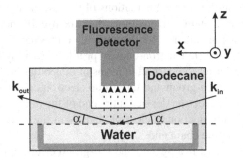

FIGURE 4.2 XFNTR and reflectivity measurement geometry along with a schematic illustration of a sample cell used for X-ray measurements from the liquid–liquid interface between an aqueous phase in a glass tray surrounded by a dodecane phase. Kinematics of X-ray reflectivity are shown: incoming X-ray wave vector $\vec{k}_{in}$; reflected wave vector $\vec{k}_{out}$, and angles of incidence and reflection, α. Dashed lines with arrows indicate fluorescence X-rays generated isotropically by the sample interface, some of which are measured by the fluorescence detector. Two other detectors that are not shown measure the incident and reflected intensity, respectively, before and after the sample cell.[22]

showing X-rays penetrating through the dodecane phase and reflecting from the liquid–liquid interface, as well as the detection of fluorescence produced at the interface.

4.2.1 AMPHIPHILE ORDERING AT LIQUID–LIQUID INTERFACES: X-RAY REFLECTIVITY

X-ray reflectivity measurements consist of the reflected X-ray intensity normalized by the incident intensity, measured for a range of reflection angles α (Figure 4.2). X-ray reflectivity probes the electron-density profile $\rho(z)$, where z is the coordinate perpendicular to the interface and the measured profile is averaged over the X-ray footprint in the x–y plane of the interface. Since the interface consists of different components—water, organic solvent, extractants, and metal ions—each with a different electron density, X-ray reflectivity measurements are sensitive to the arrangement of these components at the interface.[26]

X-ray reflectivity measurements are analyzed by fitting them to a model of the electron-density profile $\rho(z)$, which is typically a set of interfacial layers characterized by the electron density and thickness of each layer. This type of analysis yields an electron-density profile (perpendicular to the interface) with sub-nanometer spatial resolution and ±2% or better uncertainties in the electron density.[26,54] The electron-density profile is then interpreted to yield the interfacial molecular arrangement.[40,41,43–47,55]

Figure 4.3a shows the X-ray reflectivity from a self-assembled monolayer of the extractant DHDP at the interface between pure water and a dodecane solution of DHDP. This measurement revealed a monolayer covering the interface at 28°C.[22] Quantitative analysis of the X-ray reflectivity produced the electron-density profile shown by the red line in Figure 4.3b. The form of the electron-density profile is the result of the underlying molecular arrangement at the interface plus smearing due to thermal capillary-wave fluctuations of the interface. Capillary waves are long-wavelength hydrodynamic fluctuations that are usually not expected to alter the local molecular arrangement.[26] Mathematically removing the effect of capillary waves produces the *intrinsic* electron-density profile, shown by the dashed line in Figure 4.3b, which is easier to interpret in terms of molecular ordering.[56] The intrinsic profile reveals the head-group and tail-group electron densities and thicknesses, as well as a small dip in electron density at the end of the tail-group, which is attributable to either a slight reduction in electron density in the methyl group or to conformational disorder at the chain end.

The measured values of head-group and tail-group electron densities along with the measured thickness of each of these regions can be interpreted in terms of molecular configurations that are known to correspond to these values. In this case, the measured head-group electron density and thickness are quantitatively consistent with the presence of a monolayer of phosphate head-groups that completely covers the interface. Similarly, the measured tail-group electron density and thickness are quantitatively consistent with the presence of an ordered monolayer of close-packed all-trans hexadecyl chains (except at the chain end) as illustrated by the molecular cartoon in Figure 4.3c.

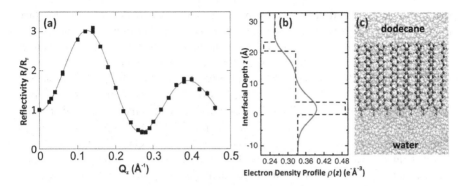

FIGURE 4.3 X-ray reflectivity data and analysis from a monolayer of extractants at the interface between pure water and 10^{-4} M DHDP in dodecane at 28°C. (a) Variation of the measured X-ray reflectivity $R(Q_z)$ (normalized by the Fresnel reflectivity $R_F(Q_z)$ that is calculated for an ideal flat interface without the monolayer) with wave vector transfer $Q_z = (4\pi/\lambda)\sin\alpha$ (where α is the angle of reflection and λ is the X-ray wavelength). The red line through the data is the best fit calculated from the electron-density profile shown in (b), from which the molecular ordering illustrated by the cartoon in (c) is determined. X-ray reflectivity measures the electron-density profile shown by the line in (b), which includes the effect of thermally induced hydrodynamic capillary waves of the interface. The dashed line in (b) is the intrinsic electron-density profile, which has the effect of capillary waves removed mathematically from the measured electron-density profile. The intrinsic profile is most easily compared to a cartoon of the molecular model in (c), though the effect of the slightly lower electron density at the methyl end of the hydrocarbon chain is not illustrated in the cartoon.[22] (Adapted with permission from *J. Phys. Chem. B*, 2014, 118(36), pp. 10662–10674. Copyright © 2014, American Chemical Society.)

4.2.2 METAL ION SURFACE DENSITY AT LIQUID–LIQUID INTERFACES: X-RAY FLUORESCENCE NEAR TOTAL REFLECTION

Although X-ray reflectivity measures the molecular-scale electron-density profile of organic–aqueous interfaces, it does not identify the presence of specific elements. XFNTR can be used to identify specific elements; for example, to identify which metal ion is at the interface.[21,57,58] XFNTR measures the interfacial density of just that type of ion, say erbium or strontium in the experiments presented below, independent of other molecules or elements that may be present at the interface.

XFNTR from the liquid–liquid interface was first used to study the time-dependent formation of interfacial crud, which is the result of third-phase formation or precipitation at the interface of metal ion–extractant complexes from the organic phase.[31] Cruds vitiate the separation and usually need to be avoided in practical separations processes. Although known to be a problem, little work has been done to chemically or structurally characterize these precipitates, which may hold important clues to organic-phase metal ion speciation.

Crud was prepared at the liquid–liquid interface between dodecane and a 0.1 M $ErCl_3$ aqueous solution by injecting 30.5 µL of HDEHP into the top (dodecane) phase, which produced an HDEHP concentration in the dodecane of about 10^{-3} M.

Figure 4.4 shows the time course of the fluorescence spectrum from the interface monitored over several hours after injection (at $t=0$ min). Since the fluorescence was produced by an X-ray beam reflected off the interface under conditions of total reflection, the fluorescence originates primarily in the interfacial region and is increasing with time. Visual observation of the sample revealed the formation of a microscopically thick precipitate at the interface. These two observations imply that the formation of interfacial crud containing Er is increasing with time.[21] Although further quantitative analysis of this set of data was not possible, more extensive experiments employing this methodology could be used to study the kinetics of crud formation.

Subsequent measurements of XFNTR at the liquid–liquid interface were analyzed quantitatively to determine metal ion densities at the interface. Spectra similar to those shown in Figure 4.4 can be analyzed and compared to spectra measured from a calibration sample to determine the number of metal ions per interfacial area. Figure 4.5 illustrates comparative results from both XFNTR and X-ray reflectivity measurements of the interfacial density of Sr ions at an interface between a dilute solution of DHDP in dodecane and a dilute solution of $SrCl_2$ in water. DHDP extractants form a monolayer at this interface, in an arrangement similar to that shown in Figure 4.3c, but with Sr ions present in the aqueous phase. Although both fluorescence and reflectivity are sensitive to the presence of Sr ions, the element-specificity

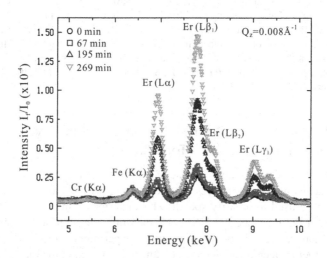

FIGURE 4.4 Time-dependent measurement of the X-ray fluorescence intensity from the liquid–liquid interface between a 10^{-3} M HDEHP dodecane solution and a 0.1 M $ErCl_3$ aqueous solution that follows the interfacial production of crud. The fluorescence intensity was measured at $Q_z=0.008$ Å^{-1} (below the critical angle for total reflection) to provide sensitivity to interfacial Er. HDEHP was injected into pure dodecane at $t=0$ min. The increase in fluorescence intensity with time represents increasing Er at the interface due to the buildup of interfacial crud. Different peaks labeled "Er (XX)" represent fluorescence from different electronic energy levels. Note that the time-invariant fluorescence from Cr and Fe was produced by extraneous scattering of the X-ray beam from a stainless-steel detector support.[21] (Reprinted from *Journal of Applied Physics*, 2011, 110, 102214, with the permission of AIP Publishing.)

of XFNTR can identify that they are strontium, whereas reflectivity records only their contribution to the electron density. Therefore, the measurement of Sr interfacial density by X-ray reflectivity is indirect and relies upon comparing measurements of samples with and without Sr, whereas XFNTR is a direct measurement of the interfacial density of Sr. As a result, we expect that the Sr interfacial density should be measured more accurately by XFNTR than by reflectivity, an expectation that is confirmed in Figure 4.5 by the smaller error bars on XFNTR measurements.

Figure 4.5 shows that the Sr interfacial density varies with the pH of the aqueous phase. Deprotonation of the DHDP phosphate head-groups is expected to occur with increasing pH, leading to binding of Sr(II) ions as well as to the creation of an electrical double layer near the negatively charged interface. These two possible locations of Sr ions, either bound to DHDP or in the electrical double layer, produced the unanticipated result that the Sr ion interfacial density measured by XFNTR is larger than that measured by X-ray reflectivity. These data provided quantitative evidence that a relatively large interfacial density of Sr ions is located immediately adjacent, presumably bound, to the phosphate head-group of the DHDP, while a smaller density of Sr ions is present in the interfacial double layer, though not bound to the head-groups.

XFNTR measured a larger ion density because it detected all Sr ions within roughly 10 nm of the interface, including ions bound to DHDP or located in the diffuse electrical double layer, whereas in this case X-ray reflectivity was sensitive primarily to the Sr bound directly to DHDP.[23] Calculations demonstrated that the

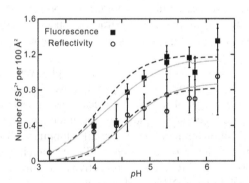

FIGURE 4.5 The variation with pH of the interfacial density of Sr(II) ions at the interface between 10^{-4} M DHDP in dodecane and 10^{-5} M $SrCl_2$ in water at 28°C. The density is given in terms of number of Sr(II) ions per 100 Å. For comparison, the number of DHDP molecules per 100 Å varies from 2.1 to 2.5 as the pH increases, indicating the presence of one Sr(II) in the interfacial region for two DHDP at higher pH. Squares represent fluorescence (XFNTR) measurements, and circles represent reflectivity measurements. Lines are published fits or calculations which demonstrated that the larger interfacial density of Sr ions measured by XFNTR is due to the sensitivity of this technique to ions that are either bound to the phosphate head-group of DHDP or in the nearby electrical double layer, whereas the reflectivity measurements are sensitive primarily to Sr ions bound to DHDP.[23] (Adapted with permission from *J. Phys. Chem. B*, 2014, 118(43), pp. 12486–12500. Copyright © 2014, American Chemical Society.)

insensitivity of X-ray reflectivity to Sr ions throughout most of the electrical double layer was the result of their relatively small density in the double layer, which provided inadequate electron-density contrast (with the bulk aqueous solution) for them to be detected. These calculations also provided evidence that the difference in ion density measured by the two techniques could be accounted for by the number of Sr(II) ions in the electrical double layer predicted by the Poisson–Boltzmann equation.[23]

4.2.3 Ion–Extractant Complexes Formed in the Midst of Extraction

The formation of interfacial ion–extractant complexes and their transport across the oil–water interface occur faster than the time required to characterize these complexes by XFNTR and X-ray reflectivity. Controlling extractant adsorption to the interface by thermal methods can arrest the extraction process and immobilize ion–extractant complexes at the interface for many hours, during which time their structure can be characterized by X-ray methods.[22] Thermal control over the extraction kinetics can be provided by an adsorption phase transition at which extractant adsorption to the oil–water interface is enhanced below a transition temperature T_o.[40,44,46,59,60] For example, cooling a sample from above T_o to below it creates a stable, condensed layer of extractants at the interface, which greatly retards the extraction of ions through the interface, whereas heating from below T_o to above it allows extractants to move freely between the interface and bulk dodecane, thereby re-establishing normal extraction. This thermal switch, which effectively turns extraction on and off, is not applicable to all extractants, though it has proven useful in the study of DHDP.[22,23,51] The choice of a long chain extractant, such as DHDP, was motivated partially by the convenience of locating the transition temperature T_o slightly above room temperature.

To take advantage of this adsorption transition, samples were prepared as follows. First, the dodecane–water sample was prepared above the transition temperature, during which time extraction of ions into dodecane was presumably facilitated by the formation of ion-extraction complexes at the interface. The sample was then cooled below the transition temperature, essentially stopping the extraction and trapping ion–extractant complexes at the interface.[22] The cartoon in Figure 4.6 illustrates the combined results of XFNTR and X-ray reflectivity measurements of the interface between dodecane solutions of DHDP and aqueous solutions of ErBr$_3$ (pH 2.5) when the sample was prepared in this way.[22] The cartoon displays an inverted bilayer consisting of supramolecular ion–extractant complexes at the interface. The complexes contained an average of three DHDP molecules for each Er(III) ion (along with a few water molecules not shown explicitly), thus leading to charge neutrality in the vicinity of each Er(III).

It is useful to discuss the analysis that underlies the drawing of the molecular cartoon in Figure 4.6 not only as an illustration of what can be done by combining the techniques of XFNTR and X-ray reflectivity but also because the inverted bilayer shown in Figure 4.6 represents a counterintuitive interfacial structure. Instead of polar head-groups located in the aqueous phase, as commonly observed for amphiphiles and illustrated for the DHDP monolayer in Figure 4.3c, the structure shown in

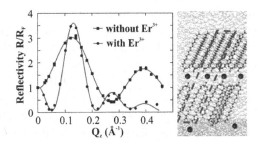

FIGURE 4.6 X-ray reflectivity from the interface between water and 10^{-4} M DHDP in dodecane with and without 5×10^{-7} M ErBr$_3$ in the water (pH 2.5 adjusted with HBr). The samples were prepared as described in the text in order to thermally arrest the extraction of Er(III) from water to dodecane and concentrate the ion–extractant complexes at the interface. The cartoon illustrates the inverted bilayer that is arrested when Er ions (green dots) are in the sample. The dashed box illustrates an ion–extractant complex: 3 DHDP molecules that coordinate with one Er(III) (water molecules coordinated with Er(III) are not shown).[22] (Adapted with permission from *J. Phys. Chem. B*, 2014, 118(36), pp. 10662–10674. Copyright © 2014, American Chemical Society.)

Figure 4.6 has DHDP head-groups separated from the water surface and DHDP tail-groups in the lower leaflet of the bilayer in contact with it. Although this structure is different from the conventional equilibrium structures of amphiphiles at interfaces, it is important to realize that this structure was prepared under non-equilibrium conditions, by thermally arresting it at the interface in the midst of extraction.

The determination of the inverted bilayer structure relied upon a quantitative analysis of the X-ray data, but also upon qualitative features of the data. Figure 4.6 shows that oscillations in the reflectivity from samples with ErBr$_3$ have half the wavelength of those from samples without ErBr$_3$, thereby revealing an interfacial film that is twice as thick as the monolayer measured in the absence of Er. Another important qualitative feature of the data is the dip in reflected intensity at the smallest values of wave vector transfer Q_z, at roughly 0.05 Å^{-1}. The importance of this feature is indicated by a general mathematical argument that constrains the interfacial structure from taking certain forms. Details of the mathematics are described elsewhere,[22] but the calculation shows that the appearance of a minimum in the reflectivity just above the critical wave vector transfer $Q_c \approx 0.01$ Å^{-1} is a signature feature of an inverted bilayer that would not be observed for more conventional interfacial layers of DHDP, such as a monolayer, bilayer, or trilayer that have at least one layer whose head-groups are in contact with the aqueous phase.

Quantitative modeling of the X-ray reflectivity data from the inverted bilayer consists of three regions, known as slabs, where the first slab (slab 1) represents the layer of hydrocarbon chains in contact with the aqueous phase at the bottom of the bilayer, slab 2 represents the head-group region where the lower and upper leaflets of the inverted bilayer meet, and slab 3 represents the hydrocarbon chains in the upper leaflet in contact with the dodecane phase. Note that the chemical components of the slabs are not assumed beforehand; their identification is produced by the analysis. For example, the large electron density of the slab 3, $\rho_3 = 0.324 \pm 0.003$ e$^-$/Å^3, is consistent with an upper leaflet of tail-groups (in contact with dodecane) that is close-packed and essentially all-trans.[61] The thickness of this layer, $d_3 = 18.5^{+2.1}_{-3.5}$ Å, is within one standard deviation of the all-trans length, 20.6 Å. Alternatively, if we assume that the difference between the layer thickness and the all-trans length is significant, then the measured electron density suggests that the deviation is due primarily to chain tilting, which can be calculated as $\cos^{-1}(d_3/L_{\text{trans}}) = 26°$ similar to values commonly observed in close-packed crystals of alkanes.[62] The area A per DHDP molecule in the upper leaflet of the bilayer is given by $A = 2A_0 L_{\text{trans}}/d_3 = 44^{+5}_{-4}$ Å^2, where $A_0 = 19.8$ Å^2 is the cross-sectional area per hydrocarbon chain in the all-trans state, and the value of the upper error bar includes the effect of placing a limit ($<35°$) on the maximum tilt angle for DHDP in the upper leaflet.[61–63]

The lower layer of tail-groups (in contact with water), represented by slab 1, has a similar thickness $d_1 = 18.9^{+0.8}_{-0.6}$ Å, but a lower electron density, $\rho_1 = 0.279 \pm 0.003$ e$^-$ Å^{-3}. The lower electron density implies a larger surface area per DHDP, which could be the result of the intermixing of dodecane into the lower leaflet, as illustrated in Figure 4.6, or it could be due to the presence of gauche conformations in the alkyl chains of DHDP or to a combination of both effects. X-ray reflectivity cannot distinguish between these possibilities because the electron density of dodecane chains is essentially identical to that of the hydrocarbon chains of DHDP.

The thickness of the middle region (slab 2) of the inverted bilayer is $d_2 = 8.5^{+3.9}_{-2.8}$ Å, and its average electron density is $\rho_2 = 0.329^{+0.013}_{-0.008}$ e$^-$ Å^{-3}. Assuming that head-groups of DHDP, Er(III) ions, and water molecules are in the middle slab, a straightforward calculation suggests that there is one Er(III) per 88 Å^2 with $3.5^{+10.5}_{-3.5}$ water molecules.[22] Note that the reflectivity measurements do not specifically probe the presence of water molecules, just their effect on the average electron density, which leads to large errors on their number in the head-group region of the inverted bilayer. Similarly, reflectivity measurements do not specifically measure the presence of Er(III) ions. Therefore, XFNTR measurements, similar to those presented in Section 4.2.2, which are ion-specific, were used to directly measure the interfacial area of 81^{+6}_{-4} Å^2 per Er ion.

The combination of results from XFNTR and X-ray reflectivity provided the basis for the model of the inverted bilayer illustrated in Figure 4.6. The interfacial area of 81^{+6}_{-4} Å^2 per Er ion from the fluorescence analysis, together with the extractant molecular area in the upper leaflet of 44^{+5}_{-4} Å^2 measured by reflectivity, suggests that each Er(III) ion binds to two phosphate head-groups from the upper leaflet of the bilayer. Greater uncertainty exists for the number of DHDP per Er(III) in the lower leaflet because X-ray measurements cannot determine whether additional DHDP chains or dodecane chains occupy the space beyond that required for one DHDP in the lower leaflet. Nevertheless, a model that includes more DHDP chains must also account for their accompanying head-groups. The data limits the fraction of head-groups in direct contact with the aqueous phase to be small, ~10%, so that most head-groups will be located in the mid-region of the inverted bilayer.

Different models were examined which considered more or less DHDP or dodecane in the lower leaflet. The simplest model of the ion–extractant complex that is consistent with both X-ray reflectivity and XFNTR data consists of three DHDP phosphate head-groups, two from the upper layer and one from the lower layer, coordinated to each Er(III) ion, as shown in the dashed box in Figure 4.6. An important outcome of this interfacial structure, along with the measured absence of Br⁻ ions at the interface (whose source might have been the ErBr$_3$ in solution), is that it leads to charge-neutral complexes. Although these interfacial measurements are not consistent with the complexing of an average of two or four DHDP molecules per Er(III) ion, they cannot exclude the presence of an additional fraction up to 0.8 DHDP per Er(III) in the lower leaflet. This indicates that a fraction of Er(III) ions in the mid-region of the inverted bilayer might be complexed to more than three DHDP molecules.

The unexpected structure of inverted bilayers at an organic solvent–water interface has been recently confirmed by X-ray studies of samples containing Y(III) cations, as well as molecular dynamics (MD) simulations that verified its short-term stability.[51] Combining these observations of inverted bilayers for Y(III) and Er(III) with the observation of a conventional monolayer for Sr(II) provides insight into the extraction mechanism. The interfacial state of Y(III) and Er(III) sandwiched between layers of DHDP extractants suggests the prompt transfer of these cations from the aqueous side of the liquid–liquid interface to a coordinated ion–extractant environment on the organic side. These ion–extractant complexes represent an intermediate state in which ions have been transported across the aqueous–organic interface but have not yet been dispersed in the organic phase. This dispersion occurred in the experiments once the temperature was raised above the adsorption transition. In contrast to this, the observation of a conventional monolayer of DHDP extractants with Sr(II) bound to DHDP head-groups, but remaining in contact with the water phase, suggests a slower kinetics of transfer of Sr(II) from water to dodecane, whose mechanism involves at least one additional step to transport the ion across the aqueous–organic interface. As suggested by a recently introduced analytical model, reverse-micelle formation at the interface may be a plausible mechanism for Sr extraction by DHDP.[51]

Comparison of the interfacial states of ion–extractant complexes for these three ions, Y(III), Er(III), and Sr(II), suggests that the oxidation state of the ion, or possibly just its charge, is the primary factor that determines the form of the intermediate state. These studies also allow for a comparison of the effect of electronic configuration on the intermediate state. We note that Sr(II) and Y(III) have the same closed-shell electronic configuration ($4p^6$) of the unreactive noble gas Kr, whereas Er(III) has a more complex $4f^{11}$ electronic configuration. It appears that the electronic configurations of these ions are not the determining factor in the structure of the intermediate state, in contrast to previous suggestions from kinetic studies of divalent ion extraction.[20]

4.2.4 Ion–Extractant Complexes Measured by Combining X-Ray and Neutron Reflectivity

Neutron reflectivity shares many of the advantages with X-ray reflectivity in the investigation of molecular structures at liquid interfaces. The two techniques are complementary because they measure different aspects of the interfacial structure. The X-ray atomic scattering amplitude is proportional to the number of electrons, whereas neutrons scatter from the nucleus. Neutron reflectivity measurements of specially prepared materials can take advantage of isotopic substitution of deuterium for hydrogen because the large difference and opposite sign of the neutron scattering cross-section from hydrogen and deuterium provides sensitivity to the hydrogen structure of materials that is missing for X-rays.[64] However, the significantly higher X-ray flux and smaller beam size and divergence give X-rays a decided advantage in many investigations.

Recent studies combined X-ray and neutron reflectivity to investigate a model solvent extraction liquid–liquid interface in the presence of diamide extractants, specifically DMDOHEMA (*N,N*'-dimethyl *N,N*'-dioctyl hexyl ethoxy malonamide) and DMDBTDMA (*N,N*'-dimethyl-*N,N*'-dibutyl-2-tetradecylmalonamide). The diamides were dissolved in dodecane at different concentrations, all below the critical aggregation concentration at which extractant complexes would form in the bulk phase. The dodecane phase was placed in contact with an aqueous phase containing neodymium nitrate salt, specifically 0.25 M of neodymium nitrate and 2 M of lithium nitrate as buffer salts to fix the activity in the aqueous phase.[24,25]

The X-ray and neutron reflectivity were each analyzed separately using standard techniques, similar to those described earlier, to produce an electron-density profile and a scattering length density profile, respectively. A Monte Carlo sampling method was then used to construct concentration profiles of each chemical component that were consistent with the measured electron density and scattering length density profiles. An example of the results is shown in Figure 4.7.[25]

Adsorption of DMDOHEMA appears to increase slightly as the critical aggregation concentration (CAC) of roughly 0.04 M is approached, whereas adsorption of DMDBTDMA appears localized near the interface at the lowest concentrations with a wider interfacial distribution at higher concentrations. Note that these concentration profiles include the effect of interfacial roughness, which is expected to be larger at higher concentrations due to the lower interfacial tension. Therefore, the wider distributions at higher extractant concentrations are due at least partially to the effect of interfacial roughness.

Differences between the samples with DMDOHEMA and DMDBTDMA are also evident in the interfacial ion adsorption. Except for the perplexing oscillations at 0.02 M DMDOHEMA, interfacial peaks in the Li and Nd ion concentrations are not visible in DMDOHEMA samples, indicating that these ions do not preferentially segregate to the interface or move into the region with extractants. Contrary to this, ion peaks are visible at the interface with DMDBTDMA, revealing an enhanced concentration of ions in the region containing extractants. In this case, ion concentration profiles overlap the extractant profile, except possibly in the sample with the lowest concentration (0.02 M DMDBTDMA). Further development of this analysis technique should provide an understanding of the statistical significance of the concentration profiles to better understand the measured differences between samples with different extractant concentrations.

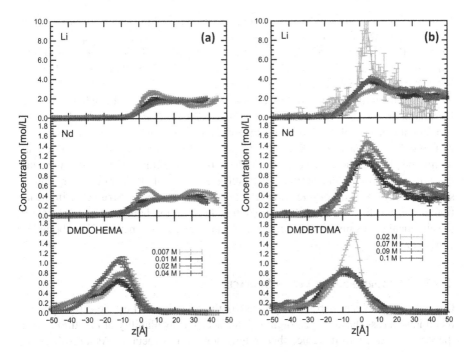

FIGURE 4.7 Monte Carlo concentration profiles determined by X-ray and neutron-reflectivity measurements from the liquid–liquid interface between aqueous solutions of neodymium nitrate salts and dodecane solutions of diamide extractants DMDOHEMA and DMDOHEMA. Concentration profiles as a function of depth z through the interface of (a) lithium and neodymium cations and extractants at four concentrations of DMDOHEMA (0.007, 0.01, 0.02, and 0.04 M), and of (b) lithium and neodymium cations and extractants at four concentrations of DMDOHEMA (0.02, 0.07, 0.09, and 0.1 M). Negative values of z correspond to the dodecane phase, whereas positive values correspond to the aqueous phase.[25] (Reprinted with permission from *Angew. Chem. Int. Ed.*, 2016, 128, pp. 9472–9476. Copyright © 2016, Wiley–VCH Verlag GmbH & Co.)

4.3 WATER–VAPOR INTERFACE STUDIES

Although solvent extraction occurs at the organic–aqueous interface, X-ray scattering studies of water–vapor interfaces for the purpose of understanding solvent extraction can be justified in two ways. First, certain aspects of interfacial ion speciation and interfacial extractant–ion interactions might be sensibly expected to be relevant to solvent extraction processes, even if measured in the absence of the organic phase. Second, there are a number of technical advantages to performing X-ray scattering measurements on water–vapor interfaces. These advantages lead to higher spatial resolution, simplicity, and ease of measurements, facile control over extractant density at the interface through the use of Langmuir trough techniques, and easier access to the interface for the purpose of placing detectors near the interface.[26]

In some cases, certain techniques can only be used at a liquid interface with vapor. For example, the measurement of electron-yield X-ray absorption spectroscopy, which will be discussed later in this section, cannot be carried out at the liquid–liquid interface because electrons ejected near the interface would be absorbed by the upper liquid phase before reaching a detector.

X-ray reflectivity and grazing-incidence X-ray diffraction have been in routine use for a number of years to characterize the interfacial structure at water surfaces with sub-nanometer spatial resolution.[26,48] Information of relevance for solvent extraction includes the interfacial distributions of ions,[65] ordering of ions against extractants within the plane of the interface,[66,67] and the out-of-plane configuration of ions bound to extractant monolayers at the water surface.[68] Recent developments in liquid-surface X-ray spectroscopy techniques reveal element-specific structural information at the interface, such as the oxidation state of an ion bound to an extractant,[69] the chemical environment of a target ion,[70,71] and the stoichiometry of ions bound to extractants.[58,72]

4.3.1 ION ORDERING AT WATER SURFACES

Critical to the solvent extraction process is the effective and efficient transport of metal ions across the aqueous–organic interface. The availability of metal ions near the water surface will influence their interaction with extractants. As a result of the variation in the chemical environment across a water–vapor interface, theory has predicted that cations are expected to be depleted from the interface, while large anions are absorbed to it due to the hydration effects[73] and polarizability.[74] These predictions are consistent with some experimental observations, including surface tension,[75] electron spectroscopy,[76] and vibrational sum-frequency generation.[77] Nevertheless, these techniques lack the spatial resolution to measure the ion density profile.

X-ray reflectivity from surfaces of aqueous $ErCl_3$ solutions measured interfacial electron-density profiles to determine the ion number-density profiles at the surface (Figure 4.8a).[65] Data analysis revealed the first observation of a nonmonotonic profile of ions decaying from the surface into the bulk water (Figure 4.8b). This profile is very different from the monotonic profiles predicted by conventional Poisson–Boltzmann (or Gouy–Chapman) theory for ion distributions in an electrical double layer, though it is consistent with recent computer simulations.[78,79] The nonmonotonic variation is evident in samples with bulk concentrations of $ErCl_3$ above 0.5 M, which exhibit the formation of a second peak in the Er ion number-density profile (Figure 4.8b). The weak second peak represents a second layer of Er ions located further from the surface with a lower concentration of ions. The separation between the first and second peaks is similar to the spacing expected between layers of Er ions, inclusive of their two hydration shells, as sketched in Figure 4.8c.[5]

As shown in Figure 4.8b, a depletion layer separates the first layer of ions from the water surface. The thickness of the depletion layer varies with bulk concentration of $ErCl_3$, from 7.8 Å at 0.2 M to 5.5 Å at 1.0 M; therefore, the subsurface layer of Er approaches the vapor interface with increasing concentration. The depletion layer thickness is comparable to the distance from the center of an Er ion to the outside

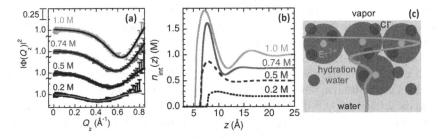

FIGURE 4.8 X-ray reflectivity measurements from ErCl$_3$ aqueous solutions at four different concentrations reveal a nonmonotonic profile of Er ions. (a) Intrinsic interfacial structure factors, $\Phi(Q_z)$, are determined from X-ray reflectivity measurements by mathematically removing the effect of thermal capillary waves. The wave vector transfer Q_z is determined by the reflection angle. Solid lines are fits to damped-sinusoidal electron-density profiles. (b) Erbium ion intrinsic number-density profiles $n_{int}(z)$ (in molar units) determined by the fits to the intrinsic structure factors shown in (a). Vapor is located at negative values of z, the aqueous solution at positive values. (c) A cartoon of ion distributions at the water surface of a 1 M ErCl$_3$ solution. The X-ray reflectivity measurements are sensitive primarily to the Er(III) ions and water, though they are consistent with the presence of Cl ions shown near the interface. The red line is the measured erbium ion intrinsic number-density profile that was also shown in panel (B).[65] (Reprinted with permission from *J. Phys. Chem. C*, 2013, 117(37), pp. 19082–19090. Copyright © 2013, American Chemical Society.)

of its second hydration shell, suggesting that Er ions maintain their two hydrations shells as they approach the water surface.

As a result of the presence of the depletion layer, as well as the nonmonotonic form of the rest of the ion density profile, the total Er ion adsorption within the entire interfacial region is negative, even though peak concentrations of Er ions within the surface layers are substantially larger than the bulk concentration. Thermodynamic analysis of the variation of surface tension with concentration has shown that the Er adsorption is negative, consistent with the X-ray measurements.[65] However, it is worthwhile mentioning that such thermodynamic measurements were previously interpreted as indicating that the ion concentration was merely depleted at the interface with a lower-than-bulk concentration of ions throughout the entire interfacial region. The measurement by Luo and coworkers demonstrates that even when ion adsorption is negative, regions of the interface can have a local ion concentration that is much larger than the bulk concentration.[65] From the point of view of solvent extraction, this measurement indicates that Er ions are readily available to interact with extractants at the interface, even though thermodynamic measurements indicate that the ion adsorption is negative.

The local hydration or solvation structure of metal ions undergoes significant restructuring upon transport from an aqueous to an organic phase. Therefore, aside from the ionic distribution, the chemical environment of metal ions at the interface is also a vital aspect for understanding the interfacial mechanism during solvent extraction. Extended X-ray absorption fine structure (EXAFS) and high energy X-ray scattering (HEXS) are well-established techniques to measure the chemical environment of metal ions in the bulk of aqueous and organic phases.

These techniques measure the identity and number of coordinating species as well as their distance to the metal ion. However, measurement of these quantities for ions near the interface remains a challenge.

Liquid-surface EXAFS from Er ions at the water surface was measured recently with the experimental setup illustrated in Figure 4.9a.[71] This measurement can be made in three different modes. In mode 1, the incident X-ray angle is set to be greater than the critical angle, which allows X-rays to penetrate through the surface into the bulk of the water. Fluorescence is then produced primarily from X-rays which probe the bulk phase coordination, in this case up to roughly 300 μm from the surface (Figure 4.9b). Modes 2 and 3 take advantage of grazing incidence for which the incident angle is set to be less than the critical angle for total reflection. X-rays are then confined to a distance within a penetration depth on the order of 5 to 10 nm from the surface. In mode 2, fluorescence is measured from this depth, which provides surface sensitivity. In mode 3, Auger electrons are collected under the same conditions of grazing incidence. In this electron-yield mode, the authors estimated the measurement to be sensitive to a depth less than 3 nm because of the short mean free path of electrons in the liquid, thus providing greater sensitivity to surface structure than the fluorescence mode 2. Figure 4.9b illustrates results from mode 1 (bulk fluorescence) and mode 3 (surface electron-yield).

EXAFS measurements with total electron yield and fluorescence detection revealed that the coordination of Er at the surface was qualitatively different from that in the bulk. The surface coordination was described by a three-shell model. The first and second shells consisted of 6 to 7 O atoms and 3 Cl atoms, respectively, suggesting the formation of a neutral $[(H_2O)_{6-7}ErCl_3]$ entity. Three Er atoms fill the third shell, indicating the presence of Er–Er correlations absent in the bulk. The proposed planar 7-mer Er cluster structure is illustrated in Figure 4.9b. This structure is very different from that

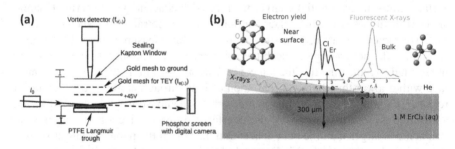

FIGURE 4.9 (a) Experimental setup for EXAFS from the surface of a 1 M ErCl$_3$ aqueous solution contained in a Langmuir trough. X-rays strike the aqueous surface above or below the critical angle for total reflection, yielding X-ray fluorescence and Auger electrons. A gold mesh for measuring the total electron yield (TEY) and a vortex detector for measuring X-ray fluorescence are placed above the surface. (b) Fourier transform of EXAFS signals from the bulk (measured with fluorescence when the incident X-rays are above the critical angle for total reflection) and at the surface (measured with electron yield for grazing-incidence X-rays) along with probable coordination models for these Er(III) environments.[71] (Adapted with permission from *J. Phys. Chem. B*, 2015, 119(28), pp. 8734–8745. Copyright © 2015, American Chemical Society.)

observed in the bulk: a simple one-shell coordination with 8 O atoms surrounding a single Er. However, the interfacial structure is consistent with the previously described nonmonotonic interfacial profile of Er measured by X-ray reflectivity.[65]

Taking advantage of the surface sensitivity provided by the short mean free path of electrons in water, which yields surface sensitivity even if the incident X-rays are not confined to the surface region, Bera and Antonio developed a simplified experimental setup that does not require a grazing-incidence geometry and can benefit from a small sample volume.[70] Figure 4.10 shows the experimental setup in which a horizontal X-ray beam strikes an inverted pendant drop of aqueous solution. Surface-EXAFS measurements from a drop of 1 M $EuCl_3$ solution revealed the presence of Eu–Cl and Eu–Eu correlations within a few nanometers of the water surface, which are absent in the bulk. The presence of ion–ion correlations is similar to the observation for Er; however, instead of the planar cluster observed for Er(III), a tetranuclear cluster was observed for Eu(III).

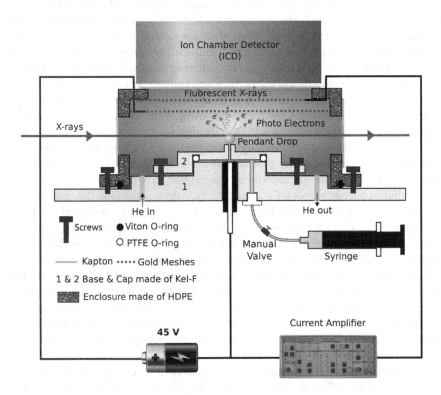

FIGURE 4.10 Schematic diagram of the inverted pendant drop setup for liquid-surface EXAFS. A horizontal X-ray beam strikes the top portion of the drop, whose diameter is 2–3 mm, yielding X-ray fluorescence and Auger electrons. Ion chamber and gold mesh collect X-ray fluorescence and ejected Auger electrons, whose penetration depth is on the order of hundreds of micrometers and a few nanometers, respectively. The inverted pendant drop is placed in a helium environmental cell, sealed by Kapton windows that pass the primary X-ray and fluorescent beams.[70] (Reprinted with permission from *Langmuir*, 2015, 31(19), pp. 5432–5439. Copyright © 2015, American Chemical Society.)

4.3.2 EXTRACTANT ORDERING AND ION–EXTRACTANT INTERACTIONS AT WATER–VAPOR INTERFACES

A key component of any solvent extraction process is the extractant, namely an amphiphilic molecule with a functional group capable of chemical interactions with the target metal ion. The ion–extractant interaction determines the selectivity of the separations process and is, therefore, critical for understanding solvent extraction. Although the entire interfacial process of solvent extraction cannot be monitored at a water–vapor interface, this interface provides a platform to study ion–extractant interactions within an interfacial aqueous environment.

Similar to studies of the liquid–liquid interface described earlier, di-hexadecyl phosphate (DHDP) has been used to model the interactions of an acidic organo-phosphorus extractant head-group (PO_4) with metal ions. The longer chain DHDP has been preferred for studies at the water–vapor interface, instead of more common acidic organophosphorus extractants like bis(2-ethylhexyl) phosphate (HDEHP), because its stronger amphiphilic character allows for greater experimental control over its surface ordering. Herein, we shall review several structural studies of ion–DHDP complexes at the water surface by liquid-surface X-ray scattering techniques.

Vaknin's group has used several X-ray scattering techniques to study the interaction of DHDP with monovalent Cs^+ at the water surface. As one example, they measured anomalous X-ray reflectivity to characterize the Cs^+ distribution under these conditions.[63] By measuring X-ray reflectivity using X-ray energies near and far from an electronic absorption edge of Cs^+ (specifically, the L_3 edge), the measured electron-density profile gains a sensitivity to the presence of Cs because its cross-section for X-ray scattering varies significantly at the selected energies (Figure 4.11a). Figure 4.11b shows that the interfacial distribution exhibits a strong dependence on the bulk concentration of Cs^+, disagreeing with calculations that assume that all DHDP molecules are fully deprotonated at neutral pH (roughly pH 6). When the effect of competitive hydronium (H_3O^+) binding to the phosphate head-group was considered in addition to Cs^+ binding the data show agreement with Poisson–Boltzmann calculations over five orders of magnitude in bulk ion concentration (Figure 4.11b shows a subset of these data). Note that the spatial resolution of these measurements, 4 Å, is well matched to the size of the head-group and bound ions. The in-plane order of DHDP molecules was also examined in this study by grazing-incidence X-ray diffraction. A close-packed hexagonal structure with a molecular area of 40 $Å^2$ was observed, yielding a maximum surface charge density of 0.4 C m^{-2} at full deprotonation.

Subsequent studies on the same system measured near-resonance X-ray spectroscopy, in which the reflected beam intensity is recorded at fixed wave vector transfers, i.e., fixed Q_z, while scanning the X-ray energy near the L_3 Cs^+ edge.[72] The measured periodic dependence on Q_z was modeled by a single Cs^+ distribution for all fixed values of Q_z, demonstrating excellent agreement with distributions obtained from anomalous reflectivity measurements. These measurements also probed the local coordination of Cs^+ at the water surface, demonstrating that the Cs ion is surrounded by eight O atoms. This coordination is similar to Cs coordination in the bulk of $CsNO_3$ solutions.

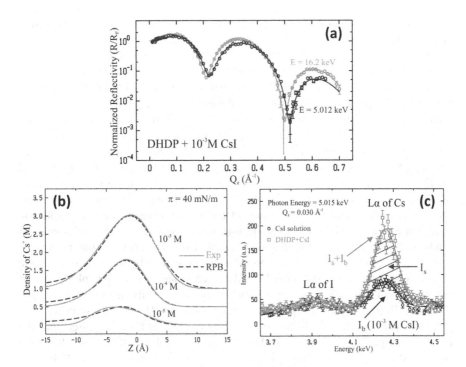

FIGURE 4.11 Studies of Cs binding to a DHDP monolayer spread on the surface of CsI aqueous solutions. (a) Normalized X-ray reflectivities measured at 16.2 keV, far from an absorption edge, and at 5.012 keV at the Cs L_3 edge. Differences in the two sets of data are the result of the different X-ray scattering cross-sections for electrons in Cs at these two X-ray energies. (b) Interfacial distributions of Cs^+ determined from measurements of the anomalous X-ray reflectivity from samples with three different bulk CsI concentrations and calculated from the Poisson–Boltzmann theory with adjusted surface charge density. (c) X-ray fluorescence spectra for 10^{-3} M CsI with and without DHDP. The shaded area represents the effect of Cs^+ ions accumulating at the surface. The data measures the effect of both Cs at the surface (I_s) and in the bulk solution (I_b) because the X-rays were incident at an angle above the critical angle for total reflection.[58,63] (Panels A and B adapted with permission from *Langmuir*, 2006, 22(13), pp. 5673–5681. Copyright © 2006, American Chemical Society. Panel C reprinted from *Journal of Applied Physics*, 2009 105, 084911, with the permission of AIP Publishing.)

Similar to studies of liquid–liquid interfaces described earlier, XFNTR was used to study Cs adsorption to DHDP monolayers on water.[58] This technique measures the surface density (number per area) of Cs^+ with high accuracy (±0.1 ion per extractant or roughly ±0.25 ions per nm^2). Figure 4.11c shows the enhanced X-ray fluorescence from the Cs Lα emission line observed in the presence of a DHDP monolayer, revealing Cs^+ adsorption to the monolayer. A quantitative analysis yielded 0.47 ± 0.09 Cs^+ per DHDP for a 10^{-3} M CsI solution, in good agreement with the result from anomalous reflectivity. These measurements also provided information on the anion: the fluorescence signal of I^- did not change in the presence of the DHDP monolayer, indicating that I^- ions were not adsorbed to the surface. This method characterizes the ion–extractant complexation by quantifying ion adsorption to the extractant head-group.

Recently, X-ray reflectivity and XFNTR were used to investigate the character of ion–extractant binding between DHDP and two trivalent ions, Fe(III) and La(III), over a wide range of pH.[68] To determine whether the binding is sensitive to surface charge or to the specificity of the extractants, arachidic acid (AA) with a carboxylic head-group was also examined. X-ray reflectivity and fluorescence measurements were used to measure the amount of bound Fe and La per molecule for AA monolayers and DHDP monolayers at different pH values. Figure 4.12a shows that pH thresholds for binding were observed for La(III) binding to AA and DHDP monolayers, whose positions are shifted slightly from the bulk pK_a values of phosphate ($pK_a = 2.1$) and carboxylic ($pK_a = 5.1$) groups. The location of these thresholds implies that the interaction of La(III) with phosphate and carboxylic groups are dominated by electrostatic interactions with the surface charge. The binding ratio saturates at about 1/3 ion per molecule at high pH, demonstrating that each La(III) binds with 3 AA or DHDP molecules, thereby maintaining surface charge neutrality. In contrast, Fe(III) binds to AA monolayers even at pH values far below the bulk pK_a (=5.1), where the surface is theoretically neutral. Specifically, Figure 4.12b illustrates a sharp binding transition near pH 2.2, consistent with the threshold pH for Fe(OH)$_3$ formation in solution. The authors suggested that the binding near pH 2.2 was due to covalent binding of Fe(OH)$_3$ to the carboxylic group of AA. Moreover, Fe exhibits a nearly pH-independent binding to DHDP. This observation, combined with the fact that the binding ratio is above 1/3, indicates the

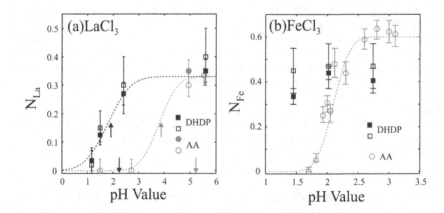

FIGURE 4.12 Studies of La(III) and Fe(III) binding to monolayers of arachidic acid (AA) and di-hexadecyl phosphate (DHDP) at the water–vapor interface. Panels (a) and (b) show the average number of La and Fe ions bound to each molecule as a function of the subphase pH. Open and filled symbols were obtained, respectively, from X-ray fluorescence near total reflection (XFNTR) and reflectivity. The two arrows pointing down in (a) indicate the bulk pK_a values of phosphate (2.1) and carboxylic (5.1) groups, respectively; the two arrows pointing up indicate the effective surface pK_a values, which are slightly lower than the bulk values for La.[68] (Reprinted with permission from *Langmuir*, 2011, 27(19), pp. 11917–11924. Copyright © 2011, American Chemical Society.)

presence of a strong covalent bond between $Fe(OH)_x^{(3-x)+}$ complexes and phosphate groups. Overall, La(III) binds electrostatically, whereas Fe(III) binding is covalent and depends upon its character in solution.

Most metal ions have different electronic states which influence the formation of solution complexes. Knowledge of the relationship between the electronic state of bulk ions and their state in the interfacial ion–extractant complex would provide a fundamental perspective for solvent extraction. It is well known that the binding energy of elements has a small chemical shift when the electronic state and chemical environment changes. Small shifts on the order of a few eV can be easily measured at modern synchrotron X-ray sources with X-ray absorption near-edge spectroscopy (XANES) technique. A recent XANES measurement of iron in its ferrous Fe(II) and ferric Fe(III) states in bulk aqueous solutions and as Fe–DHDP complexes at the water–vapor interface observed the expected 5 eV chemical shift in the bulk, but the state of the Fe–DHDP complexes at the interface appeared to be independent of the bulk state.[69] Note that these experiments were done using a reflection geometry which enabled easy switching between bulk and surface sensitive measurements on the same sample, thereby allowing for a direct comparison of bulk and surface states. The measurement suggested that ferrous and ferric ions coexist at the interface with a uniform fractional valence (i.e., $Fe^{[2+n]+}$, $0 < n < 1$). Moreover, the presence of a pre-edge in the XANES spectra of the Fe–DHDP interfacial complex indicated a broken inversion-symmetry environment, where one side of the iron is bound to a phosphate head-group and the other side is surrounded by water molecules.

In a recent investigation of platinum anion complexes, Uysal and coworkers studied the adsorption of $PtCl_6^{2-}$ complexes to a positively charged monolayer of 1,2-dipalmitoyl-3-trimethylammonium-propane (DPTAP) on the surface of aqueous solutions containing an excess of Cl^- ions.[80] A combination of anomalous X-ray reflectivity and XFNTR measurements determined the concentration of $PtCl_6^{2-}$ in the Stern and diffuse layers. Results demonstrated a strong dependence on $PtCl_6^{2-}$ concentration: $PtCl_6^{2-}$ concentration in the Stern layer is negligible at lower concentrations; while $PtCl_6^{2-}$ adsorbs predominantly in the Stern layer at higher concentrations (see Figure 4.13). The authors suggested a two-step absorption process. At low concentration, $PtCl_6^{2-}$ adsorbs to the diffuse layer due to its strong hydration. With increasing concentration, electrostatic interactions and ion–ion correlations are able to overcome hydration and move $PtCl_6^{2-}$ ions into the Stern layer. Note that liquid-surface X-ray scattering provides a unique tool to identify the location of adsorbed ions interacting with DPTAP head-groups.

Ion–extractant complexes can exhibit long-range ordering within the plane of the surface. Dutta and coworkers used grazing-incidence X-ray scattering (GIXD) to investigate the in-plane structure of DHDP–Er complexes at the water–vapor interface.[66] In this technique, an evanescent X-ray wave confined to within a few nanometers of the surface is used to diffract from the in-plane order. As anticipated, GIXD measured long-range order of DHDP molecules within the monolayer, but also revealed a thin Er lattice formed under the monolayer (see Figure 4.14a). Surprisingly, the DHDP lattice and the ionic lattice are incommensurate, suggesting that ions are not chemically bonded to the charged monolayer. Instead, the ordering is produced by electrostatic interactions between charges in the counterlayer.

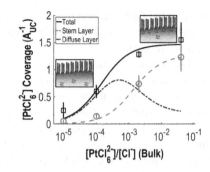

FIGURE 4.13 Measurements of $PtCl_6^{2-}$ adsorption to a monolayer of DPTAP on the surface of water. Coverage of $PtCl_6^{2-}$ ions (in units of number of $PtCl_6^{2-}$ ions per two DPTAP molecules) as a function of bulk concentration. The total and Stern layer coverages are calculated from X-ray fluorescence near total reflection and anomalous X-ray reflectivity measurements, respectively. The diffuse layer coverage is the difference between these two.[80] (Reprinted with permission from *J. Phys. Chem. C*, 2017, 121(45), pp. 25377–25383. Copyright © 2017, American Chemical Society.)

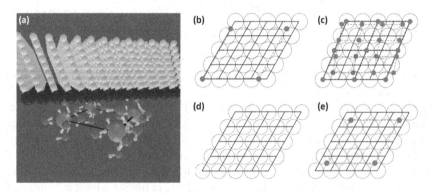

FIGURE 4.14 Studies of lanthanides below monolayers of ODPA and DHDP on the surface of aqueous solutions. (a) Schematic diagram of the ionic lattice formed under a floating monolayer. (b–e) Real-space lattice structures formed in (b) ODPA with light lanthanides, (c) ODPA with heavy lanthanides, (d) DHDP with light lanthanides, and (e) DHDP with heavy lanthanides. Black circles and red dots represent the chain lattice and ionic lattices, respectively.[67] (Reprinted with permission from *Langmuir*, 2017, 33(6), pp. 1412–1418. Copyright © 2017, American Chemical Society.)

An extension of this study included a range of lanthanide ions and use of a single-chain extractant, octadecylphosphonic acid (ODPA).[67] Compared with DHDP molecules, ODPA doubles the surface charge density when head-groups are fully dissociated. The measured ionic lattice structures exhibited a strong dependence on the type of monolayer molecule and on the atomic number of the lanthanides. Both commensurate and incommensurate lattices were formed under ODPA monolayers; only incommensurate lattices were detected with DHDP. Moreover, a sharp structural transition from light to heavy lanthanides was observed around Gd–Tb regardless of monolayer molecule. It is well known that the number of water molecules in the first hydration shell gradually changes from nine in lighter lanthanides to eight in heavier lanthanides. This behavior, combined with the similar chemical properties of all lanthanides, suggests that the hydration shell plays an important role in the interfacial structure with a possible impact on extraction efficiency.

4.4 OUTLOOK

Basic questions on solvent extraction persist: where and how do ion–extractant complexes form, how are ions transported across the liquid–liquid interface, and from a practical point of view, how will this information be used to improve the kinetics and efficiency of separations? The X-ray and neutron scattering studies reviewed in this chapter suggest a new opportunity for investigating these questions that focus on the role of the liquid interface. Detailed information on the position within the interface of metal ions, extractants, and ion–extractant complexes, as well as their chemical state and structural ordering can now be obtained with molecular-scale resolution. These experiments clarify that the interface is not just a boundary that must be crossed by diffusion. It is a region where ions and extractants order and interact in ways that are different than in the bulk of the liquids, ways that influence the ion transport that underlies solvent extraction.

It is likely that the experiments reviewed here represent just the beginning of our use of these techniques to understand the role of the interface in solvent extraction. This is not only because the experiments that have been reported to date have investigated a narrow selection of chemical systems, but because the character and static nature of the systems will be broadened. Most of the experiments reported to date have investigated strongly amphiphilic extractants that are not representative of the majority of extractants in commercial use which are much weaker amphiphiles with greater solubility in the aqueous phase. Many of the published experiments have been a more natural extension of the earliest liquid-surface X-ray studies of Langmuir monolayers of surfactants that originally established many of the X-ray techniques. Nevertheless, the study of extractants like HDEHP, tributyl phosphate (TBP), and many others will be necessary to make a connection to their practical application in separations processes. Their study will not be a trivial extension of current experiments because of their lower scattering contrast and higher aqueous solubility, but we expect that the powerful analytical capabilities of X-ray and neutron interface-sensitive techniques will reveal their interfacial ordering and chemistry. Finally, most of the studies have probed equilibrium states, yet the transport of ions across the interface is a dynamic process. Recent investigations of non-equilibrium states of ion–extractant complexes at the liquid–liquid interface are providing a guide to investigating this dynamic process,[51] but many more investigations and the application of different scattering techniques will be required to characterize them.

4.5 SUMMARY

This chapter has reviewed recent X-ray studies of liquid–liquid interfaces and liquid–vapor interfaces that investigated the molecular ordering and binding of ions and extractants. The goal of characterizing the complexation of ions, extractants, and solvents at the liquid–liquid interface while measuring the mechanisms of ion transport for the separation of critical elements is far from being accomplished, but important early steps have been taken. Measurements of the surface of aqueous solutions have probed the distribution of Er ions in the absence of extractants,[65] as well as the coordination changes that Er and Eu ions undergo as they move from the bulk to the interface.[70,71] These measurements characterized the availability of ions in the interfacial region to interact with extractants. Adding model extractants to the aqueous electrolyte solution has led to a detailed characterization of ion–extractant interactions with phosphoric acid head-groups, including ion distributions of Cs^+ as well as a direct measurement of its preferential adsorption over I^- for the extractant-covered interface.[58,63] The importance of electrostatic vs. covalent nature of binding of Fe(III) and La(III) to a phosphoric acid head-group was investigated by varying pH and liquid-surface X-ray measurements.[68] Measurements of $PtCl_6^{2-}$ suggested a two-step adsorption process as the ion concentration is increased that enables the chlorometalate ion to partially shed its hydration shell.[80] Finally, the well-known variation from nine to eight water molecules in the first hydration shell with increasing lanthanide element mass has been shown to play a role in interactions with phosphoric acid and phosphatidic acid moieties.[67]

Measurements of the liquid–vapor interface have enabled high-resolution studies of ions and ion–extractant interactions, but cannot address the role of the organic phase or the process of transport across the organic–aqueous interface. Recent measurements of a dodecane–water interface with diamide extractants and Nd(III) ions used a combination of X-ray and neutron reflectivity measurements along with a Monte Carlo analysis of chemical component distributions to discover an unexpected variation in the interaction of Nd(III) with different diamides at the interface.[24,25] These observations have direct implications for solvent extraction processes. Experiments with a model extractant have explored the state of ion–extractant complexes formed during solvent extraction, identifying very different structures that form with either trivalent ions, Er(III) and Y(III), or divalent ions, Sr(II), that likely affect both their extraction kinetics and efficiency.[22,23,51]

This chapter was written to provide the reader with an understanding of these accomplishments, as well as an introduction to the capabilities of synchrotron X-ray techniques for studying liquid interfaces.[26] These techniques include scattering measurements such as reflectivity and grazing-incidence diffraction that measure the electron density variation both within and perpendicular to the interfacial plane, thereby providing a basis to understand ionic and molecular ordering at the interface. They also include interface-sensitive X-ray spectroscopy and fluorescence that can identify elements within the interface and near-interfacial region, as well as characterize their interfacial density, oxidation state, and coordination to neighboring atoms. Although a nearly complete coverage has been provided of the techniques in use today for investigating liquid interfacial aspects of solvent extraction, applications of other synchrotron X-ray techniques are ongoing, and new capabilities can be expected to provide further insight into the role of interfaces in solvent extraction.

ACKNOWLEDGMENTS

MLS acknowledges support from U.S. DOE OBES, Chemical Sciences, Geosciences, and Biosciences Division DE-SC0018200, and thanks Dr. Lynda Soderholm for introducing him to solvent extraction and to the expertise of her group in this area, as well as initial support under a subcontract from U.S. DOE OBES Contract No. DE-AC02-06CH11357. NSF's ChemMatCARS Sector 15 is principally supported by the Divisions of Chemistry (CHE) and Materials Research (DMR), National Science Foundation, under grant number NSF/CHE-1834750. Use of the Advanced Photon Source, an Office of Science User Facility operated for the U.S. Department of Energy Office of Science by Argonne National Laboratory, was supported by the U.S. DOE under Contract No. DE-AC02-06CH11357.

REFERENCES

1. Tasker, P. A.; Plieger, P. G.; West, L. C. Metal Complexes for Hydrometallurgy and Extraction. In *Comprehensive Coordination Chemistry II: From Biology to Nanotechnology*; McCleverty, J. A., Meyer, T. J., Eds.; Elsevier: Oxford, 2004; Vol. 9; pp 759–808.
2. Stevens, G. W.; Perera, J. M.; Grieser, F. Interfacial Aspects of Metal Ion Extraction in Liquid–Liquid Systems. *Rev. Chem. Eng.* 2001, *17*, 87–110.
3. Tondre, C.; Hebrant, M.; Watarai, H. Rate of Interfacial Reactions Compared to Bulk Reactions in Liquid–Liquid and Micellar Processes: An Attempt to Clarify a Confusing Situation. *J. Colloid Interface Sci.* 2001, *243*, 1–10.
4. Benay, G.; Wipff, G. Liquid–Liquid Extraction of Uranyl by TBP: The TBP and Ions Models and Related Interfacial Features Revisited by Md and PMF Simulations. *J. Phys. Chem. B* 2014, *118*, 3133–3149.
5. Soderholm, L.; Skanthakumar, S.; Wilson, R. E. Structures and Energetics of Erbium Chloride Complexes in Aqueous Solution. *J. Phys. Chem. A* 2009, *113*, 6391–6397.
6. Wilson, R. E.; Skanthakumar, S.; Cahill, C. L.; Soderholm, L. Structural Studies Coupling X-Ray Diffraction and High-Energy X-Ray Scattering in the UO_2^{2+}-HBr$_{aq}$ System. *Inorg. Chem.* 2011, *50*, 10748–10754.
7. Knope, K. E.; Skanthakumar, S.; Soderholm, L. Two Dihydroxo-Bridged Plutonium(IV) Nitrate Dimers and Their Relevance to Trends in Tetravalent Ion Hydrolysis and Condensation. *Inorg. Chem.* 2015, *54*, 10192–10196.
8. Steytler, D. C.; Jenta, T. R.; Robinson, B. H.; Eastoe, J.; Heenan, R. K. Structure of Reversed Micelles Formed by Metal Salts of Bis(Ethylhexyl) Phosphoric Acid. *Langmuir* 1996, *12*, 1483–1489.
9. Ibrahim, T. H.; Neuman, R. D. Nanostructure of Open Water-Channel Reversed Micelles. I. H-1 NMR Spectroscopy and Molecular Modeling. *Langmuir* 2004, *20*, 3114–3122.
10. Gannaz, B.; Antonio, M. R.; Chiarizia, R.; Hill, C.; Cote, G. Structural Study of Trivalent Lanthanide and Actinide Complexes Formed Upon Solvent Extraction. *Dalton Trans.* 2006, *38*, 4553–4562.
11. Meridiano, Y.; Berthon, L.; Crozes, X.; Sorel, C.; Dannus, P.; Antonio, M. R.; Chiarizia, R.; Zemb, T. Aggregation in Organic Solutions of Malonamides: Consequences for Water Extraction. *Solvent Extr. Ion Exch.* 2009, *27*, 607–637.
12. Ellis, R. J.; Anderson, T. L.; Antonio, M. R.; Braatz, A.; Nilsson, M. A SAXS Study of Aggregation in the Synergistic TBP–HDBP Solvent Extraction System. *J. Phys. Chem. B* 2013, *117*, 5916–5924.

13. Danesi, P. R.; Chiarizia, R.; Coleman, C. F. The Kinetics of Metal Solvent Extraction. *Crit. Rev. Anal. Chem.* 1980, *10*, 1–126.

14. Hughes, M. A.; Rod, V. A General Model to Account for the Liquid/Liquid Kinetics of Extraction of Metals by Organic Acids. *Faraday Discuss. Chem. Soc.* 1984, *77*, 75–84.

15. Qiao, B.; Muntean, J. V.; Olvera de la Cruz, M.; Ellis, R. J. Ion Transport Mechanisms in Liquid–Liquid Interface. *Langmuir* 2017, *33*, 6135–6142.

16. Hunt, E. C. The Interaction of Alkyl Phosphate Monolayers with Metal Ions. *J. Colloid Interface Sci.* 1969, *29*, 105–115.

17. Szymanowski, J. Kinetics and Interfacial Phenomena. *Solvent. Extr. Ion Exch.* 2000, *18*, 729–751.

18. Testard, F.; Berthon, L.; Zemb, T. Liquid–Liquid Extraction: An Adsorption Isotherm at Divided Interface? *C. R. Chim.* 2007, *10*, 1034–1041.

19. Watarai, H. Interfacial Molecular Aggregation in Solvent Extraction Systems. In *Ion Extraction and Solvent Extraction: A Series of Advances*; Moyer, B. A., Ed.; CRC Press: Boca Raton, FL, 2014; Vol. 21; pp 159–196.

20. Plucinski, P.; Nitsch, W. Kinetics of the Interfacial Ion Exchange in Winsor II Microemulsion Systems. *J. Colloid Interface Sci.* 1992, *154*, 104–112.

21. Bu, W.; Hou, B.; Mihaylov, M.; Kuzmenko, I.; Lin, B.; Meron, M.; Soderholm, L.; Luo, G.; Schlossman, M. L. X-Ray Fluorescence from a Model Liquid/Liquid Solvent Extraction System. *J. Appl. Phys.* 2011, *110*, 102214-102211–102214-102216.

22. Bu, W.; Yu, H.; Luo, G.; Bera, M. K.; Hou, B.; Schuman, A. W.; Lin, B.; Meron, M.; Kuzmenko, I.; Antonio, M. R.; Soderholm, L.; Schlossman, M. L. Observation of a Rare Earth Ion–Extractant Complex Arrested at the Oil–Water Interface During Solvent Extraction. *J. Phys. Chem. B* 2014, *118*, 10662–10674.

23. Bu, W.; Mihaylov, M.; Amoanu, D.; Lin, B.; Meron, M.; Kuzmenko, I.; Soderholm, L.; Schlossman, M. L. X-Ray Studies of Interfacial Strontium–Extractant Complexes in a Model Solvent Extraction System. *J. Phys. Chem. B* 2014, *118*, 12486–12500.

24. Scoppola, E.; Watkins, E.; Li Destri, G. L.; Porcar, L.; Campbell, R. A.; Konovalov, O.; Fragneto, G.; Diat, O. Structure of a Liquid/Liquid Interface during Solvent Extraction Combining X-Ray and Neutron Reflectivity Measurements. *Phys. Chem. Chem. Phys.* 2015, *17*, 15093–15097.

25. Scoppola, E.; Watkins, E. B.; Campbell, R. A.; Konovalov, O.; Girard, L.; Dufrêche, J. F.; Ferru, G.; Fragneto, G.; Diat, O. Solvent Extraction: Structure of the Liquid–Liquid Interface Containing a Diamide Ligand. *Angew. Chem. Int. Ed. Engl.* 2016, 55 (32), 9326–9330.

26. Pershan, P. S.; Schlossman, M. L. *Liquid Surfaces and Interfaces: Synchrotron X-Ray Methods*; Cambridge University Press: Cambridge, 2012.

27. Walker, D. S.; Brown, M. G.; McFearin, C. L.; Richmond, G. L. Evidence for a Diffuse Interfacial Region at the Dichloroethane/Water Interface. *J. Phys. Chem. B* 2004, *108*, 2111–2114.

28. Rowlinson, J. S.; Widom, B. *Molecular Theory of Capillarity*; Clarendon Press: Oxford, 1982.

29. Mitrinovic, D. M.; Zhang, Z.; Williams, S. M.; Huang, Z.; Schlossman, M. L. X-Ray Reflectivity Study of the Water-Hexane Interface. *J. Phys. Chem. B* 1999, *103*, 1779–1782.

30. Mitrinovic, D. M.; Tikhonov, A. M.; Li, M.; Huang, Z.; Schlossman, M. L. Noncapillary-Wave Structure at the Water–Alkane Interface. *Phys. Rev. Lett.* 2000, *85*, 582–585.

31. Luo, G.; Malkova, S.; Pingali, S. V.; Schultz, D. G.; Lin, B.; Meron, M.; Benjamin, I.; Vanysek, P.; Schlossman, M. L. Structure of the Interface between Two Polar Liquids: Nitrobenzene and Water. *J. Phys. Chem. B* 2006, *110*, 4527–4530.

32. Luo, G.; Malkova, S.; Pingali, S. V.; Schultz, D. G.; Lin, B.; Meron, M.; Graber, T. J.; Gebhardt, J.; Vanysek, P.; Schlossman, M. L. The Width of the Water/2-Heptanone Liquid–Liquid Interface. *Electrochem. Comm.* 2005, *7*, 627–630.

33. Tikhonov, A. M.; Mitrinovic, D. M.; Li, M.; Huang, Z.; Schlossman, M. L. An X-Ray Reflectivity Study of the Water–Docosane Interface. *J. Phys. Chem. B* 2000, *104*, 6336–6339.

34. Benjamin, I. Reactivity and Dynamics at Liquid Interfaces. In *Reviews in Computational Chemistry*; Lipkowitz, K. B., Ed.; John Wiley & Sons: Hoboken, NJ, 2013; Vol. 28; pp 205–313.

35. Zarbakhsh, A.; Bowers, J.; Webster, J. R. P. Width of the Hexadecane–Water Interface: A Discrepancy Resolved. *Langmuir* 2005, *21*, 11596–11598.

36. Davies, J. T.; Rideal, E. K. *Interfacial Phenomena*, 2nd ed.; Academic Press: New York, London, 1963.

37. Davies, J. T. The Distribution of Ions under a Charged Monolayer, and a Surface Equation of State for Charged Films. *Proc. Roy. Soc. A* 1951, *208*, 224.

38. Yue, B. Y.; Jackson, C. M.; Taylor, J. A. G.; Mingins, J.; Pethica, B. A. Phospholipid Monolayers at Non-Polar Oil/Water Interfaces Part I. Phase Transitions in Distearoyl-Lecithin Films at the N-Heptane Aqueous Sodium Chloride Interface. *J. Chem. Soc. Faraday Trans. I* 1976, *72*, 2685.

39. Taylor, J. A. G.; Mingins, J.; Pethica, B. A. Phospholipid Monolayers at Non-Polar Oil/Water Interfaces Part 2. Dilute Monolayers of Saturated 1,2-Diacyl-Lecithins and a-Cephalins. *J. Chem. Soc. Faraday Trans. I* 1976, *72*, 2694.

40. Schlossman, M. L.; Tikhonov, A. M. Molecular Ordering and Phase Behavior of Surfactants at Water–Oil Interfaces as Probed by X-Ray Surface Scattering. *Annu. Rev. Phys. Chem.* 2008, *59*, 153–177.

41. Tikhonov, A. M.; Schlossman, M. L. Vaporization and Layering of Alkanols at the Oil/Water Interface. *J. Phys. Condens. Matter* 2007, *19*, 375101.

42. Tikhonov, A. M.; Patel, H.; Garde, S.; Schlossman, M. L. Tail Ordering Due to Headgroup Hydrogen Bonding Interactions in Surfactant Monolayers at the Water–Oil Interface. *J. Phys. Chem. B* 2006, *110*, 19093–19096.

43. Pingali, S. V.; Takiue, T.; Luo, G.; Tikhonov, A. M.; Ikeda, N.; Aratono, M.; Schlossman, M. L. X-Ray Reflectivity and Interfacial Tension Study of the Structure and Phase Behavior of the Interface between Water and Mixed Surfactant Solutions of $CH_3(CH_2)_{19}OH$ and $CF_3(CF_2)_7(CH_2)_2OH$ in Hexane. *J. Phys. Chem. B* 2005, *109*, 1210–1225.

44. Tikhonov, A. M.; Pingali, S. V.; Schlossman, M. L. Molecular Ordering and Phase Transitions in Alkanol Monolayers at the Water–Hexane Interface. *J. Chem. Phys.* 2004, *120*, 11822–11838.

45. Tikhonov, A. M.; Schlossman, M. L. Surfactant and Water Ordering in Triacontanol Monolayers at the Water–Hexane Interface. *J. Phys. Chem. B* 2003, *107*, 3344–3347.

46. Li, M.; Tikhonov, A. M.; Schlossman, M. L. An X-Ray Diffuse Scattering Study of Domains in $F(CF_2)_{10}(CH_2)_2OH$ Monolayers at the Hexane–Water Interface. *Europhys. Lett.* 2002, *58*, 80–86.

47. Tikhonov, A. M.; Li, M.; Schlossman, M. L. Phase Transition Behavior of Fluorinated Monolayers at the Water–Hexane Interface. *J. Phys. Chem. B* 2001, *105*, 8065–8068.

48. Bu, W.; Schlossman, M. L. Synchrotron X-Ray Scattering from Liquid Surfaces and Interfaces. In *Synchrotron Light Sources and Free-Electron Lasers: Accelerator Physics, Instrumentation and Science Applications*; Jaeschke, E., Khan, S., Schneider, J. R., Hastings, J. B., Eds.; Springer: Cham, Switzerland, 2016; pp 1–33.

49. Bera, M. K.; Bu, W.; Uysal, A. Liquid Surface X-Ray Scattering. In *Physical Chemistry of Gas–Liquid Interfaces*; Faust, J. A., House, J. E., Eds.; Elsevier, Inc.: Amsterdam, Netherlands, 2018; pp 167–194.

50. Nash, K. L.; Madic, C.; Mathur, J. N.; Lacquement, J. Actinide Separation Science and Technology. In *The Chemistry of the Actinide and Transactinide Elements*, 3rd ed.; Morss, L. R., Jean Fuger, N. M. E., Ed.; Springer: Dordrecht, 2006; Vol. 4; pp 2622–2798.

51. Liang, Z.; Bu, W.; Schweighofer, K. J.; Walwark, D. J., Jr.; Harvey, J. S.; Hanlon, G. R.; Amoanu, D.; Erol, C.; Benjamin, I.; Schlossman, M. L. Nanoscale View of Assisted Ion Transport across the Liquid–Liquid Interface. *Proc. Natl. Acad. Sci. U.S.A.* 2018, DOI:10.1073/pnas.1701389115.

52. Nash, K.; Lumetta, G. J. *Advanced Separation Techniques for Nuclear Fuel Reprocessing and Radioactive Waste Treatment*; Woodhead Publishing: Philadelphia, PA, 2011.

53. Tian, Q. Z.; Hughes, M. A. Synthesis and Characterisation of Diamide Extractants for the Extraction of Neodymium. *Hydrometallurgy* 1994, *36*, 79–94.

54. Als-Nielsen, J.; McMorrow, D. *Elements of Modern X-Ray Physics*; Wiley: Hoboken, 2001.

55. Zhang, Z.; Mitrinovic, D. M.; Williams, S. M.; Huang, Z.; Schlossman, M. L. X-Ray Scattering from Monolayers of $F(CF_2)_{10}(CH_2)_2OH$ at the Water–(Hexane Solution) and Water–Vapor Interfaces. *J. Chem. Phys.* 1999, *110*, 7421–7432.

56. Percus, J. K.; Williams, G. O. The Intrinsic Interface. In *Fluid Interfacial Phenomena*; Croxton, C. A., Ed.; Wiley: Chichester, 1986; p 1.

57. Yun, W. B.; Bloch, J. M. X-Ray near Total External Fluorescence Method – Experiment and Analysis. *J. Appl. Phys.* 1990, *68*, 1421–1428.

58. Bu, W.; Vaknin, D. X-Ray Fluorescence Spectroscopy from Ions at Charged Vapor/Water Interfaces. *J. Appl. Phys.* 2009, *105*, 084911–084916.

59. Matubayasi, N.; Motomura, K.; Aratono, M.; Matuura, R. Thermodynamic Study on the Adsorption of 1-Octadecanol at Hexane/Water Interface. *Bull. Chem. Soc. Jpn.* 1978, *51*, 2800–2803.

60. Takiue, T.; Uemura, A.; Ikeda, N.; Motomura, K.; Aratono, M. Thermodynamic Study on Phase Transition in Adsorbed Film of Fluoroalkanol at the Hexane/Water Interface. 3. Temperature Effect on the Adsorption of 1,1,2,2-Tetrahydroheptadecafluorodecanol. *J. Phys. Chem. B* 1998, *102*, 3724.

61. Small, D. M. *The Physical Chemistry of Lipids*; Plenum: New York, 1986.

62. Kitaigorodskii, A. I. *Organic Chemical Crystallography*; Consultants Bureau: New York, 1961.

63. Bu, W.; Vaknin, D.; Travesset, A. How Accurate Is Poisson-Boltzmann Theory for Monovalent Ions near Highly Charged Interfaces? *Langmuir* 2006, *22*, 5673–5681.

64. Crowley, T. J.; Lee, E. M.; Simister, E. A.; Thomas, R. K. The Use of Contrast Variation in the Specular Reflection of Neutrons from Interfaces. *Phys. B Condens. Matter.* 1991, *173*, 143–156.

65. Luo, G.; Bu, W.; Mihaylov, M.; Kuzmenko, I.; Schlossman, M. L.; Soderholm, L. X-Ray Reflectivity Reveals a Non-Monotonic Ion-Density Profile Perpendicular to the Surface of $ErCl_3$ Aqueous Solutions. *J. Phys. Chem. C* 2013, *117*, 19082–19090.

66. Miller, M.; Chu, M.; Lin, B.; Meron, M.; Dutta, P. Observation of Ordered Structures in Counterion Layers near Wet Charged Surfaces: A Potential Mechanism for Charge Inversion. *Langmuir* 2016, *32*, 73–77.

67. Miller, M.; Chu, M.; Lin, B.; Bu, W.; Dutta, P. Atomic Number Dependent "Structural Transitions" in Ordered Lanthanide Monolayers: Role of the Hydration Shell. *Langmuir* 2017, *33*, 1412–1418.

68. Wang, W. J.; Park, R. Y.; Meyer, D. H.; Travesset, A.; Vaknin, D. Ionic Specificity in pH Regulated Charged Interfaces: Fe^{3+} Versus La^{3+}. *Langmuir* 2011, *27*, 11917–11924.

69. Wang, W. J.; Kuzmenko, I.; Vaknin, D. Iron near Absorption Edge X-Ray Spectroscopy at Aqueous–Membrane Interfaces. *Phys. Chem. Chem. Phys.* 2014, *16*, 13517–13522.

70. Bera, M. K.; Antonio, M. R. Polynuclear Speciation of Trivalent Cations near the Surface of an Electrolyte Solution. *Langmuir* 2015, *31*, 5432–5439.

71. Bera, M. K.; Luo, G.; Schlossman, M. L.; Soderholm, L.; Lee, S.; Antonio, M. R. Erbium(III) Coordination at the Surface of an Aqueous Electrolyte. *J. Phys. Chem. B* 2015, *119*, 8734–8745.

72. Bu, W.; Ryan, P. J.; Vaknin, D. Ion Distributions at Charged Aqueous Surfaces by near-Resonance X-Ray Spectroscopy. *J. Synchrotron Radiat.* 2006, *13*, 459–463.

73. Ruckenstein, E.; Manciu, M. Specific Ion Effects via Ion Hydration: II. Double Layer Interaction. *Adv. Colloid Interface Sci.* 2003, *105*, 177–200.

74. Levin, Y. Polarizable Ions at Interfaces. *Phys. Rev. Lett.* 2009, *102*, 147803.

75. Matubayasi, N.; Tsunetomo, K.; Sato, I.; Akizuki, R.; Morishita, T.; Matuzawa, A.; Natsukari, Y. Thermodynamic Quantities of Surface Formation of Aqueous Electrolyte Solutions – IV. Sodium Halides, Anion Mixtures, and Sea Water. *J. Colloid Interface Sci.* 2001, *243*, 444–456.

76. Ghosal, S.; Hemminger, J. C.; Bluhm, H.; Mun, B. S.; Hebenstreit, E. L. D.; Ketteler, G.; Ogletree, D. F.; Requejo, F. G.; Salmeron, M. Electron Spectroscopy of Aqueous Solution Interfaces Reveals Surface Enhancement of Halides. *Science* 2005, *307*, 563–566.

77. Raymond, E. A.; Richmond, G. L. Probing the Molecular Structure and Bonding of the Surface of Aqueous Salt Solutions. *J. Phys. Chem. B* 2004, *108*, 5051–5059.

78. Jungwirth, P.; Tobias, D. J. Molecular Structure of Salt Solutions: A New View of the Interface with Implications for Heterogeneous Atmospheric Chemistry. *J. Phys. Chem. B* 2001, *105*, 10468–10472.

79. Casillas-Ituarte, N. N.; Callahan, K. M.; Tang, C. Y.; Chen, X. K.; Roeselová, M.; Tobias, D. J.; Allen, H. C. Surface Organization of Aqueous $MgCl_2$ and Application to Atmospheric Marine Aerosol Chemistry. *Proc. Natl. Acad. Sci. U.S.A.* 2010, *107*, 6616–6621.

80. Uysal, A.; Rock, W.; Qiao, B.; Bu, W.; Lin, B. Two-Step Adsorption of $PtCl_6^{2-}$ Complexes at a Charged Langmuir Monolayer: Role of Hydration and Ion Correlations. *J. Phys. Chem. C* 2017, *121*, 25377–25383.

5 Solvent Extraction through the Lens of Advanced Modeling and Simulation

Aurora E. Clark, Michael J. Servis, Zhu Liu,
Ernesto Martinez-Baez, Jing Su, Enrique R. Batista,
Ping Yang, Andrew Wildman, Torin Stetina,
Xiaosong Li, Ken Newcomb, Edward J. Maginn,
Jochen Autschbach, and David A. Dixon

CONTENTS

5.1   INTRODUCTION

The last three decades has seen tremendous growth of modeling and simulation applied to solvent extraction systems. Historically, much success has been observed in gas-phase calculations of the electronic structure and bonding of metals with extracting ligands. The level of accuracy of binding energetics, and the ability to understand trends in ligand selectivity based upon the relative strength of electrostatic and bonding interactions, have now enabled data-driven approaches that are at the forefront of ligand-design strategies. However, inherent to ligand design is a robust understanding of aqueous- and organic-phase speciation. Determining the thermodynamic ensemble of aqueous species has been a major topic of research for the last decade, with advances seen in reliable aqueous thermochemistry, *ab initio* molecular dynamics, and new force fields for use with classical methods—all to understand the prevalence and distribution of solvation and coordination environments in aqueous solutions. Those solutions potentially contain complexing acid anions, where ion-pairing phenomena may be relevant or where multiple miscible solvents may come into play. Although the organic phase is generally considered from the perspective of ideal metal–ligand complexes, accumulating experimental data has now spurred computational chemists to study the self-assembly and aggregation phenomena of extractants or surface-active molecules that may or may not contain a myriad of ion configurations. This has challenged computational chemists, as these aggregates may be relatively minor species at low metal loading but grow in importance with increasing concentrations relevant to processing conditions. The ability of these aggregates to potentially contain water hampers industrial flowsheets and alters ligand-design strategies, as the coordination environment of the aggregate may not be able to leverage the appropriate bonding environments to engender selectivity. From a simulations perspective, a growing body of work has been dedicated toward understanding aggregation phenomena; however, much is to be learned about the fundamental driving forces behind the formation of such species.

Aside from the thermodynamic distributions in aqueous or organic phases, there is also growing interest in the mechanisms of solvent extraction itself. So-called diffusion vs. kinetic regimes of solvent extraction are in direct relation to the organization and dynamics of the interface itself, the reacting metal ions and extracting ligands that must encounter one another, and the transport process that may require significant reorganization of the interfacial solvent and surface-active molecules. The sheer complexity of available reaction pathways and ensembles of species at the interface has meant that our appreciation for the mechanisms and governing factors of transport across a range of solution compositions are still in their infancy. That being said, new tools are being developed for interrogating the structure of the interface, and simulations (with experimental collaboration) are beginning to reveal its secrets.

Within this chapter, we discuss, with some rigor, the computational methods that can be employed to study speciation of molecular configurations, environments, and complexes in the aqueous phase and how these change from the bulk to the interfacial region. This includes the ability to study the structure and dynamics of the interface at the nm scale, the mechanisms and energetics of transport across the interface, and the bulk organic speciation. Some emphasis is given to separations processes relevant to heavy metals, with examples demonstrating both the successes and failures of different methodologies as applied to certain applications. This being the first chapter in this book series to treat modeling and simulation, we are excited to share these approaches and examples with the broader solvent extraction community.

5.2 A LANDSCAPE OF COMPUTATIONAL METHODS

5.2.1 Electronic-Structure (Quantum-Chemistry) Techniques

Molecular-scale (time-independent) electronic-structure methods have played an essential role in understanding many of the important nuances within solvent extraction chemistry. This includes the prediction of speciation in complex mixtures, the spectroscopic signatures that enable solution-phase characterization, the mechanisms of chemical transformations, and the fundamental design principles of extractant ligands. This section provides a brief background regarding the fundamental methods of quantum chemistry and considerations of important computational details if chemically meaningful information is to be obtained.

5.2.1.1 Essentials of Methods and Basis Sets

Within the last three decades, numerous "black box" software programs have emerged that enable both beginner and expert to calculate a wide variety of molecular properties with relative ease. However, as with all science, there are many cautionary tales—where preconceptions, or a missing facet of essential physics can lead to an unproductive and even false scientific path. For better or worse, the speed at which modern calculations can be performed accelerates both sound and unsound computational studies. All applications of quantum-chemistry[1,2] methods need to balance the desired accuracy of the calculation and the associated computational resources that are required. There are some fairly general rules to keep in mind when it comes to the chemical application of *ab initio* ("from first-principles") quantum-chemistry electronic-structure methods. Every rule has exceptions, of course, but we focus on typical cases. One rule is "Size matters." The size of a system means here, roughly, the number of atoms, or the number of electrons, or the number B of one-particle basis functions. One-center basis functions in quantum-chemistry calculations usually resemble atomic orbitals (AOs), although one should be careful to distinguish between AO-like basis functions and the orbitals of atoms.[3] The computational effort of Hartree-Fock (HF) and density-functional theory (DFT, consisting of Kohn–Sham and generalized Kohn–Sham approximations)[4] with hybrid functionals scales *formally* as B^4, non-hybrid DFT as B^3, and wave-function theory (WFT) with explicit treatment of the electron correlation as B^n with n starting at 5 and increasing systematically with an improved treatment of electron correlation. The scaling of the computational effort limits the size of a system that can be treated meaningfully by computations. A calculation that is practical for a given extracting ligand, for instance, may be completely impractical for a metal complex with several of these ligands.

Recommendations about basis sets[5–7] are difficult to provide in a few sentences, but as a general rule, the basis-set first and foremost needs to reflect where the electrons "want to be." Atomic shells of the same element in different bonding environments have different optimal radial extensions, requiring at the very least split-valence (double-ζ) basis sets. Bound atoms are polarized. Calculations with non-polarized basis sets (save for hydrogen in large organic molecules) are not state of the art. The description of diffuse electronic states (for example, anions) requires diffuse basis functions. Angular momenta and radial function exponents that are very different from those needed to describe the atomic shells and chemical bonds may be required to describe, for instance, how a system responds to external fields or the "electron cusps" that form in the wavefunction[8] when the electron correlation is treated increasingly accurately. Another rule, applying in particular to WFT without explicit correlation, is "Improving the accuracy of a calculation systematically requires a (formally) higher scaling method *and* more basis functions" (of the right kind). However, even very accurate methods rely to some degree on error cancellation, such that one cannot expect any improvements—likely the opposite—simply by treating the aforementioned ligand at a high-accuracy level and the bigger complex at a less accurate one to obtain the complexation energy, for example.

In DFT, the true energy functional remains elusive, and all applications require approximations beyond the basis-set as well. There is the notion of a "ladder" of functionals[9] with increasing overall accuracy, without necessarily going to very large basis sets. For instance, when stepping up—at increasing computational cost—the first few rungs from a local-density approximation (LDA) to a generalized gradient approximation (GGA) to a hybrid GGA there is generally an improvement, albeit not systematically, in the electronic structure and the associated properties such as energies, molecular structures, and spectroscopic properties. Beyond this, a general, widely accepted, and practically proven *ab initio* way toward improved properties in DFT calculations has yet to emerge. For the time being, DFT with widely available approximate functionals is suitable for many applications and rightly considered as the workhorse of quantum chemistry. Coupled-cluster (CC) theory[10] with singles, doubles, and perturbative triples, that is, CCSD(T), is able to deliver chemical accuracy (1 kcal/mol) for relative energies of single-configurational systems (see below), but it formally scales as B^7. Reduced-scaling CC methods[11] and approximate variants such as CC2[12] are increasingly used for larger systems, however.

Many solvent extraction systems involve the separations of metals, generally those later in the periodic table. In this case, relativistic effects[13–16] become increasingly important. Relativistic effects in atomic valence shells, as a percentage of a corresponding nonrelativistic property such as the energy, scale in leading order as Z^2/c^2, where Z is the nuclear charge of the element and c is the speed of light. Besides the overall order Z^2/c^2, indicating that relativistic effects are stronger for heavier elements, there are important variations depending on the element and the property of interest.[17,18] For systems with light(er) elements, chemical accuracy for relative energies, or structure optimizations, do not usually demand relativistic corrections. Heats of formation from atomization energies may require such corrections, however, even for lighter elements. Ultimately, whether relativistic effects need to be considered or not depends on the desired accuracy of the calculation. Starting roughly at Cs ($Z=55$), relativity noticeably impacts the physicochemical properties, and in the third transition metal row and beyond, it impacts the chemistry even qualitatively. Therefore, calculations involving heavy metals always need to include relativistic effects. The most important relativistic effects for heavy element compounds are obtained by replacing the one-electron operators of the Schrodinger (S) Hamiltonian by the Dirac (D) operator

$$\hat{h}^S = \frac{1}{2m_e} \hat{p} \cdot \hat{p} + V \rightarrow \hat{h}^D = \begin{bmatrix} V & c\sigma \cdot \hat{p} \\ c\sigma \cdot \hat{p} & V - 2m_e c^2 \end{bmatrix} \tag{5.1}$$

Here, $\hat{p} = -\hbar \vec{\nabla}$ is the electron-momentum operator, σ is a 3-vector of the 2×2 Pauli spin matrices, V is the electron-nucleus attractive potential energy, and $V=0$ coincides in $\hat{h}^D$ with the electron rest-mass energy $m_e c^2$. The Dirac operator is a 4×4 object, since each entry in the "split" notation above is a 2×2 block. Therefore, the orbitals that $\hat{h}^D$ acts upon must have four components and are called 4-spinors. Corresponding relativistic quantum-chemical calculations are referred to as 4-component. The 4-spinors are composed of two coupled 2-spinors, which carry redundant information. It is possible to transform $\hat{h}^D$ and the 4-spinors to obtain a fully relativistic 2-component scheme based only on 2-spinors. It has taken many years of research to find practical methods to do this for the one-electron $\hat{h}^D$ operator without introducing approximations, but the problem was eventually solved.[19,20] A commonly used subset of these methods is referred to as eXact 2-component, or X2C. When the decoupling of the Dirac spinors is done approximately, the scheme is called quasi-relativistic. Popular variationally stable quasi-relativistic methods are the zeroth-order regular approximation (ZORA)[21,22] and low-order Douglas–Kroll–Hess (DKH).[23,24] The aforementioned relativistic Hamiltonians are meant to be used in all-electron calculations and require different basis sets as well as careful treatments of which electrons are correlated in WFT. The treatment of relativistic effects on the two-electron interaction and molecular properties other than the energy remains an active area of research.

Compared to $\hat{h}^S$, relativistic Hamiltonians furnish so-called scalar relativistic (SR) corrections to the kinetic and potential energy, and they introduce a coupling between the electron spin and position degrees of freedom called spin-orbit (SO) coupling. In 2-component schemes, it is usually straightforward to split the operator into separate SR and SO parts. When the SO interaction is discarded in a 2-component scheme, an operator is left that, like $\hat{h}^S$, is not spin-dependent and acts on (1-component) spatial orbitals. Therefore, many traditional nonrelativistic quantum-chemistry programs have been retrofitted with SR-ZORA, SR-DKH, or SR-X2C. A notation such as SO-X2C etc. usually means that both SR and SO effects are included.

All-electron relativistic calculations require specifically designed and usually large basis sets, especially for the heavy elements in a system. For general molecules, when the SO interaction is included, the orbitals become complex and have two or four components, which increases the computational cost further. Relativistic effects on heavy-atom valence shells can alternatively be treated via relativistic pseudo-potentials (RPPs)[25] designed specifically to mimic relativistic effects on the valence orbitals in otherwise nonrelativistic calculations. The most common RPPs are SR, but there are also variants to be used in 2-component calculations that have parameters for the SO interaction. RPPs have the advantage that potentially a great number of core shells are removed from a calculation that is effectively nonrelativistic. However, the relativistic effects are treated approximately. Furthermore, the core nodal structure of the valence orbitals is not generated, leaving properties inaccessible that depend on this structure, such as nuclear magnetic resonance (NMR) shielding and J-coupling, electron-nucleus hyperfine coupling, and electric-field gradients for atoms treated by RPPs.

The type of complexes that are relevant in transition metal (TM), lanthanide (Ln), and actinide (An) separations may have open-shell metal centers, namely unpaired electrons that are formally localized at one or several metal centers. The electronic structure may, in this case, require two or more electron configurations, for even a qualitatively correct description. This is referred to as a multi-configuration (MC) or multi-reference case (not to be confused with multiple resonance structures), or static correlation as opposed to the dynamic correlation, which describes the explicit avoidance of two electrons in the wave function. The approximate correlation functionals in DFT and low-order truncated CC WFT, for instance, are good at describing the dynamic correlation of single-configuration systems, but they can have severe difficulties with MC cases. The full CC (or full configuration interaction, CI) wave function treats both static and dynamic correlation, but in practice, as usual, a compromise must be made. For an MC system, the first priority is to get its description qualitatively correct, that is, to treat the static correlation. Among the more frequently applied methods for MC systems is complete active space[26] (CAS) WFT and its variants. In a CAS calculation, an active space of orbitals and electrons, comprising the open shells and often additional orbitals, is selected, and a full CI calculation in this active space is performed, usually with simultaneous orbital optimization. Factorial scaling with the active space size severely limits these calculations, but in recent years CAS variants based on electronic states from density-matrix renormalization group (DMRG) calculations with polynomial scaling are showing much promise for large MC problems.[27–29] CAS-type calculations can be performed with relativistic spinors,[30] also with DMRG,[31] but often only SR effects are treated variationally while the SO interaction is introduced via a CI-like state interaction.[32,33]

The dynamic correlation in CAS-type calculations is usually treated approximately by perturbation theory (PT).[34,35] There is also truncated multi-reference configuration interaction (MR-CI),[36] which remains in use in particular for calculations of electronic spectra and to introduce dynamic correlation in a MC ground state. Promising singlet-paired CC methods[37,38] and multi-configuration pair-density-functional methods[39] have also been developed in recent years that may allow routine calculations of MC systems with the inclusion of dynamic correlation, and progress in MC–CC theory[40] has been reported.

For truly MC systems, DFT calculations cannot generally be recommended if the open shells participate in the properties of interest (bonding, spectroscopic, or magnetic). The KS approach is based on a single configuration. The static correlation is in principle taken care of by the elusive functional, but the available approximations treat MC systems rather poorly. However, there are scenarios when details of the open shell do not play an important role and can be treated as an "average of configurations" (AOC) in DFT calculations. An example would be a structure optimization of a metal complex with unpaired electrons only in non-bonding d or f orbitals. Among the successful variants of AOC approaches are all-electron DFT calculations with fractional occupations for a non-bonding open shell (e.g., six electrons distributed evenly over the seven $4f$ orbitals in a Eu(III) complex). For f-elements, there are also f-in-core RPPs[25] for use in DFT and WFT calculations where the partially filled f-shell is not treated explicitly. The latter approach is particularly suitable for Ln complexes as the $4f$ shell usually participates in bonding only weakly, yet care must be taken to ensure this is indeed the case.

In summary, and moving forward, these methods form the underlying methodology that serves as a starting point for predicting the behavior in solution, the characteristic chemical signatures of different species (coupling theoretical predictions with experiment), and for the elucidating the underlying chemical principles that stabilize specific molecular configurations or enable ligand selectivity.

5.2.1.2 Accounting for the Solution Phase Using a Continuum Approximation

The solution phase imparts a number of new interactions to a solute that can significantly alter the preferred coordination environment, shift spectroscopic observables, or result in perturbed reactivity. To account for solution-phase effects, there are two major approaches based upon the relative importance of the specific solute–solvent configurations, versus the bulk physical properties of a solvent. Building upon Section 2.1.1, the former approach solves the SE Hamiltonian for a discreet solution-phase system model that includes all degrees of freedom and involves the full description of the solute, solvent, and the solute–solvent and solvent–solvent interactions. Generally, an ensemble of solvation environments is also needed to depict and understand the average solute–solvent interactions and the sensitivity of those interactions to local effects.

Alternatively, it may also be appropriate to simplify the solute and solvent into two different models, where the solute is fully described at a high quantum-mechanical (QM) level, and the solvent is simplified and approximated as a continuum medium that is generally isotropic and solely described by its dielectric

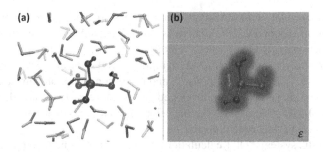

FIGURE 5.1   Illustration of (a) discreet and (b) continuum solvation models.

response function (see Figure 5.1). This simplification is warranted if the average solute–solvent interaction is responsible for the desired property to be calculated. As an example, the free energy of solvation is often independent of a specific configuration of solute–solvent interactions, like the exact hydrogen-bond configurations of water molecules that may be solvating an ion. The realm of models that use a continuum medium to describe the solvation environment around a solute of interest are called continuum solvation models (CSMs). As noted in extensive reviews[41], continuum approaches were introduced by Rivail et al.[42,43] in the early 1970s and have been extensively modified to determine a number of solvent effects, from the calculation of solvation free energies to shifts in UV–Vis spectra. A CSM framework employs an effective Hamiltonian can then be written as:

$$H_{eff}^{tot}\left(r_m\right)= H^M\left(r_M\right)+ H^{MS}\left(r_M\right) \qquad (5.2)$$

where H^M is the "effective" Hamiltonian in a vacuum, in the selected QM formalism (i.e., HF or DFT), and H^{MS} comprises the description of the interaction between the solute and the solvent molecules. H^M expands upon the pioneering ideas of Born,[44] further developed by Kirkwood[45] and Onsager,[46] which redefined the polarization of the solvent as a "reaction field" (RF). Since higher electrical multipoles of the solute are affected by the RF of the solvent, continuum solvation models need to include effects of back-polarization of the solute by the solvent in a self-consistent way. Models exploiting the self-consistent RF (SCRF) approach for the dielectric response are implemented in classical CMSs and are naturally incorporated into H^{MS} in Eq. 5.2 in the form of QM Fock-like operators. Instead of including an atomistic description of the solvent, H^{MS} is generally expressed as an appropriate solvent response function $Q(r,r')$ based on an averaged continuous solvent distribution, where r is not the whole set of solvent coordinates but only a couple of position vectors (r,r'). The nature of $Q(r,r')$ can be separated into different terms each related to an interaction with a different physical origin. The number of different interactions that are included in $Q(r,r')$ and the mathematical expressions and algorithms to account for solvation effects define the main characteristics of distinctive CSMs formulations.

   Among the most popular CSMs are those introducing a cavity (i.e., encapsulating the solute molecule) that is geometrically dependent on the solute molecular shape and projecting the dielectric response function on the surface of such cavity.

This is the case of the polarizable-continuum models (PCMs)[47] and the conductor-like screening model (COSMO).[48] In the PCM approach, each type of solute–solvent interaction can be associated with an independent operator that contributes to the solvation energy, to get:

$$G(M) = G_{cav} + G_{el} + G_{dis} + G_{rep} + G_{tm} \qquad (5.3)$$

where G_{cav}, G_{el}, G_{dis}, G_{rep}, and G_{tm} are the cavitation, electrostatic, dispersion, repulsion, and thermal-motion contributions to the solvation energy. The cavitation energy results from the spatial redistribution of the solvent molecules accommodating the solute. There are different available cavity models for the solute in different CSM implementations, namely, spherical, ellipsoidal, and molecular-shaped cavities.[44–46,49,50] The best compromise between accuracy and computational cost is based on the use of cavity contours reproducing the molecular shape of the solute using interlocked superposition of atomic spheres with the van der Waals radii. The GEPOL[51–53] algorithms (used in PCM) and other similar approaches are the most widely used methods for CSM cavity construction.

The electrostatic interaction between a solute and its surrounding dielectric-continuum environment is represented by G_{el}, and it is, in many cases, the dominant part of the solute–solvent interactions. Computing G_{el} is done by solving the Poisson equation, nested within a QM framework, for the polarization function $\vec{P}$ of the dielectric medium:

$$\vec{P} = \frac{\varepsilon - 1}{4\pi} \vec{E} \qquad (5.4)$$

where ε is the dielectric constant in the solvent and $\vec{E}$ is the electric field. In practice, the polarization of the medium is described by the surface polarization charge density σ on the solute cavity surface (apparent surface charge (ASC) approach). The ASC defines a "reaction potential" that is expressed in its integral form as:

$$V_\sigma = \int \frac{\sigma(s)}{|r - s|} ds. \qquad (5.5)$$

To compute V_σ, the previously generated cavity surface Σ is tessellated and thus approximated in terms of a set of small, discreet triangular elements (tesserae) with areas A_i, and centered at positions t_i, where σ_i is considered to be constant. Eq. 5.5 is transformed into a finite sum over the apparent surface elements ($q = A_i \sigma_i$)

$$V_\sigma = \sum \frac{\sigma A_k}{|r - s|} = \sum \frac{q}{|r - s|}. \qquad (5.6)$$

Both the PCM and the COSMO models are ASC methods distinguished by the mathematical formulation used to calculate σ (for details see review by Tomasi et al.[41]). Additional contributions to the solution free energy from non-electrostatic solute–solvent interactions (i.e., $G_{dis-rep}$ and G_{tm} in Eq. 5.3) are generally considered as corrections to the dominant electrostatic term (especially in polar solvents) or even ignored. Recent CSMs include dispersion and repulsion effects either in a semiclassical (MST,[54] SMD,[55] and early implementations of PCMs[56-58]) way or included in the QM–SCRF continuum formalism (i.e., the integral equation formalism (IEF) PCM model[59-61]). The self-consistent method developed by Amovilli and Mennuci[62] and the more recent Pomogaeva and Chipman[63] model are examples of attempts to include these non-electrostatic interactions in PCMs. Many publications have used the CSMs formalism to study chemical reactions and compute accurate values of free energies of solvation. Solvation free energies (G_{solv}) are calculated as the change in Gibbs free energy of the solute in the gas and in the solution phase. The solvation free energy is defined generally as:

$$\Delta G_{solv} = \left(E_{soln} + G_{nes} \right) - E_{gas} \tag{5.7}$$

where E_{soln} and E_{gas} refer to the electronic energy of the solute in presence and absence of the continuum solvent field respectively, and G_{nes} includes all non-electrostatic contributions (i.e., cavitation energy, dispersion–repulsion interactions, and thermal corrections). It is important to note that the SCRF models described above provide a free energy of solution at 298 K, but explicit temperature dependence of the solvation energy has been included in more recent CSMs implementations (SM8T[64] or COSMO-RS[65]).

5.2.1.3 Predictive Spectroscopy

5.2.1.3.1 UV–Vis

The utility of optical electronic spectroscopy for investigating chemical systems is well known. It is often one of the first methods used to interrogate components of solutions, be it in research or analytical laboratories, and to analyze kinetics, structure, and composition of chemical species and reactions.[66,67] Optical electronic spectroscopy corresponds to excitations of the valence electrons to an excited state. These transitions can be significantly modified by the presence of a solvent,[68] and can even be used to probe intermolecular interactions.[69] Despite the common use of this technique in solvated environments, accurately describing the effects of the solvent on the absorption of molecules remains a theoretical challenge. Some current approaches are discussed below.

From a theoretical point of view, a few methods exist to model electronic transitions. Spectra can be accessed by solving the time-dependent Schrodinger equation for a molecular system, either in the real-time[70,71] or frequency domain.[72,73] Due to its favorable computational cost and sufficient accuracy, one of the most widely employed methods is time-dependent density-functional theory (TDDFT) within the linear-response formalism.[72,74] While this method is sufficiently accurate in most cases, more expensive, yet more accurate methods may be employed, such as the equation of motion coupled-cluster (EOM-CC).[73,75] The linear-response formalism of TDDFT results in the following eigenvalue equation:[72]

$$\begin{pmatrix} A & B \\ B^* & A^* \end{pmatrix} \begin{pmatrix} X \\ Y \end{pmatrix} = \omega \begin{pmatrix} 1 & 0 \\ 0 & -1 \end{pmatrix} \begin{pmatrix} X \\ Y \end{pmatrix} \tag{5.8}$$

where the A and B matrix elements are

$$A_{ia,jb} = \delta_{ij}\delta_{ab}\left(E_a - E_i\right) + \left(ia \mid jb\right) - \alpha\left(ij \mid ab\right) + \left(1 - \alpha\right)\left(ia \mid f_{xc} \mid jb\right) \tag{5.9}$$

$$B_{ia,jb} = \left(ia \mid bj\right) - \alpha\left(ib \mid aj\right) + \left(1 - \alpha\right)\left(ia \mid f_{xc} \mid bj\right) \tag{5.10}$$

capturing the excitation matrix and the de-excitation matrix, respectively. The term $(ia|jb)$ is a two-electron repulsion integral in Mulliken notation, where i, j are occupied orbitals and a, b are unoccupied virtual orbitals. The orbital energies are defined as E_i and E_a for the occupied and virtual orbitals respectively. Further, the term $(ia|f_{xc}|jb)$ represents the second functional derivative of the exchange-correlation energy, which is also known as the xc kernel. The matrices X and Y are the first-order electron-density responses for the given excitations obtained in the eigenvalue problem. The parameter α is the fraction of Hartree–Fock exchange in the hybrid exchange-correlation functional, where i and j are occupied molecular orbitals, a and b are virtual molecular orbitals, ε is the orbital energy.

Solvation can be taken into account using continuum models (Section 2.1.2) or through an embedded model where the method to treat a chromophore is at a high, quantum-mechanical level of theory, and the solvent at a lower, often classical level of theory (e.g., a mixed quantum-mechanical/molecular-mechanical (QM/MM) method),[76–78] where the molecules of the solvent are treated explicitly with a force field. Both of these approaches affect the TDDFT equation simply by adding a perturbation to the excitation and de-excitation operators.

$$A_{ia,jb} = \delta_{ij}\delta_{ab}\left(E_a - E_i\right) + \left(ia \mid jb\right) - \alpha\left(ij \mid ab\right) + \left(1 - \alpha\right)\left(ia \mid f_{xc} \mid jb\right) + V_{ia,jb}^{\text{solvent}} \tag{5.11}$$

$$B_{ia,jb} = \left(ia \mid bj\right) - \alpha\left(ib \mid aj\right) + \left(1 - \alpha\right)\left(ia \mid f_{xc} \mid bj\right) + V_{ia,jb}^{\text{solvent}} \tag{5.12}$$

In the case of a polarizable-continuum model (PCM), the perturbation term is given by[79]

$$V_{ia,jb}^{\text{solvent}} = \sum_{I,J} \phi_{ia}(R_I) Q_{I,J}\left(\varepsilon_{\text{solvent}}\right)\phi_{jb}(R_J) \tag{5.13}$$

The indices I and J run over PCM tesserae (discrete tiles on the solvent cavity surface), R are the coordinates of the tesserae, φ is the electrostatic potential, E is the dielectric constant of the solvent, and Q is the PCM response function. Further details can be found in Cammi et al.[79] In the case of QM/MM, only polarizable models have direct contributions to the TDDFT equations, which has the following form in the induced-dipole formulation:[76]

$$V_{ia,jb}^{\text{solvent}} = \sum_{I} \mu_I \cdot E(R_I) \tag{5.14}$$

where I runs over the MM polarizable sites, μ is the induced dipole at those sites, R is the coordinate of the dipole, and E is the electric field. Further computational details can be found in Curutchet et al.[76]

Due to the underlying assumptions, care must be taken when applying the previously mentioned solvation models. The traditional formulation of PCM only accounts for solvent polarization in an average manner and neglects specific solvent–solute interactions, such as hydrogen bonding. It is worth mentioning that extensions to PCM have been made to model specific interactions, such as supplementing the PCM potential with an additional repulsion potential.[80,81] However, the authors of these extensions note that they have focused only on reproducing the energy of the hydrogen-bonding interaction, and this extended model may not be sufficient to describe other properties, such as absorption energies. Polarizable MM, on the other hand, has an explicit, atomistic description of the solvent, so specific solvent–solute interactions can be accounted for. With this explicit solvent representation, however, comes the computational expense of needing to perform calculations over an ensemble of solvent configurations.

In addition to these considerations for choosing the solvent model, additional thought must be dedicated to the UV–Vis spectroscopy of metals relevant to solvent separation. In particular, heavy metals can have large relativistic effects, namely contraction of the inner orbitals whose electrons are moving a significant fraction of the speed of light, and spin-orbit coupling. One can treat these effects phenomenologically through the zero-order relativistic approximation (ZORA) and a spin-orbit perturbation term. Alternatively, one can treat these effects directly from relativistic theory and use a relativistic Hamiltonian, such as the exact two components (X2C) Hamiltonian (see Section 2.1.1). The effects of both solvation and relativistic effects on the optical spectroscopy of metal complexes have been widely reported in the literature.[82–87]

5.2.1.3.2   X-Ray absorption spectroscopy (XAS)

XAS is an element-specific spectroscopic technique that can probe the electronic structure and the local geometry of molecular systems and materials. With recent advances in synchrotron technology and bench-top X-ray laser light sources, XAS has become a routine experimental technique with high resolution.[88–92] X-ray spectra arise from core electronic excitations and are typically split into two distinct sections known as the near-edge X-ray absorption fine structure (NEXAFS) that lies below the ionization edge of the absorbing atom, and the extended X-ray absorption fine structure (EXAFS) that is higher in energy than the ionization absorption edge.[93] In addition, the NEXAFS is also sometimes called the X-ray absorption near-edge structure (XANES). Since the NEXAFS corresponds to bound state excitations, it is primarily used to obtain information about local geometric structure. On the other hand, the EXAFS contains information about ionized electrons that scatter off of neighboring atoms in order to determine the extended structure of periodic crystals.

As mentioned above, an important aspect of XAS is that core excitations from different elements are highly separated in energy. This allows for the isolation of specific elements in X-ray spectra. In addition to this element specificity, XAS can be broken down further. Experimental investigations using XAS typically come in three distinct types for a select element, known as the K-edge, L-edge, and M-edge.

These edges correspond to different core orbitals of a selected principal quantum number being excited and are typically highly separated in energy for a specific element. The K-edge contains the excitations from the 1s orbital, L-edge contains the excitations from the 2s and 2p orbitals, and the M-edge contains excitations from the 3s, 3p, and 3d orbitals.

Since the NEXAFS region of the spectrum involves core-to-bound state electronic excitations, the information from this region provides information about the local geometric structure. In addition, the core hole lifetimes of X-ray excitations are on the order of femtoseconds, so nuclear geometries can be considered frozen during a single absorption event.[93] These properties of XAS are important when probing the physical properties of molecular geometries in solvated systems. Multi-reference WFT calculations of XANES spectra of actinide complexes have been reported recently, with very promising results regarding the interpretation of experimental spectra in terms of metal–ligand chemical bonding.[94] However, the most commonly employed technique to model such transitions, employ TDDFT using atom-centered Gaussian basis sets in order to solve for the local molecular-orbital picture of the absorption.

Eq. 5.8 is identical for solving all electronic spectra, and the full matrix contains all of the information about both UV–Vis and X-ray absorption. Therefore, one can solve for all of the absorption roots, UV–Vis to X-ray, if the full TDDFT matrix is diagonalized. However, in practice, this is usually computationally intractable, as the whole matrix is too large to store for even moderate-sized molecular systems. To remedy this problem, one can use iterative diagonalization algorithms to solve for a smaller window of absorption frequencies.[95,96] In order to solve for the X-ray absorption eigenvalues, energy-specific methods (ES-TDDFT) of iterative diagonalization can be used to compute the spectral interior eigenvalues of the TDDFT matrix.[97]

XAS is particularly useful for studying the electronic structure of molecules containing heavy metal atoms, due to the high resolution of core atomic transitions. Typically, when solving for excitations using TDDFT, it is assumed that the wavelength of incident light is much longer than the size of the quantum system, so the photon can be treated as a uniform field. This is commonly known as the electric-dipole approximation (EDA) and is sufficient for spectroscopies such as UV–Vis. However, for high-energy X-ray absorption, the wavelength of the incoming photon can be sub-nanometer, causing the EDA to break down, and higher-order multipole terms become significant to the point where the EDA gives a qualitatively incorrect line shape. Classical light, interacting with a quantum-molecular system, can be modeled as a planewave potential perturbing a molecular Hamiltonian. Neglecting time-dependence, the planewave can be expanded in a Taylor series such as:

$$\exp\left(i\boldsymbol{k}\cdot\boldsymbol{r}\right) = 1 + i\boldsymbol{k}\cdot\boldsymbol{r} + \frac{1}{2}\left(i\boldsymbol{k}\cdot\boldsymbol{r}\right)^2 + \ldots \tag{5.15}$$

where $\boldsymbol{k}$ is the wave vector describing the direction of propagation for the planewave, and $\boldsymbol{r}$ is the spatial coordinate vector. The EDA truncates the Taylor expansion at the first term, but in order to describe XAS beyond the EDA, more terms in the expansion must be included. By including more terms, we can describe electric-quadrupole contributions, magnetic-dipole terms, etc. (Figure 5.2).

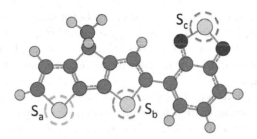

FIGURE 5.2 Single unit of a PCPDTBT polymer, highlighting 3 different sulfur atoms, Sa-c. Since sulfur is a relatively light atom, the electric-dipole approximation is sufficient, and the difference between $f^{(0)}$ and $f^{(2)}$ is negligible. (Reprinted with permission from Lestrange et al. (2015). Copyright © 2015, American Institute of Physics.)

However, arbitrary truncation of the multipole expansion can introduce origin-dependence of oscillator strengths, which can affect the qualitative result of an XAS calculation. Specifically, one extension to the EDA would be to truncate the planewave expansion at second order. This would give the electric-dipole term (E1), the magnetic-dipole term (M1), and the electric-quadrupole term (E2) for the oscillator strength, f which is proportional to the multipole moment squared. We can denote this second-order truncation of the oscillator strength as $f^{(E1+M1+E2)}$, and the EDA as $f^{(0)}$. Truncating the oscillator strength expansion at the M1 and E2 terms to match the planewave truncation gives an inconsistent treatment of second-order terms, and the electric-octopole (E3) and magnetic-quadrupole (M2) terms are necessary to cancel out origin dependence in the oscillator strength.[99] We can denote this origin-independent correction of the oscillator strength with E3 and M2 to the second order as f^2. This phenomenon is explained in detail in Lestrange et al.,[98] where the sulfur K-edge is calculated for a single polymer unit of poly[2.6-(4,4-bis-(2-ethylhexyl)-4H-cyclopenta[2,1-b:3,4-b']dithiophene)-alt-4,7(2,1,3-benzothiadiazole)] (PCPDTBT), shown in Figure 5.2. In Figure 5.3, it is shown that using the naive second-order truncation, $f^{(E1+M1+E2)}$, the X-ray absorption spectrum changes drastically depending on the origin **r**, while the $f^{(0)}$ and f^2 oscillator strengths are both origin-independent.

In heavy metal atoms, the breakdown of the EDA is even more pronounced due to shorter wavelengths of the absorbed X-rays. For metal systems, $f^{(0)}$ and f^2 will differ greatly. In order to properly describe X-ray transitions in the framework of TDDFT, one needs to carefully select the truncation of the planewave expansion when describing oscillator strengths beyond the EDA, and avoid expansions that introduce nonphysical origin dependence (Figure 5.3).

5.2.1.3.3 Nuclear magnetic resonance (NMR)

As essential as optical spectroscopy, NMR is equally important to elucidating complex solution-phase speciation. Herein, we focus upon calculations for metal complexes. For closed-shell (no unpaired electrons) systems, there are a number of DFT and WFT methods available for calculating the two basic solution-phase NMR parameters, viz. nuclear magnetic shielding (σ) and indirect nuclear spin-spin coupling (J-coupling). These are static magnetic-response properties of the electronic

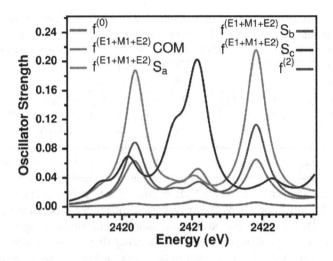

FIGURE 5.3 Computed sulfur K-edge X-ray absorption spectra for different origins in the PCPDTBT molecule. S_{a-c} are the specific sulfur atoms denoted in Figure 5.2, and COM is the center of mass as the origin for calculating oscillator strengths. (Reprinted with permission from Lestrange et al. (2015). Copyright © 2015, American Institute of Physics.)

system, which means that the calculation needs to determine how the electronic structure responds in first order to the presence of the external magnetic field from the spectrometer or the inhomogeneous magnetic fields associated with the nuclear spins. Moreover, NMR parameters are famously sensitive to relativistic effects. For the foreseeable future, NMR calculations on closed-shell systems that are of relevance in separations science will most likely utilize DFT,[100] combined with an all-electron relativistic Hamiltonian if heavy metals are involved.[101] In order to treat solvent effects, it is advisable to use some flavor of molecular dynamics to generate solvent–solute clusters with explicit solvation at the quantum-mechanical level for the NMR calculations,[100,102–106] in particular when the metal has open coordination sites accessible by the solvent. Molecular dynamics (MD) simulations are also useful in order to study NMR relaxation processes,[107] for instance, due to paramagnetic[108] or quadrupolar[109,110] effects. Chemical shifts of light atoms in the vicinity of a heavy metal can be strongly impacted by SO coupling. For example, ^{29}Si shifts in a series of M–Si-bonded paddle-wheel complexes with M = Ni, Pd, Pt showed Si to be ~ 40 ppm more shielded in the Pt than in the Pd compound because of SO coupling.[111] Carbon-13 shifts in diamagnetic U(VI) compounds can reach several hundred ppm because of the strong SO coupling at the actinide.[112,113] The SO effect is particularly effective if the metal–ligand bonding orbital(s) have large s character at the light atom and may, therefore, become spectacularly large (dozens of ppm or more) for hydride complexes.[112,114] Usually, such effects are calculated with all-electron relativistic methods, but SO-RPPs are also suitable.[115] Relativistic effects on J-coupling between light ligands in a complex can often be treated at the SR level, all-electron or via RPPs even if the coupling path involves the metal shell.[116–118] For J-coupling with the metal, or J-coupling among two metal centers, it is mandatory to use an all-electron relativistic method. For example, one-bond J-coupling between Hg and light ligands may increase by a factor of 3 relativistically.[101] SO effects are often secondary for J-coupling, even when the coupling involves one or two heavy atoms, with the notable exception of heavy p-block elements.[119,120]

Ab initio computational NMR capabilities for open-shell paramagnetic complexes are far less well developed than their diamagnetic counterparts.[121–123] The best success has so far been achieved with DFT for ligand paramagnetic NMR (pNMR) shifts in systems with spin states that can be described reasonably well with a single configuration, such as 3d metallocenes,[124,125] Ru complexes,[126–128] and others.[129–131] In this case, the paramagnetic effects on the shift can be approximated from a product of the system's g-tensor and the hyperfine tensor of the ligand nucleus.[132] Other examples are discussed in refs 121–123. Recently, a SO-CAS-level implementation of the full pNMR shift expression[123,133] was reported, along with successful calculations of ligand pNMR shifts for selected actinide complexes in which the ligand shifts are dominated by the orbital angular momentum rather than unpaired spins.[134,135] The CAS approach does not treat the spin polarization in the ligands fully, and other authors have opted for treating the ligand hyperfine terms with DFT in a hybrid DFT-CAS scheme for cases where the spin magnetization is more important.[136] For lanthanides, the crystal-field theory for ligand pNMR shifts of Bleaney and others[137–139] remains in use, but DFT-based calculations have also been attempted.[127,140]

5.2.1.4 From Gas-Phase Electronic Structure to Solution-Phase Energetics

The above overview of the essential considerations and terminology of electronic-structure methods, as well as the manner in which molecular species are characterized, provide the basis for now considering specific applications of practical importance to the solvent extraction community: How do we combine these methods to predict the energetic favorability of distinct molecular configurations in solution? That is, what is the speciation of solutes we are separating, and what are the fundamental physical underpinnings of the speciation ensemble behind such energetic ordering? An accurate description of the electronic structure and bonding of aqueous metal ions and their extractant–metal–ligand complexes is critical for theoretical modeling of their thermodynamic properties in solution. The latter is of essential importance toward understanding which species will exist in solution and how to use such information to control speciation and thus solvent extraction efficiency.

From a QM perspective, all approaches toward solution-phase thermochemistry begin with an understanding of the gas-phase analog. Indeed, much can be learned from studying the trends in gas-phase physical properties—be it bonding or energetics. It is now possible to calculate thermodynamic properties, including bond-dissociation energies, atomization energies and heats of formation, electron affinities and ionization (redox) potentials, Lewis acidities and proton affinities (Brønsted acidity and basicity) of gas-phase molecules for all of the elements through $Z = 103$, to chemical accuracy (<1 kcal/mol for main group compounds and <3 kcal/mol for compounds containing metals) by using high-level theoretical approaches, for example, by the Feller–Peterson–Dixon (FPD) composite approach.[141–144] This includes the careful treatment of all important effects including core-valence correlation, relativistic effects, and the zero-point energy. Such calculations are made possible, in part, by the development of correlation-consistent basis sets for the entire periodic table.[145–148]

Focusing upon the behavior of the heaviest elements, the lanthanides and actinides, many important features should be noted regarding studies associated with solvent extraction. However, discussion of the practical methods and computational protocols often hold for studies of lighter metals. The lanthanide and actinide elements have both f and d valence shells. As the $4f$ orbitals are core-like, the chemistry of the lanthanide ions is in general very similar and largely governed by $5d$ orbitals. The most common oxidation state for all lanthanide elements is +3 in solution and in the solid state. Compared to lanthanides, the electronic structure of actinides is significantly more nuanced. This complication is mainly due to $5f$ orbitals. In the early actinides, the $5f$ orbitals are more extended and very close in energy to the $6d$ orbitals. As the atomic number increases, the $5f$ orbitals become more localized and progressively lower in energy, while the $6d$ energies remain relatively constant. Therefore, starting from Am, the late actinides can behave similarly to lanthanides. The similar chemical properties of lanthanides and late actinides as it pertains to ionic radii, oxidation state and coordination structure are the underlying source of challenge for separating lanthanides (e.g., Eu) and minor actinides (Am and Cm) in used nuclear fuel reprocessing.

For actinides, the best current approach for calculating accurate energies is to use the new all-electron Douglas–Kroll[23,149,150] correlation-consistent basis sets[151] developed for the actinides by Peterson.[146,147] These basis sets have been shown to have smaller errors than pseudopotential-based basis sets, and the additional computational cost is not that large, as only the Hartree–Fock level is really affected. One can then use highly correlated MO theory at the coupled-cluster theory CCSD(T) level[152–158] including the use of improved orbitals[159,160] to make reliable predictions about the potential-energy surfaces for different An oxidation states. Multi-reference (MR) methods, including the complete active space self-consistent field (CASSCF) method[161] and CASSCF with second-order perturbation theory CASPT2,[162,163] can be used to test the appropriateness of using a correlated single-reference wave-function approach such as CCSD(T). An issue with such MR methods is the appropriate choice of the space for the open-shell electrons. In addition, if one wants to build databases of results, one has to define a consistent active space, for example, if one is to treat all of the actinides. Assuming that bonding has a significant ionic component, the active space for the actinides can be considered to be the $5f$ electrons on the An. This choice needs to be checked to see if the active space needs to be expanded to include the $6d$ and $7s$ orbitals on the An and possibly any ligand active electrons. In most cases, we have found that the $6d$ orbitals on the actinide are doubly occupied, and there is very little $7s$ character present, based on a natural bond order (NBO[164,165])/ natural population analysis (NPA).[166–168] It is important to use the most recent version of the software,[169] especially if effective core potentials are being used. The effects of core-valence interactions need to be considered, and the all-electron basis sets are useful for dealing with such interactions. However, one has to carefully investigate the orbitals that are being correlated due to the interleaving of the frozen core and correlated orbitals. Such CCSD(T) and MR-CI calculations including scalar relativistic effects can be carried out with MOLPRO,[170] NWChem,[171-173] and DIRAC software programs.[174,175]

Spin-orbit corrections can be incredibly important to determining the correct ground electronic state of An and influencing spectroscopic assignments. They can be included in different ways depending on the software program employed. For example, via (i) spin-orbit DFT (SO-DFT) approaches developed in NWChem with appropriately chosen functionals and basis sets;[176] or (ii) using the ADF code[177,178] at the ZORA-spin orbit level[179-183] with the BLYP functional[184,185] and the TZ2P basis-set. This level has provided useful results for the AnO_2^+.[186] A third approach is the average-of-configuration open-shell Hartree-Fock approach as implemented in DIRAC. This method was developed because the spin-orbit interaction couples the spin and spatial components, so it is not straightforward to employ a restricted open-shell Hartree–Fock method. As implemented in DIRAC, the energy of a selected set of open-shell states is calculated to provide an estimate of spin-orbit effects. The separation into scalar and spin-orbit effects can be checked by performing 2-component and full 4-component calculations. Calculations on the size of systems that are of interest to solvent separations cannot be done currently in any routine way—but advances in massive parallelization are anticipated that include the use of GPUs within DIRAC[187] and are being developed for use at the Oak Ridge Leadership Computing Facility at Oak Ridge National Laboratory. There are a variety of modules available in DIRAC for such calculations including the KRCI open-shell module. KRCI uses Dirac's Kramers pairs for both two- and four-component relativistic calculations and it is based on a Generalized Active Space approach. One can also use the Kramers unrestricted coupled-cluster method[188,189] in DIRAC as well for testing the approach given above. For the 4-component calculations, one can use the basis sets developed by Dyall.[190-192]

Using these approaches in addition to DFT can provide not only accurate gas-phase energetics but also detailed information regarding the chemical bonding that differentiates the reactivity of various metals, including actinides from one another and from lanthanides. As an example, recent theoretical and experimental studies indicate partial covalency of 5f orbitals in late actinide complexes,[193-195] distinct from the ionic bonding character of lanthanide 4f orbitals. One representative example is the 5f covalency in $AmCl_6^{3-}$ vs. the 4f covalency in valence isoelectronic $EuCl_6^{3-}$.[195] Relativistic DFT calculations with the inclusion of spin-orbit coupling were performed on these two hexachloride compounds to simulate their Cl K-edge X-ray absorption spectra. Theoretical results indicate that the amount of Cl 3p-mixing with Am(III) 5f orbitals is more pronounced than that with Eu(III) 4f orbitals, which is highly consistent with experimental data (see Figure 5.4). This appears to confirm Seaborg's 1954 hypothesis[196] that Am(III) 5f-orbital covalency was more substantial than 4f-orbital mixing for Eu(III).

In the absence of learning about the statistical ensemble of solvation environments about metal ions and ML complexes, we are generally concerned about solution-phase thermochemistry. The effect of solvent on actinide chemistry can be predicted by using micro-solvation/self-consistent reaction field (SCRF) techniques (Section 2.1.2).[197-205] A hybrid supermolecule-continuum approach with the continuum

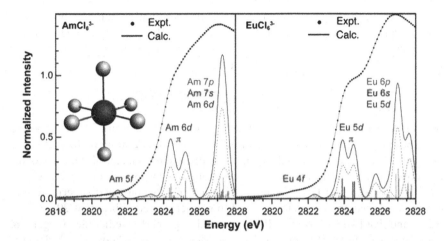

FIGURE 5.4 Experimental Cl K-edge XAS spectrum (dotted) and DFT calculations (solid black trace) of $AmCl_6^{3-}$ and $EuCl_6^{3-}$. The orange, green, purple, gray bars, and dashed traces represent the energy and oscillator strength for the calculated transitions involving $5f/4f$, $6d/5d$, $7s/6s$, and $7p/6p$ final states in the $AmCl_6^{3-}/EuCl_6^{3-}$ complex, respectively. (Reprinted with permission from Cross et al. (2017). Copyright © 2017, American Chemical Society.)

represented by a SCRF with appropriate parameters[206,207] can be used to predict free energies of solvation. One such approach is the COSMO continuum solvation model which has been implemented in NWChem,[171] ADF,[178] and Gaussian-03/09/16.[208] Other solvation models include the solvation model based on density (SMD) approaches.[209–211] Continuum-based solvation models such as COSMO partition the solute–solvent environment into two distinct regimes, one of which is a cavity containing the solute molecule, and the other is the solvent medium surrounding the cavity defined solely by its dielectric constant. Induced charges on the surface of the cavity due to solute–solvent polarization allow one to obtain the free energy of solution of the solute under study. Issues include the need for the correct radii for atoms, the inclusion of non-bonded parameters for all atoms, the potential appearance of cavities inside a molecule, the choice of the number of explicit solvent molecules, and issues with charged species, especially charged species with charge magnitudes >1. In addition, it is to be noted that current SCRF approaches yield a free energy of solvation, $\Delta G_{solvation}(298\ K)$. This needs to be combined with a gas-phase free energy, $\Delta G_{gas}(298\ K)$, to give the free energy in solution, $\Delta G_{solution}(298\ K)$, given by

$$\Delta G_{solution}\left(298\,K\right) = \Delta G_{gas}\left(298\,K\right) + \Delta G_{solvation}\left(298\,K\right) \tag{5.16}$$

5.2.1.4.1 A Classic Example: Elucidating the First Solvation Shell for UO_2^{2+}

The majority of experimental measurements in solution show that the $UO_2(H_2O)_5^{2+}$ is the dominant species in aqueous solution. There is also evidence that $UO_2(H_2O)_2^{4+}$ may be present as a minor component of these systems, rather than the usual value of five equatorial ligands that is often cited. Both NMR and EXAFS studies have indicated the presence of structures with coordination numbers below five. The interpretation of high-energy X-ray scattering (HEXS)[212,213] of the uranyl ion with a perchlorate counter-ion indicate an equilibrium exists in solution between the four- and five-coordinate uranyl (reaction (r1)).

$$UO_2\left(H_2O\right)_4^{2+} + H_2O \leftrightarrow UO_2\left(H_2O\right)_5^{2+} \tag{r1}$$

Integration of the HEXS peak at 2.420(1) Å yielded 46.1 electrons, 3.9 electrons less than the value of 50 electrons needed for five waters coordinating in the first solvation shell. This data can be analyzed to show that there is an equilibrium where the five-coordinate uranyl is favored over four-coordinate by $86 \pm 7\%$ to $14 \pm 7\%$, giving $\Delta G(298\ K) = -1.19 \pm 0.42$ kcal/mol for reaction (r1). A combined supermolecule-continuum solvation approach was used to predict this equilibrium at 298 K.[214] It was found that two solvation shells are needed to properly predict the energetic of reaction (r1) as the positive charge form the UO_2^{2+} can "leak" out of (polarize) the second solvation shell. A second solvation shell of 10 H_2O molecules was added to $UO_2(H_2O)_5^{2+}$ so that each H atom on a first solvation shell water molecule has one hydrogen bond. The geometries were optimized with DFT using the B3LYP exchange and correlation functionals, and single point energy calculations were performed at the MP2 level with large basis sets due to the presence of hydrogen bonding.[215] Even though the DFT errors are small per hydrogen bond, they accumulate and the reaction energy for reaction (r1) cannot be predicted properly. The calculated free energies of reaction are very sensitive to the choice of O and H radii and isodensity values in the PCM and self-consistent isodensity polarizable-continuum model (SCIPCM) solvation models, respectively. Rearrangement reactions based on an intramolecular isomerization leading to a redistribution of water in the two shells provide good values in comparison to experiment with values of $\Delta G_{exchange}$ from -2.2 to -0.5 kcal/mol, meaning the inclusion of a second hydration sphere accounts for most solvation effects. One must also carefully consider the appropriate standard state corrections, for example, the release of a H_2O molecule requires a correction of $+4.3$ kcal/mol. The best-calculated value for $\Delta G_{reaction}$ is -2.0 kcal/mol in comparison with the value of -1.2 ± 0.4 kcal/mol from the HEXS experiment. The predicted free energy of solvation of UO_2^{2+} at 298 K is 411 ± 5 kcal/mol in excellent agreement with the best experimental estimate of 421 ± 15 kcal/mol. Surprisingly, the free energy of solvation is nearly converged with one solvent shell embedded in a continuum. More recent work[216] has examined the contributing factors to the solution-phase correction to the free energy of molecular clusters of U(III–VI) as a function of the number of explicit waters (up to 77 H_2O), cavity type, and continuum model. Therein, it was observed that the free-energy correction in solution does not smoothly converge to zero as the number of explicitly solvating water molecules approaches the bulk limit, and the convergence behavior varies significantly with cavity and model. This is likely the source of previously reported discrepancies in predicted free energies of solvation for metal ions when using different PCM cavities and/or models. Thus, the cancellation of errors when using a molecular cluster must be carefully considered based upon the size of the cluster, cavity type and PCM model employed.

Section Summary. While the methods described above illustrate the capabilities of quantum-mechanical methods to determine energetic favorability of chemical species, their potential reactivity, and characteristic chemical signatures, one concept that is lacking is that of the ensemble of all potential species and reactions at thermal equilibrium. Indeed, to truly understand many aspects of solvent extraction, a statistical-mechanics and statistical-thermodynamics approach is warranted. This is largely due to the complex chemical composition of most solvent extraction systems, which impart a myriad of molecular configurations, and intact species that may act as reactants or products of the extraction. Further, such solutions may not be homogeneous, and in particular, significant variations in composition and reactivity may be encountered near the liquid interface. Thus, the QM methods applied above are only as useful as they are informed by the most relevant species in solution and available reaction and transport pathways. Toward this end, simulations that leverage statistical mechanics—specifically Monte Carlo (MC) and molecular dynamics (MD)—are needed to provide insight into the equilibrium ensemble of species and processes.

5.2.2 STATISTICAL-MECHANICS METHODS

5.2.2.1 Representations of Potential-Energy Functions

Statistical-mechanics methods often employ classical descriptions of inter- and intramolecular interactions to calculate the energy of a given molecular configuration. The combination of a potential-energy function with associated parameters is known as a force field. Typically, that potential-energy function is parametrized by empirical fitting to experimental or first-principles calculated data. This classical description typically only considers nuclear degrees of freedom with motions of the electronic structure treated implicitly. The quality of an MD or MC simulation is determined by the ability of its force field to reproduce the potential energy U accurately. To meet this goal, several force fields have been developed over the last decades including all kinds of energy potential terms, such as Optimized Potentials for Liquid Simulations (OPLS),[217] Assisted Model Building with Energy Refinement (AMBER),[218] Generalized AMBER Force Field (GAFF),[219] Chemistry at HARvard Macromolecular Mechanics (CHARMM),[220] etc. A typical force field[221,222] consists of covalent (bonded) interactions, van der Waals, and electrostatic (Coulomb), with an overall expression given by:

$$U_{\text{total}} = \sum_{\text{bonds}} \frac{1}{2} k^l (l - l^0)^2 + \sum_{\text{angles}} \frac{1}{2} k^\theta (\theta - \theta^0)^2 + \sum_{\text{torsions}} \frac{1}{2} k^\varphi{}_n [1 - \cos(n\varphi)]$$

$$+ \sum_{i<j}^{N} 4\varepsilon ij \left[\left(\frac{\sigma_{ij}}{r_{ij}} \right)^{12} - \left(\frac{\sigma_{ij}}{r_{ij}} \right)^6 \right] + \sum_{i<j} \frac{q_i q_j}{4\pi\varepsilon_0 \varepsilon_r r_{ij}}$$

(5.17)

The first three terms represent intramolecular interactions, including harmonic potentials that capture bond stretching and angle bending, and a torsional potential represented by a cosine series depending on the dihedral angle. The constants k and equilibrium positions l^0 and θ^0 are uniquely determined for each intramolecular interaction type. The last two terms are intermolecular interactions, namely Lennard–Jones (LJ) and Coulomb potentials. The LJ potential, as an empirical potential of pairwise van der Waals interactions, consisting of a short-ranged repulsion to represent orbital overlap and a long-range attractive force due to electronic dispersion. The Coulomb potential represents classical electrostatic interactions between charged species.

5.2.2.1.1 Force-Field Development

The quality of dynamic and thermodynamic properties obtained from simulation is crucially dependent on the ability of the molecular-mechanical force field to accurately describe the interactions present in the simulated system. Classical force fields are often parameterized to reproduce experimental data. This approach was taken by Guilbaud and Wipff,[223] who conducted the first MD simulations on the uranyl ion in water. They tried various sets of partial charges to determine the effect on structural parameters (bond length and angle), hydration number, and hydration free energy. Using this approach, they studied the complexation between the uranyl ion and 18-crown-6 and successfully reproduced the second-sphere coordination complex inferred from kinetic studies. Wipff and coworkers[224] applied their empirically tuned force field to the study of uranyl extraction by tri-*n*-butylphosphate (TBP), one of the key components of the Plutonium-URanium Reduction EXtraction (PUREX) process.[225,226] Using MD, they studied the complexation of uranyl nitrate and TBP, and elucidated TBP's role as a co-solvent in the extraction process.[227] In a later work, they calculated the free energy of transfer of uranyl nitrate across the interface. These examples illustrate how a properly tuned classical force-field-based simulation can provide molecular-level information that would be difficult or expensive to obtain experimentally.

Unfortunately, parameterization based on experiments is not suitable for systems where experimental data are limited, as is often the case for transuranics. An alternative approach to using experimental data to develop a force field is the use of highly accurate QM calculations on a small, representative gas-phase system, and tuning the classical force field to reproduce the QM potential-energy surface. Liu et al.[228] adopt this approach to develop non-polarizable and polarizable force-field models for a class of chlorinated hydrocarbons to reproduce their correct dielectric constants. Hagberg and coworkers[229] explored the QM potential-energy surface of a uranyl–water dimer and used the resulting force-field parameters to study the actinyl ion in the aqueous phase. Their model was also able to reproduce structural quantities such as the radial distribution function (RDF) and coordination number in the first solvation shell, in agreement with X-ray scattering data.

While force-field fitting from the QM data has the advantage of not relying upon already existing experimental data, it does have its limitations. In particular, the use of a pair of molecules in the QM calculation has been shown to fail in the case of highly charged ions (such as Cu(III) and Th(IV)), where gas-phase electron transfer can take place. The solvation of highly charged ions stabilizes the system and suppresses electron transfer. Thus, any force-field parameterized from QM energy calculations of such highly charged species will be inaccurate. One widely explored solution to this problem is the use of multiple water molecules to solvate the bare gas-phase actinyl ions.[230–232] Maginn and coworkers[233] explored the minimum number of water molecules in a QM calculation necessary to produce a high-quality classical force field for the uranyl–water system. It was demonstrated that many-body solvation effects are non-negligible, contributing to a 35–40% reduction in binding energy. They also explored the effect of various classical water models on thermodynamic and structural properties. The classical model developed by this technique was able to correctly reproduce hydration free energy, which constitutes a severe test of the underlying potential. The methodology was extended to the study of the transuranic ions, namely NpO_2^{2+}, PuO_2^{2+}, and AmO_2^{n+} $(n = 1, 2)$.[234]

In addition to fitting classical potentials to QM energies, "force matching" methods have been developed wherein classical potentials are instead fit to the gradient of the QM potential-energy surface (PES), hence the term force matching. The fitting quantum PES can be determined through gradients calculated across sets of geometries from pairwise dissociation curves, QM trajectories (i.e., from *ab initio* molecular dynamics (AIMD) simulation) or QM/MM calculations on classical trajectories where interactions of interest receive a quantum treatment and the remainder of the trajectory is treated classically. In these methods, typically an objective function relating the difference between the gradients of a classical and quantum PES from AIMD simulation is minimized. This yields optimized parameters for the given functional form of the classical potential. An example objective function, χ^2, given by Li and Wang[235] has the form:

$$\chi^2 = \sum_i w_i \left(F_i^{\text{ref}} - F_i^{\text{fit}} \right)^2 + \sum_i w_i \left(\tau_1^{\text{ref}} - \tau_i^{\text{fit}} \right)^2 + \lambda \sum_\mu \left(\sum_v n_v q_{\mu v} \right)^2 \quad (5.18)$$

Here, the sum of the square of the difference in forces on atoms, F, and torques, τ, between the reference PES gradient and the fitted gradient are weighted by w_i and minimized. Additionally, a weight λ is applied to a constraint on the fitted charges, q. Parameter values can be taken from an average over all sites and snapshots. This approach is generalizable to any force-field analytical form.

The first fitting PES gradient method based on *ab initio* trajectories was suggested by Ercolessi and Adams[236], where they fit a "glue" potential for solid-phase aluminum. Parrinello and Voth et al. similarly developed three-site water models from Car–Parrinello molecular dynamics (CPMD) trajectories.[237] Their models were able to accurately capture the structural and self-diffusion properties in close quantitative agreement with both experimental measurements and *ab initio* trajectories. Koziol et al. similarly developed force matched classical water potentials for extreme pressure and temperature conditions.[238] Maple et al.[239] and Doemer et al.[240] have demonstrated the development of classical potentials from a quantum PES gradient for organic molecules. Wang and coworkers developed a methodology called "adaptive force matching," wherein classical potentials are fit iteratively through comparing the classical PES gradient from a classical trajectory to a QM/MM PES calculated for that same trajectory. Parameters of the classical potential are then optimized to the objective function. The fitted classical potential is then used to re-run the trajectory and is iteratively improved until it converges on the QM/MM PES gradient.[235,241–243] This method was applied by Huang and Tsai to demonstrate the differing dependence of first shell and second shell water on fitted potentials, including explicit polarization.[244] Insofar as a static classical force field can describe a molecular system, these methods are valuable for *ab initio* force-field development.

5.2.2.1.2   *Polarizable vs. Non-polarizable Force Fields*

MD simulations with non-polarizable potentials often incorporate induced polarization by means of applying atomic partial charges. This can be capable of quantitatively capturing single-ion properties, such as free energies of solvation. However, the non-polarizable models can only account for changing the electronic structure in response to different local environments in an averaged manner.[245] This fails to describe concentrated ionic solutions, where long-ranged electrostatic interactions have been demonstrated to be a significant contributor to the dynamic properties of actinyl ions. Spezia and coworkers recently developed polarizable and non-polarizable force fields for An(III) ions that captures both the RDF while giving different mean residence times for water. In addition to the dearth of experimental data, there was not a consistent trend in the water residence times between the polarizable and non-polarizable force fields.[246] It is also important to note that more complex solution compositions also require the capability of the force field to capture nonpolar, polar, and hydrogen-bonding environments in a consistent fashion, as is demonstrated by multipole expansion-based polarizable force fields,[247] like AMOEBA (Atomic Multipole Optimized Energetics for Biomolecular Applications).[248] Recognizing this fact, recent works have greatly expanded the suite of AMOEBA potentials for various ions with water, including not only monovalent ions, but also the trivalent lanthanide and actinide cations La, Eu, Gd, Ac, Am, and Cm.[249] Having a breadth of force fields that account for polarization relevant to different solution conditions and at different computational expenses are a continuing area of needed development.

5.2.2.1.3   Ab Initio *Implementations in Statistical Mechanics Methods*

Within both MD and MC, the energies and forces can be evaluated, not with analytical functions, but using the quantum-mechanical methods generally described in Section 5.2.1.1. Most applications utilize periodic boundary conditions, and therein density-functional theory is most often employed to calculate energy. As in classical dynamics, the use of a periodic boundary eliminates surface effects that may occur within a cluster model of a chemical system; however, due to the increased computational cost associated with the QM energy determination, the simulation size is generally on the order of hundreds of atoms, and as such only phenomena whose characteristic correlation length is fairly small can be studied.

The ensemble derived from a quantum-mechanical potential-energy surface enables the making and breaking of chemical bonds, and for other changes in electronic structure that are challenging using an analytical approximation to the energy. AIMD methods are typically of two major classes: Born–Oppenheimer (BOMD) and Car–Parinello (CPMD). In the former, the lowest wave function, energy, and forces are calculated at every timestep in the dynamics, the system is propagated upon a single electronic state (usually the ground state), and Gaussian basis sets are employed. Very small time steps are generally employed in BOMD, so that large variations in the wave function do not occur from one timestep to the next. However, this necessarily limits the total simulation (and sampling) time accessible. Thus, typically only 10's of ps of total time are generally produced during a typical BOMD production run. This is in contrast to the 100's of ns that are routinely available through the many different classical MD software programs.

The determination of accurate forces for the equations of motion is very computationally time-consuming, and this is a motivation for instead performing CPMD which allows the electronic degrees of freedom to evolve at the same time as the nuclear degrees of freedom. This method usually employs planewave basis sets and pseudo-potentials for metals. For metal-containing systems, basis-set and pseudo-potential development is an active area of research. For example, norm-conserving pseudo-potentials (for the planewave basis sets) can be challenging to study the late lanthanides because they require very large cutoff energies. Other challenges associated with CPMD include ensuring that the system propagates on the ground electronic state. Due to the way in which the electronic degrees of freedom evolve, the wave function can hop to an excited state, and the resulting dynamic properties will be entirely incorrect. Although there is a decrease in computational cost when using CPMD relative to BOMD, there are potential tradeoffs in accuracy, and in the end only modestly longer timescales are accessible relative to BOMD.

Many successful BOMD and CPMD studies have probed local configurational environments about solutes relevant to solvent extraction. Some notable examples include 1) understanding solvent exchange mechanisms about a solute (which is relevant to ligand complexation),[250–253] 2) using the velocity auto-correlation function to obtain highly accurate IR/Raman spectra that inherently account for solvation effects and anharmonicity,[254] and 3) using thermodynamic integration to determine pKa values.[255,256]

5.2.2.2 Obtaining an Equilibrium Ensemble

For a given classical force field, either MC or MD simulations can be used to obtain an equilibrium ensemble of the system. In MC simulation, configurations are generated proportional to their Boltzmann weight through trial moves. After sufficient sampling, the ensemble of configurations for a given system is approximated. Alternatively, in MD simulation, a configuration is propagated in time according to Newton's equations of motion with forces on atomic sites determined from the potential-energy surface given by the force field. As a result of the ergodic hypothesis, after a sufficiently long simulation, the time average of the observed configurations is equivalent to the ensemble average. MC simulation methods that employ the Gibbs ensemble are capable of predicting phase-composition equilibria, which can be of interest for liquid–liquid extraction systems.[257] Alternatively, for a known phase composition, MD has the benefit of providing dynamic information. Prior to generating an equilibrium ensemble of configurations, it is necessary to verify a given system is in fact at equilibrium. This often involves verifying convergence-of-state variables such as total energy or, for ensembles without fixed volume, density. Awareness of possible kinetic barriers is essential to determining whether sufficient configurations are generated to adequately sample the equilibrium configurational ensemble with statistical accuracy. In MD simulations of solvent extraction applications, for example, the lifetimes of metal–ligand solvate dissociation may be of the same order or larger than accessible simulation times, preventing the system from readily sampling the equilibrium ensemble of metal–ligand complexes. The applicability of MD potential-energy descriptions depends on their computational cost relative to the equilibration and sampling simulation time required by a given system. The order of computational efficiency of potential-energy descriptions is classical additive potentials > classical non-additive potentials for explicit polarization >> fully quantum-mechanical energies/forces. This corresponds to accessible simulation times of 100–200 ns,[258,259] up to 10 ns,[260,261] and 10s of ps (with substantially smaller simulation sizes),[262] respectively.

Before choosing a simulation approach, the correlation lengths and sampling times of the properties of interest in the chosen system must be considered. In the aqueous phase, water exchange rates about highly charged ions is quite slow. The time until a water molecule reaches a 50% probability of leaving the first coordination shell of the aqueous U(VI) dioxocation was computed to be on the order of 10s of ns.[261,263] This indicates that, for potential-energy descriptions more computationally intensive than additive classical potentials, studying water exchange in statistically significant quantities requires biased sampling methods. In the organic phase, there is no doubt that the correlation lengths and equilibration and sampling times exceed those accessible to AIMD. If a full ternary surfactant–water–oil system is to be studied with MD, the equilibration from a mixed or partially mixed initial configuration (surfactant dispersed in one phase) requires up to 100 ns,[259] precluding even the application of non-additive potentials. Care must be taken to apply potentials which facilitate slow equilibration processes balanced against choosing accurate potentials which capture intermolecular interactions. One approach that is increasingly seen in the literature is to use equilibrated classical MD trajectories as an initial configuration when conducted AIMD on relaxed liquid phases, thus accelerating the equilibration of the quantum potential.[254] While this circumvents long equilibration times, the statistical sampling time is still limited by the nature of the interaction potential.

5.2.2.3 Subensemble Analysis

Subensemble analysis consists of approaches that allow the assignment of contributions from specific chemical species of interest within a total ensemble of species to a macroscopically measured quantity. Though typical spatial or temporal correlation functions, like radial distribution functions, are incredibly useful for understanding average structural organization, they "hide" valuable information about structure or correlations which may contribute significantly and differently to the macroscopic property of interest. This approach could relate the spectroscopic signatures of specific chemical species to macroscopically obtained spectra (via QM methods),[264] or break down the unique contributions from microscopically heterogeneous complex solutions or liquid interfaces to averaged properties. Once species are effectively classified by their contributions to a measurable quantity, and the populations of those species are obtained, the total signal of a measurable quantity can be constructed. This obviates having to measure this quantity for all configurations while also providing insight into how the different classes of species or structures contribute to that quantity.

Central to subensemble analysis is effectively collapsing a vast configurational space to a tractable number of states. Doing so can require combining metrics, such as the Cartesian position and intermolecular interaction topology, to describe the high dimensional configurational space inherent to many solvent extraction-related structures. Then, subensemble analysis links those states to macroscopically measurable quantities, reducing the computational effort required to connect simulations to experiment. Those observable properties are often sensitive to complex correlations: orientation at a fluctuating liquid–liquid interface or the relationship between solvents in the same solvation environment of a metal ion. In the first example, coupling a molecule's presence in the interfacial layer of water with its position relative to the time-averaged interface provides insight into its orientation and contributions to measurements of the net molecular dipole of an interface. For the second example, the metal solvent distance of a single solvent may be coupled to the total solvation environment of the metal ion. Treating the metal and solvating molecules with graph theory, where molecules are defined as nodes and direct interactions as edges, the complex solvation can be described as a graph topology. The methods for reducing structural dimensionality are numerous, and the resulting division of simulation ensembles into experimentally distinct subensembles can complement interpretation of experimental data.

Subensemble sampling has been applied to solvent extraction-related systems to assist with the interpretation of numerous spectroscopic techniques. In the context of aqueous chemistry, Graham et al. used subensemble sampling to investigate the effect of sodium ions on aluminate structure measured by NMR chemical shifts relevant to gibbsite crystallization of solvent extraction legacy waste processing.[265] While also investigating aluminate and sodium hydroxide solutions, Wildman et al. compute XANES spectra with TDDFT to identify a pre-edge shoulder corresponding to aluminate-sodium ion–ion interaction.[266] Palmer et al. implemented a methodology wherein they compiled configurations from classical MD to compute average EXAFS spectra, testing the approach on strontium chloride solutions.[267] Perez-Conesa et al. compared calculated EXAFS spectra from classical MD configurations of UO_2^{2+} and solvating water to experimental data. Decomposing the spectral contributions from different scattering paths, they attributed differences in experimental and computed spectra to single scattering path superposition.[268] Later, they extended this approach to aqueous americium ions to demonstrate a mixed composition of $Am^{3+} + AmO_2^{2+}$ ions reproduces experimentally measured EXAFS spectra.[269]

Near interfaces, subensemble approaches have been applied to vibrational sum-frequency generation (VSFG) of aqueous interfaces applicable to solvent extraction. Nagata et al. and Walker et al. used classical MD simulation to decompose an experimentally determined VSFG spectrum of a water–vapor interface, assigning peaks to groups of interfacial water with different orientations relative to the interface and different numbers of hydrogen-bond donations.[264,270] For ions near the interface, trends in interfacial behavior have been identified based on certain ion properties, such as polarizability and hydration enthalpy, which can be correlated to the empirical Hofmeister series.[271] One such trend is that weaker hydration leads to stronger adsorption at charged aqueous interfaces. Recent work that combined VSFG and MD revealed that heavy anionic chlorometalate complexes are both strongly hydrated and strongly adsorbed to a charged Langmuir monolayer. After adsorption, these heavy anions retain part of their hydration sphere and induce a unique interfacial water structure. These results may have important implications on the relation of interfacial water structure and hydration enthalpy to the general understanding of specific ion effects.[271]

5.2.2.4   Sampling to Obtain Free-Energy Differences

One aspect that makes classical molecular simulation attractive is the ability to compute free-energy changes between thermodynamic states. In theory, the free-energy difference between two states A and B can be computed from the Zwanzig relationship:[272]

$$\Delta G_{A \to B} = -k_B T \left\langle e^{-\frac{\Delta U}{k_B T}} \right\rangle \tag{5.19}$$

However, the two states of interest are often connected by a free-energy barrier. Since classical simulations can only achieve a finite amount of sampling, without proper care, thermodynamic averages will be inaccurate and highly dependent upon initial conditions. In addition, many important chemical properties have an exponential dependence on the free energy. Small errors in the free-energy calculation can, therefore, have a large effect on the accuracy of the computed properties. Special techniques have been developed to overcome this limitation. One widely used technique is the Multistate Bennett Acceptance Ratio (MBAR) method.[273] MBAR allows for a free-energy difference to be computed by splitting the process into various simulation windows, run in parallel. The algorithm uses energy information from all of the windows to calculate the overall free-energy change. This provides a vastly improved estimate of the free energy.

Free-energy calculations are often used to calculate partition coefficients, which characterize the thermodynamic preference of a solute molecule for one solvent over another. The octanol–water partition coefficient is of particular interest—it is calculated as a measure of the solute's hydrophobicity, which is important to characterize due to its correlation with a molecule's bioaccumulation potential.[274] In addition, experimental measurements of octanol–water partition coefficients are plentiful and thus serve as a good test of force-field accuracy. Mobley and coworkers recently calculated the partition coefficients of small organic solutes in octanol–water and cyclohexane–water environments using MBAR.[275] With the widely used GAFF, they achieved an acceptable agreement with experiment. Such techniques would be quite useful in studying actinide extraction processes but have yet to be implemented for these systems.

In addition to partition coefficients, the solubility of a solute in a pure solvent is another property of interest. In theory, one could simulate a pure crystalline solute in direct coexistence with a pure solvent and wait to reach equilibrium. However, such simulations are plagued by long timescales of dissociation. Vega and coworkers[276] performed a 500 ns simulation of NaCl in H_2O and failed to witness a single-ion dissolution event. Joung and Cheatham[277] approached the problem somewhat differently; they started with a supersaturated ionic solution in contact with the crystal.

They were able to see re-precipitation and determine the solubility, but extremely long (several µs) simulations were required. Therefore, the direct simulation of solubility in this manner is challenging and computationally expensive, and a cheaper method would be highly desirable.

In theory, the solubility limit of a crystalline species can be calculated by determining the absolute chemical potential of the species in the crystalline phase and then finding the concentration dependence of the chemical potential for the species in solution. The solubility limit is the concentration where the solute chemical potential is the same in the two phases. To calculate the absolute chemical potential of the solute in the crystalline or solution phases, a reference state is used, whose chemical potential is analytically known. The Einstein crystal is a common reference state for the solid phase, while the ideal gas, Lennard–Jones, and hard-sphere models are used as solution-phase reference states. However, the use of a reference state becomes a challenge when complex molecules with many intramolecular degrees of freedom are involved.

Maginn and coworkers[278] developed the pseudo-supercritical path integration (PSPI) method, which avoids the need to use absolute reference states altogether. Rather than computing differences in free energy between two different reference states, the PSPI method determines the free-energy difference between a common reference state and the species in the crystalline and solution phases. A weakly interacting fluid phase is used, where the strength of interactions between the molecules is gradually reduced in such a way that the free energy is continuous along the path. Singularities in the Born-Mayer-Huggins-Tosi-Fumi (BMHTF) potential required an analytical reference state for the crystalline phase, but no reference state is required if continuous potentials are used. The method is illustrated in Figure 5.5. By computing $\Delta\mu_1^\alpha, \Delta\mu_1^\beta$, and $\Delta\mu_1^\gamma$, one can compute the difference in chemical potential difference $\mu_1^c - \mu_1^s$. This method was used to study the solubility limit of NaCl in water and light alcohols. This method has only been used for simple systems like NaCl but is applicable to actinides.

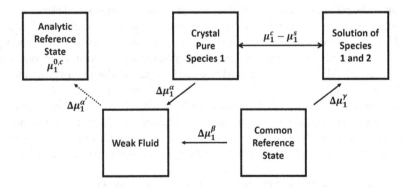

FIGURE 5.5 Schematic showing the steps of the PSPI method. (Adapted from Paluch et al. (2010).)

5.2.2.5   Rare-Event Sampling for Reaction-Pathway Energetics

Rare-event sampling methods are implemented in molecular simulation to investigate states which are inaccessible through statistical sampling of feasible simulation time and length scales. One such method which is often applied to study of solvent extraction relevant systems is the potential of mean force. The potential of mean force (PMF) technique can be used to determine the free-energy surface along a coordinate. For studies on solvent extraction systems, a common coordinate is the transfer of a metal ion, extractant or extractant–metal complex across the aqueous–organic interface. While that transfer event may be sufficiently energetically unfavorable or inhibited by kinetic barriers so as to elude the nanometer and nanosecond length and time scales of simulation methods, the free-energy profile of that event can be studied with a PMF. For a discussion on constructing a potential of mean force and the techniques used to do so, such as umbrella sampling[279], see ref 280.

As an example of a rare-event commonplace in solvent extraction, the transport of ions, extractants and their complexes across immiscible aqueous–organic interfaces is often studied with PMFs to understand the free-energy pathways of ion disengagement from the aqueous surface. That process is essential to describing the dynamics of diffusion from the interface. Schweighofer and Benjamin[281,282] identified a phase-transfer free-energy barrier for an ion traversing the water–nonpolar phase boundary starting in either phase. The barrier resulted from the hydrogen bonding of the interfacial water, either in breaking hydrogen bonds with water when transferring from aqueous to organic or in breaking water–water hydrogen bonds to form hydrogen bonds when the ion approaches from the organic phase. Dang[283] later demonstrated that for an equilibrium transfer process of an ion across a water–dichloromethane interface, there is no activation barrier present for transfer in either direction, and the previously reported barrier resulted from not sampling equilibrated trajectories along the reaction coordinate. More complicated processes necessitate the correct selection of the reaction coordinate of relevance, which may not be obvious *a priori*. Further, many reaction mechanisms may facilitate transport, and thus it remains a challenge to develop techniques capable of identifying and sampling all relevant reaction pathways.

Section Summary. The statistical-mechanics methods described here have been demonstrated to be powerful tools for elucidating the structure, dynamics, and physical properties of complex fluids, and the speciation of solutes dissolved therein. Subensemble sampling is demonstrated to be essential to understand the unique spectroscopic signatures of chemical species and for identifying correlations of structure and reactivity within solutions. Methods to understand free-energy changes that do or do not include description of the barriers are also essential to understand the driving forces behind transport in solvent extraction. With these tools in hand, we now consider practical applications of both quantum- and statistical-mechanics techniques to learn about the fundamental interactions of metals with extracting ligands, and the three major chemical environments of interest: the bulk aqueous and organic phases and the interfacial region.

5.3   SOLVENT EXTRACTION—A COMPUTATIONAL CHEMIST'S POINT OF VIEW

Liquid–liquid extraction of a target solute from an initial phase to a final phase typically consists of changing the solubility of that solute from favoring the initial phase to favoring the final phase. This often occurs through the formation of solvates consisting of the target solute and a molecule soluble in the final phase, typically referred to as an extractant, such that the solute–extractant complex favors the final phase. The difference in solvation free energy of the complex between phases is the driving force of the extraction process. However, the initial speciation of the reacting solutes must be known in order to determine the metal complexation energetics, as well as the energetically viable distribution of metal–ligand products (which may or may not be monomeric). At the same time, mechanistic information from the reactions at the interface are essential to impart insight regarding the kinetics of extraction. Within this section, we discuss important lessons learned from modeling and simulation in each of these domains, alongside motivation for further research where there still exist substantial gaps in understanding.

5.3.1 THE AQUEOUS SOLUTION

Before considering extracted organic-phase metal-containing species, it is important to first understand the molecular speciation of solutes in the aqueous solution. We focus here upon metal ion solutes, considering the solvation environment, then in complex solutions with acid anion ligands that may coordinate to the metal. This, in turn, represents the metal species that may react with the extracting ligand.

Much work has been done on methods of accurately calculating solution thermochemistry. As mentioned in Section 2.1.2 and 2.1.4, explicit clusters of solvated ions embedded in continuum methods, properly benchmarked, can reliably predict the most favorable number of coordinating solvent molecules. Further, either MC or MD simulations can predict the ensemble distribution of energetically similar solvation configurations, accounting for more complex aqueous conditions that include the presence of complexing anions, the formation of extended ion–ion interactions, and the presence of mixed solvents. It is important to recognize that competitive forces exist in multicomponent electrolyte solutions which consist of both ion–solvent and solvent–solvent interactions. These can influence a myriad of processes, including ligand complexation. As desolvation is often the initial step in metal complexation, even small changes in solvation structure can potentially perturb the mechanism, thermodynamics, and kinetics of complexation.[284–287] One excellent example comes from recent work (using classical MD, AIMD, and MP2 cluster calculations) that studied the changes to both the ion–solvent and solvent–solvent interactions in mixed composition water–alcohol solutions.[288–290] There, the changes to the hydrogen-bond dynamics of the solution were demonstrated to kinetically restrict solvent exchange about Cm^{3+}, despite the fact that the presence of the MeOH in the first solvation shell significantly decreased the ion–solvent binding energy for both Cm^{3+}–H_2O and Cm^{3+}–MeOH.

While the solution-phase conditions are essential to understand speciation, so too is the potential aqueous phase reactivity. This is a much more challenging task computationally, particularly for metals that are prone to hydrolysis, which includes the actinide elements. The first step in any actinide hydrolysis process is the loss of a proton from the inner solvent shell as shown in reaction (r2) with $h = 1$.

$$M\left(OH_2\right)_m^{n+} \rightarrow M\left(OH_2\right)_{m-h}\left(OH\right)_h^{(n-h)+} + hH^+ \tag{r2}$$

Hybrid supermolecule-continuum approaches for the prediction of free energies of solvation and pK_a's in aqueous solution have been used to reliably establish the free energies of solvation of the proton and hydroxide anion, for example, as well as provided good estimates of the pK_a's of very strong acids and of actinide-containing acids.[197–204,291] To obtain the solution-phase acidity, the energy of the various species can be calculated by using a SCRF to model the free energy in aqueous solution. The acidity can be directly calculated from reaction (r2) using the appropriate estimate for the free energy of solvation of the proton.[197] The pK_a values can then often be improved by using a known standard, and we can calculate the energy of reaction (r3)

$$M\left(H_2O\right)_n^{m+} + M'\left(H_2O\right)_{n-1}\left(OH\right)^{(m-1)+} \rightarrow M\left(H_2O\right)_{n-1}\left(OH\right)^{(m-1)+} + M'\left(H_2O\right)_n^{m+} \quad (r3)$$

where the acidity of $M'(H_2O)_n{}^{m+}$ is known, to obtain the relative pK_a. This can usually provide pK_a's within 1 to 3 pK_a units for hydrolysis of various species including Group 2 and transition metal ions in the +II oxidation state.[204] As further protons are removed, there is also likely to be a loss of H_2O molecules as has previously shown happens $M^{3+}(H_2O)_6$ complexes.[292,293]

We provide an example for the prediction of the pK_a of uranium in water in three different oxidation states.[159] As described above, it is necessary to establish the number of H_2O molecules in the first solvation shell and, as discussed above, this is not a simple task even for $UO_2{}^{2+}$.[291] The pK_a for U(IV), U(V)$O_2{}^+$, and U(VI)$O_2{}^{2+}$ in aqueous solution have been predicted using density-functional theory at the B3LYP level with different numbers of water molecules as shown in Table 5.1. The experimental pK_a's are 0.54 ± 0.06 for U(IV) and 5.25 ± 0.24 for U(VI)$_2{}^{2+}$. Bylaska and coworkers have used AIMD to predict a value of 8.5 for U(V)$O_2{}^+$.[250] The values for a single solvation shell for U(IV) are not in good agreement with the experimental value. We illustrate this for 8 and 9 H_2O molecules in the first solvation shell as shown in Table 5.1. With the use of a single solvation shell (reactions (r4) and (r5)), the calculated values are far from experiment, and there is a significant difference in the values for 8 or 9 H_2O molecules. One can improve the agreement with a single solvation shell if one includes water hydrolysis as shown in reaction (r6) for 8 H_2O molecules, but this value is still far from experiment. There is a dramatic improvement when one includes beyond a single solvation shell with results for 8 or 9 H_2O molecules now in reasonable agreement with experimental results for U(IV) as shown in reactions (r7) and (r8) as well as with each other. As there are 8 or 9 water molecules in the first solvation shell that are not exactly the same when there are multiple solvent shells present, a proton was removed from each of the 8 or 9 water molecules in the first solvation shell, which leads to different pK_a's ranging from −4.0 to −0.2. for 8 and −3.6 to −1.5 for 9 (see above for a discussion of sampling issues). Not all of the structures generated after removing a proton had converged wavefunctions which is why there are less than eight or nine pK_a values for reactions (r7) and (r8). The average values of −2.0 (r8) and −2.7 for (r8). Both are within ~3 pK_a units of experiment and predict values that are too acidic. This value could be improved by performing calculations with correlated molecular-orbital-theory levels, but the results do show reasonable agreement with experiment given the computational cost. For $UO_2{}^+$, the average pK_a value of 11.4 with 5 H_2O molecules in the first solvation shell and more than one solvation shell is again within 3 pK_a units of the Bylaska value. For $UO_2{}^{2+}$,

TABLE 5.1

Acidity Free Energies in the Gas Phase and Aqueous Solution in kcal/mol at 298 K and pK$_a$ values for U in Different Oxidation States with Average Values in Parentheses[a]

Reaction	Reaction	ΔG_{gas}	pK$_a$
$U(H_2O)_8^{4+} \rightarrow U(OH)(H_2O)_7^{3+} + H^+$	(r4)	−62.1	−22.0
$U(H_2O)_9^{4+} \rightarrow U(OH)(H_2O)_8^{3+} + H^+$	(r5)	−45.4	−13.0
$H_2O + U(H_2O)_8^{4+} \rightarrow U(OH)(H_2O)_8^{3+} + H^+$	(r6)	−61.8	−8.8
$U(H_2O)_8(H_2O)_{22}^{4+} \rightarrow U(OH)(H_2O)_7(H_2O)_{22}^{3+} + H^+$	(r7)	71.4, 71.8, 72.9, 74.3, 74.4, 74.4, 79.9 (74.2)	−0.2, −0.7, −0.7, −2.5, −2.8, −3.4, −4.0 (−2.0)
$U(H_2O)_9(H_2O)_{20}^{+4} \rightarrow U(OH)(H_2O)_8(H_2O)_{19}^{3+} + H^+$	(r8)	74.3, 76.4, 78.8, 79.0, 79.5 (77.6)	−1.5, −1.8, −3.1, −3.5, −3.6, (−2.7)
$UO_2(H_2O)_5(H_2O)_{25}^+ \rightarrow UO_2(OH)(H_2O)_4(H_2O)_{25}^+ + H^+$	(r9)	254.8, 256.5, 267.9 (259.7)	9.3, 10.6, 14.3, (11.4)
$UO_2(H_2O)_5(H_2O)_{25}^{2+} \rightarrow UO_2(OH)(H_2O)_4(H_2O)_{25}^+ + H^+$	(r10)	195.3, 199.8, 202.9, 204.8, (200.7)	3.9, 4.8, 8.1, 9.0, (6.5)
$UO_2(H_2O)_4(H_2O)_{26}^{2+} \rightarrow UO_2(OH)(H_2O)_3(H_2O)_{26}^+ + H^+$	(r11)	197.6, 198.4, 200.3 (198.7)	2.9, 4.6, 4.8 (4.1)

Source: M. Vasiliu, D. Xia, and D. A. Dixon, to be published, 2019.

[a] U(IV) is a triplet, U(V)O$_2$ is a doublet and U(VI)O$_2$ is a singlet.

the average value with 5 H$_2$O molecules in the first solvation shell is about 1.2 pK$_a$ units too large (not acidic enough) compared to experiment whereas for 4 H$_2$O molecules in the first solvation shell (about 1.2 pK$_a$ units too acidic).

5.3.2 METAL–LIGAND COMPLEXATION

5.3.2.1 Extractant Design—Electronic-Structure Considerations

Assuming that the aforementioned methods accurately determine the correct reacting species, the energetics of ligands that are either available to bind in the aqueous phase, or ligands that serve as the extractant, can be considered. Coordinating ligands in aqueous solution derive from multiple sources, including the acid anions used to set solution pH. Such ligands are generally not selective for a specific metal ion but do alter the molecular speciation that reacts with an extracting ligand. Extractant ligand design is focused upon both selectivity and efficiency (e.g., transport into the organic phase).

Selectivity can be based upon multiple electronic and structural elements of the ligand. In the former, 5*f* covalency has been implied to be prominent for heavy actinides, which is driven by orbital energy degeneracy between the metal and ligand. Specifically, DFT calculations have indicated that the modest increase in measured actinide–dipicolinate stability constants, which is well reproduced theoretically, is coincident with a significant increase in An 5*f* energy degeneracy with the dipicolinate molecular orbitals for Bk and Cf relative to Am and Cm (see Figure 5.6).[296] Although the interactions in the actinide–dipicolinate complex are largely ionic, the decrease in 5*f* orbital energy across the series manifests in orbital mixing and hence covalency driven by energy degeneracy. The same 5*f* covalency driven by energy degeneracy explained that the complexation of both An(III) and An(IV) ions with 3,4,-LI(1,2-HOPO) ligand generally becomes more favorable for heavier actinides.[297] Besides, the ability of this HOPO ligand to either accept or donate electron density as needed from its pyridine rings is found to be key to its extraordinary stability across the actinide series.

Although the late actinides from Am to Es imply a certain 5*f* covalency, the actinide–ligand bonding is primarily of ionic character, which is similar to lanthanide–ligand bonding. This is reflected in the recent theoretical and experimental results of bond lengths of M–X bonds (M = Am–Es, Nd–Gd, and X = O, N, S).[298,299] The trend of M–X bond distances generally follows that of the ionic radii of the metal ions. However, it is found that whether water coordinates to the metal or not can appreciably influence the M–O bond distance, which may perturb this trend. Additionally, theoretically predicted redox potentials indicate that although +3 is the most common for late actinides, while +4 may be accessible for Bk and may be stabilized by a strongly binding ligand like HOPO.[297] Compared with late actinides, the early actinides show complicated behavior as it pertains to their oxidation state due because the 5*f* orbitals are less contracted. For example, theoretical calculation and experimental crystal structures reveal that both +3 and +4 oxidation states of the plutonium ion can coexist in one diplutonium complex.[300]

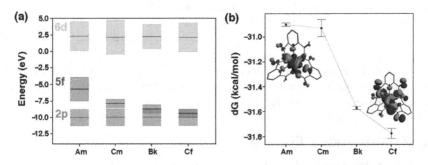

FIGURE 5.6   (a) Average energy levels of molecular orbitals composed of An 6d (red), An 5*f* (blue), and oxygen 2*p* (gray) orbitals, weighted by MO composition. (b) Experimental Δ*G* for the formation of the 1:3 An–dipicolinate complex. Note: Energies of the 5*f* orbitals decrease across the series, becoming degenerate with the ligand orbitals at Bk in (a). This corresponds to a change in 5*f* molecular orbitals from highly localized (Am, MO image on the left) to delocalized (Cf, MO image on the right) in (b). (Reprinted with permission from Kelley et al. (2017). Copyright © 2017, American Chemical Society.)

5.3.2.2 Calculation of Stability Constants—Ligand-Binding pK Values

In considering selectivity further, two aspects can be relevant: the absolute stability constant for the formation of a specific ML complex and the relative binding energy differences between different metals or ligands for a single metal. In the former, the calculation of absolute ligand-binding pK values, there are several examples where theory has contributed significantly. As part of an effort to design ligands for the extraction of UO_2^{2+} from seawater,[301–305] the aqueous stability constants for UO_2^{2+} complexes with negative oxygen–donor ligands were calculated using an approach similar to the one described above using DFT/B3LYP and an SCRF using the integral equation formalism polarizable-continuum model (IEF-PCM).[306] Additional functionals and the solvation model based on density (SMD) approach were also tested. The goal is to calculate the energy of reaction (r12)

$$UO_2\left(H_2O\right)_5^{2+} + L^{z-} \rightarrow UO_{2L}\left(H_2O\right)_{5-m}\left(OH\right)^{2-z} + m\left(H_2O\right) \qquad \text{(r12)}$$

where an additional correction factor of 1.83 kcal/mol at 298 K is needed for the ligand for the conversion from 1 atm to 1 M. As expected on the basis of the pK_a results for a single solvent shell, the actual values are not in very good agreement with experiment as shown in Table 5.2. Clearly, there is very poor agreement between the experimental and computational binding energies. The computed log K (r12) values are far too large as compared to experiment. The errors tend to be larger for the dianions than for the monoanions. This is due, in part, to the fact that the anion solvation energies with no solvent molecules present are not large enough, and formation of the ion pair will over-stabilize the system. However, when the calculated log K (r12) values are plotted against the experimental values, two linear fits are found as shown in Figure 5.7 with the anions and dianions falling on different linear fits. These linear correlations can then be used to substantially improve the predicted log K (r12) values and can be used to predict the pK (r12) of new ligands that are not in the training set shown in Table 5.2. This suggests that such correlations could be developed from limited experimental data and then used with computational results to predict pK (r12) values to within about 2–3 pK units for unknown ligands. Of course, it will be necessary to have computational results for the ligand binding, but these are much cheaper to obtain than the experiments and fits can be generated for each type of cation and a set of anions for separation system design. This approach has been expanded recently[307] to develop a predictive modeling approach for the binding of uranium to amidoxime ligands grafted onto fibers to extract uranium from seawater. By combining the computational ligand-binding model with experimental measurements and process adsorption modeling, it was shown that it should be possible to design ligands to separate uranium from vanadium in seawater using certain types of fibers as the support.

5.3.2.3 Ligand Selectivity—Leveraging Error Cancellation to Determine Accurate Free-Energy Changes

Extractant ligand selectivity in liquid–liquid extraction is closely related to the distribution ratios of metal ions. For a specific metal ion, its distribution ratio (D) is

TABLE 5.2

Comparison of log K (r12) Values for Binding Anion Ligands to UO_2^{2+} in Aqueous Solution.[a] Experimental Data From Kim et al., 2013.

Ligand	Expt'l	Calc'd	Fit
	3.1[b]	20.6	4.3
	7.3[b]	30.8	6.5
	7.7[b]	30.1	7.3
	5.6[b]	28.5	5.4
	13.0[b]	42.6	12.0
	16.8[b]	51.1	16.1
OH^-	8.8[c]	34.7	8.7
NO_3^-	0.0[c]	7.5	0.3
CO_3^{2-}	9.7[c]	41.5	11.6
SO_4^{2-}	3.0[c]	23.1	2.8
HPO_4^{2-}	7.2[c]	34.5	8.2
$H_2PO_4^-$	3.3[b]	15.9	2.9
ClO_4^-	0.5[d]	6.5	0.0
	11.1[e]		10.0
	19.1[f]		20.0
	16.8[g]		14.9

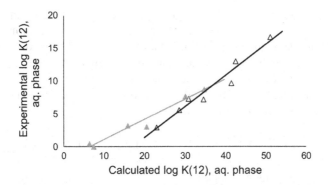

FIGURE 5.7 Plot of experimental log K (r12) vs. experimental log K (r12) in aqueous solution. The red points are for monoanions, and the black points are for dianions. Note that the two linear fits are different. (Data from Kim et al. (2013).)

the ratio at equilibrium of the concentration of this metal ion in the organic to that in the aqueous phase, which measures the relative affinity of this metal ion toward the ligand in the organic phase compared to its environment in the aqueous phase. The selectivity of a ligand (also called separation factor SF) between two metal ions is defined as the ratio of the distribution ratios of these two metal ions: $SF_{Am/Eu} = D_{Am}/D_{Eu}$. In principle, the SF and therefore efficacy of a ligand can be calculated using methods based on first-principles. Specifically, for the example of extracting an M^{3+} ion by a neutral solvating extractant from aqueous solutions containing nitric acid to form a neutral metal complex in the organic phase, the corresponding free energy of extraction, ΔG_{ext}, can be calculated using the following reaction pathway (r13):

$$M^{3+}_{(aq)} + 3NO^-_{3(aq)} + nL_{(org)} \xrightarrow{\Delta G_{ext}} M(NO_3)_3(L)_{n(org)} \qquad (r13)$$

In a rough approximation (neglecting the activity coefficients), the distribution ratio of M^{3+}, D_M, can be derived from the ΔG_{ext}:

$$\ln(D_M) = n\ln[L] + 3\ln[NO_3^-] - \Delta G_{ext}/RT \qquad (5.20)$$

Alternatively, a metal ion in the aqueous phase can be extracted by an acid (HA) present in the organic phase via a cation-exchange mechanism given in reaction (r14):

$$M^{3+}_{(aq)} + 3HA_{(org)} \xrightarrow{\Delta G_{ext}} M(A)_{3(org)} + 3H^+_{(aq)} \qquad (r14)$$

Similarly, the D_M can be approximately obtained from the ΔG_{ext}:

$$\ln(D_M) = 3\ln[HA] - 3\ln[H^+] - \Delta G_{ext}/RT. \qquad (5.21)$$

The ΔG_{ext} can be calculated using a Born–Haber thermodynamic cycle. However, it is very challenging to accurately calculate the solvation free energies of each species.

For example, using different solvation models to account for the solvent effects, the solvation free energy of M^{3+} (M = Am, Eu) can differ by 17–165 kcal/mol.[308] Using the assumption that lanthanides and actinides form the same type of complexes in the same environment because the metal ions have similar ionic radii and coordination properties, one can instead consider the difference of extraction free energies of two metal ions ($\Delta\Delta G_{ext}$). The resulting reaction could result in a more accurate calculated value due to error cancellation:

$$An^{3+}_{(aq)} + Ln(NO_3)_3(L)_{n(org)} \xrightarrow{\Delta\Delta G_{ext}} Ln^{3+}_{(aq)} + An(NO_3)_3(L)_{n(org)} \qquad (r15)$$

corresponding to solvation extraction and

$$An^{3+}_{(aq)} + Ln(A)_{3(org)} \xrightarrow{\Delta\Delta G_{ext}} Ln^{3+}_{(aq)} + An(A)_{3(org)} \qquad (r16)$$

corresponding to a cation-exchange extraction. For both reactions (r15) and (r16), $\Delta\Delta G_{ext} = \Delta G_{ext}(An) - \Delta G_{ext}(Ln)$. Furthermore, the separation factor SF between An^{3+} and Ln^{3+} for a particular ligand can be derived from $\Delta\Delta G_{ext}$:

$$SF_{An/Ln} = D_{An}/D_{Ln} = \exp(-\Delta\Delta G_{ext}/RT) \qquad (5.22)$$

Therefore, $\Delta\Delta G_{ext}$ can directly correspond to the selectivity of An^{3+} over the Ln^{3+} for a particular ligand. Many recent reports use this method when comparing the selectivity of one ligand over multiple metal ions. The ligands being studied include diethylenetriamine-pentaacetic acid (DTPA),[309] diglycolaminde ligands,[310,311] sulfur–, nitrogen–, and oxygen–donor ligands,[312] heterocyclic N-donor ligands,[313] phenanthroline derived-ligands,[314,315] and mixed N,O-ligands,[316,317] and isostructural diamide and dithioamide ligands.[318]

As an extension of this concept, to predict the relative selectivity of multiple extractants, an even better cancellation of errors can be achieved by considering the competition reaction among all metal–ligand complexes, as proposed in an Am(III) and Eu(III) separation study, by calculating $\Delta\Delta\Delta G_{ext}$.[308] Taking solvating extraction of An^{3+} and Ln^{3+} by two neutral ligands L^a and L^b as an example to illustrate in Figure 5.8 (and similarly extended for cation-exchange extractants), this method cancels out the potential large uncertainties introduced by the solvation effects for highly charged species, such as the trivalent metal ions. Besides, the $\Delta\Delta\Delta G_{ext}$ can be directly related to the difference of experimental $SF_{Am/Eu}$ values between two ligands:

$$\Delta\Delta\Delta G_{ext} = \Delta\Delta G_{ext}(L^b) - \Delta\Delta G_{ext}(L^a) = -RT\ln(SF_{L^b}/SF_{L^a}). \qquad (5.23)$$

Therefore, $\Delta\Delta\Delta G_{ext}$ corresponds to the difference in the selectivity of An^{3+} over Ln^{3+} between two ligands. This method provides a convenient way to differentiate between the selectivities of different extractants and pave the way for the high-throughput ligand design. Recent publications applied this method to study the separation of Am(III) and Eu(III) using a series of N,O-donor ligands[316] or mixed N,O-donor and N,S-donor ligands.[310,318]

$$An^{3+}_{(aq)} + 3NO_3^-{}_{(aq)} + nL^b{}_{(org)} \xrightarrow{\Delta G_{ext}} An(NO_3)_3(L^b)_{n\ (org)}$$

$$-\left(An^{3+}_{(aq)} + 3NO_3^-{}_{(aq)} + nL^a{}_{(org)} \xrightarrow{\Delta G_{ext}} An(NO_3)_3(L^a)_{n\ (org)} \right)$$

$$+\left(Ln^{3+}_{(aq)} + 3NO_3^-{}_{(aq)} + nL^a{}_{(org)} \xrightarrow{\Delta G_{ext}} Ln(NO_3)_3(L^a)_{n\ (org)} \right)$$

$$-\left(Ln^{3+}_{(aq)} + 3NO_3^-{}_{(aq)} + nL^b{}_{(org)} \xrightarrow{\Delta G_{ext}} Ln(NO_3)_3(L^b)_{n\ (org)} \right)$$

$$An(NO_3)_3(L^a)_{n\ (org)} + Ln(NO_3)_3(L^b)_{n\ (org)} \xrightarrow{\Delta\Delta\Delta G_{ext}} Ln(NO_3)_3(L^a)_{n\ (org)} + An(NO_3)_3(L^b)_{n\ (org)}$$

FIGURE 5.8 Competition reactions between two ligands and two metal ions. $\Delta\Delta\Delta G_{ext}$ corresponds to the selectivity difference between two ligands and two metal ions.

Although we have successfully constructed the appropriate thermodynamic cycles to maximize error cancellation, the challenge still remains to accurately reproduce and predict the experimental *SF* values for actinide and lanthanide separation. This is important for two reasons. The first is that for typical N-donor ligands (except some BTBP type ligands) and mixed N,O-donor ligands, such as 2.2:6,2-terpyridine (TERPY), 2,4,6-tri-(2-pyridyl)-1,3,5-triazine (TPTZ), *N,N,N,N*-tetrakis[(2-pyridyl)methyl]-ethylenediamine (TPEN), pyridine-2,6-dicarboxamides, or 1,10-phenanthroline dipicolinamides,[319–322] the $SF_{Am/Eu}$ values are 5–100 at 298 K, giving the experimental ΔG_{ext} in a narrow range of 0.95–2.73 kcal/mol. This reaches the boundary of the accuracy of DFT methods, which is typically larger than 1–2 kcal/mol. The second challenging aspect is the accurate treatment of the electronic structure of the actinide and lanthanide complex itself. Due to the energetic near-degeneracy of the $4f/5f$ orbitals, the ground state is of multi-reference character. Additionally, there exist low-lying electronic excited states, which are typically 1–8 kcal/mol above the ground state.[323,324] For example, the lowest three excited states of Eu^{3+} in $EuCl_6^{3-}$ is 1–3 kcal/mol higher than the ground state.[323] Thus, great caution is needed in the DFT calculations to avoid converging to an incorrect ground state that may lead to the opposite sign for the $\Delta\Delta G_{ext}$ and $\Delta\Delta\Delta G_{ext}$. It is also well known that technical choices in the calculation; for example, the exchange-correlation functionals, basis sets, and pseudo-potentials, may have significant influence on the geometric structure, ligand affinity, and selectivity toward metal ions. For example, GGA type functionals with small-core pseudo-potentials for the metal do not yield the correct trend of Am–S vs. Eu–S bond length in metal complexes with bis(2,4,4-trimethylpentyl)dithiophosphinic acid (the main component of Cyanex 301).[325] Besides, using Pople style 6-31G** and 6-311G** type basis sets without diffuse functions cannot accurately describe ligand affinity for Ln(III) and An(III).[316] Thus, in addition to the appropriate thermodynamic cycles, choosing proper calculation parameters is also crucial to obtain accurate thermodynamic data in the theoretical study of lanthanide and actinide separation. Generally, hybrid functionals with large basis sets are recommended in the DFT calculations.

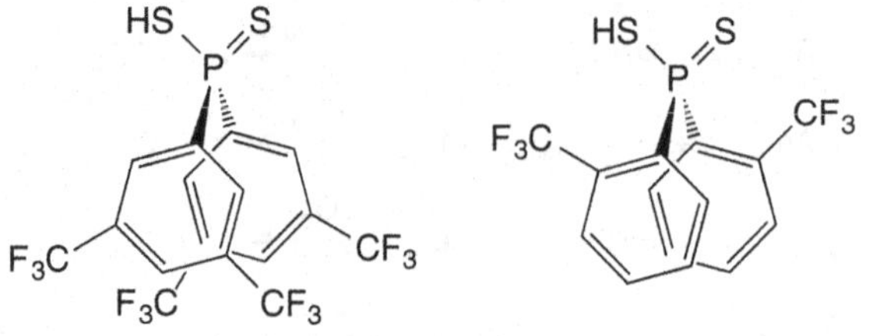

	L1		L2
$SF_{Am/Eu}$	20		100,000
$\Delta\Delta G_{ext}$	−1.8		−6.8
$\Delta\Delta\Delta G_{ext}$		−5.0	

FIGURE 5.9 Dithiophosphinic acids L1 and L2 with experimental separation factors and corresponding free-energy differences at room temperature, 298 K. Energies in kcal/mol. (Reprinted with permission from Keith and Batista (2012). Copyright © 2012, American Chemical Society.)

As an example of a successful study that combines the appropriate thermodynamic cycles, let us consider a thermodynamic examination of the relative selectivity of Am^{3+} over Eu^{3+} between two dithiophosphinic acids (L1 and L2, Figure 5.9) using $\Delta\Delta G_{ext}$ and DFT/B3LYP calculations.[308] The experimental $SF_{Am/Eu}$ values along with the corresponding $\Delta\Delta\Delta G_{ext}$ for L1 and L2 are listed in Figure 5.9.[326] As described above, the formulation of $\Delta\Delta\Delta G_{ext}$ is shown in Figure 5.10, and the ΔG_{ext} was calculated to be 3.09 kcal/mol. Compared with ~5.0 kcal/mol from experiment, theoretical $\Delta\Delta\Delta G_{ext}$ with a negative value correctly demonstrates that L2 should act as a better selective extractant for Am than L1.

$$
\begin{array}{rl}
 & Am^{3+}_{(aq)} + 3HL2_{(org)} \xrightarrow{\Delta G_{ext}} Am(L2)_{3\,(org)} + 3H^{+}_{(aq)} \\
- & \left(Am^{3+}_{(aq)} + 3HL1_{(org)} \xrightarrow{\Delta G_{ext}} Am(L1)_{3\,(org)} + 3H^{+}_{(aq)} \right) \\
+ & \left(Eu^{3+}_{(aq)} + 3HL1_{(org)} \xrightarrow{\Delta G_{ext}} Eu(L1)_{3\,(org)} + 3H^{+}_{(aq)} \right) \\
- & \left(Eu^{3+}_{(aq)} + 3HL2_{(org)} \xrightarrow{\Delta G_{ext}} Eu(L2)_{3\,(org)} + 3H^{+}_{(aq)} \right) \\
\hline
 & Am(L1)_{3\,(org)} + Eu(L2)_{3\,(org)} \xrightleftharpoons{\Delta\Delta\Delta G_{ext}} Am(L2)_{3\,(org)} + Eu(L1)_{3\,(org)}
\end{array}
$$

FIGURE 5.10 Formulation of $\Delta\Delta\Delta G_{ext}$. (Reprinted with permission from Keith and Batista (2012). Copyright © 2012, American Chemical Society.)

5.3.3   THE ORGANIC SOLUTION

It is challenging to understand organic-phase speciation using experimental methods alone, which often rely on assumptions of ideal solution behavior found only under dilute conditions. While the extraction of solutes originally in the aqueous phase may proceed through the formation of a limited number of discrete complexes under dilute conditions, numerous species with unique stoichiometries and structural topologies may become relevant in concentrated systems. Characterizing these species is often impossible using experimental methods due to the complexity of the concentrated system, which precludes the explicit enumeration of possible species required my many experimental characterization techniques. This can be exacerbated by the fact that, in addition to metals and counterions, extractants concurrently extract water and acid as separate charge-neutral adducts or through incorporation into the extracted metal solvates themselves. Changes in solvation resulting from organic-phase extractant–solute aggregation can impact solute distribution and engender phenomena including synergism[327] and third-phase formation.[328] Taken together, non-ideal behavior and the diversity of extracted species highlights the complexity of concentrated organic phase solutions and the value of molecular simulation to assist in their identification and quantification. Herein, recent molecular-simulation and related quantum-mechanical computational efforts undertaken to improve the understanding of organic-phase speciation and aggregation will be discussed.

The amphiphilic character of extractants results in self-association in nonpolar solvents. Understanding this organization and the physical and thermodynamic properties of binary extractant–solvent mixtures serves as a baseline for organic-phase structure in solvent extraction as well as a means of force-field benchmarking and validation. Cui et al. investigated the role of partial atomic charges and Lennard–Jones parameters on physicochemical properties pure tri-*n*-butyl (TBP) solutions to develop an appropriate force-field description,[329] which they later expanded to binary TBP–alkane solutions.[330] Similar efforts to reparameterize TBP with different force-field descriptions followed[331–333] with additional considerations to the dimerization constant and small-angle X-ray scattering (SAXS) profile. Vo et al. investigated self-association of the extractant TBP in an aliphatic solvent[331,334] using 2D PMFs to compute the configurational free energy landscapes of TBP dimers[331] and trimers[334]. Having reparameterized TBP force fields for use in alkane solvents, the self-association and mixing energetics of TBP in a range of organic solvents[334,335] and other similar solvating organophosphorus extractants[335] have been investigated.

Co-extraction of water and acid impact the distribution of metal ions and yet are spectroscopically challenging to characterize. They have, therefore, been the subject of numerous simulation studies. Ye et al. described the time evolution of water concentration in a biphasic simulation of the TBP–water–*n*-hexane interface.[336] Servis et al. measured the speciation of extracted TBP–water–nitric acid clusters in *n*-dodecane with a graph-theoretic approach, finding that the observed hydrogen bonding was well described by an equilibrium binding-constant model. It was further determined that the third-phase formation phase transition in that system without metal ions resulted from the formation of a percolating hydrogen-bonded network of extracted solutes.[337] While the existing scattering literature typically described third-phase formation as a result of reverse-micelle formation, even in the absence of metal ions,[338] this alternative phase-transition mechanism was corroborated by Ivanov et al.[339] using MD simulation combined with NMR and infrared (IR) spectroscopies.

Computational approaches have been applied to elucidate phenomenological trends in extraction equilibrium constants between metal ions and acidic anions. Qiao et al. determined changes to lanthanide–ligand energetics upon phase transfer for a malonamide extractant.[340] Observed trends in tetraoctyl diglycolamide (TODGA) selectivity across the lanthanides have been investigated with computational methods to understand the roles of ligand strain resulting from changing ionic radii,[341] outer-sphere anion coordination[342] and the association of water with those outer-sphere anions.[343] These studies have illustrated the complex relationship between the structure and composition of metal–ligand species and metal distribution energetics.

For higher extracted aqueous solute concentrations, the formation of large solute–extractant aggregates are commonly studied by combining experimental and simulation methods. Experimental X-ray and neutron scattering data are commonly used to validate simulation results, while simulation can provide detailed molecular descriptions of the ensemble of structures which result in a given measured scattering profile. Mu et al. demonstrated that SAXS data agree with simulations showing an absence of reverse-micelle like structures using Gibbs ensemble MC to generate the heavy and light organic phases which result under third-phase formation conditions.[257] Combining MD simulation with X-ray and neutron scattering, Qiao et al. found that morphologies of diamide extractant aggregation with water in n-heptane is impacted by extractant alkyl tails and posited that certain resulting morphologies more readily accommodate extracted water.[259] Simulation and corresponding SAXS and EXAFS data showed that spherical reverse-micelle-like aggregates of europium nitrate with a malonamide extractant expanded into long ellipsoids upon incorporation of water and nitric acid into the polar core, forming an extended hydrogen-bonding network of polar solutes.[344]

Uranyl nitrate complexes extracted by TBP in alkane solvents have been described in the experimental literature as reverse micelles, modeled by hard-sphere scattering form factors with surface adhesion.[345,346] Simulation studies have recently attempted to refine this description of the structure of and interactions between these complexes. Guilbaud et al. observed growth of uranyl nitrate aggregates under high metal concentrations resulting from "bridging" of the uranyl cations by nitrate anions, forming extended linear aggregates with up to four uranyl ions.[260] This is contrasted by a different study by Servis et al., where $UO_2(NO_3)_2(TBP)_2$ solvates formed closely packed dimers, stabilized by metal–ligand interactions between pairs of complexes, while also having long-range isotropic correlation[258] more analogous to the hard-sphere description given in the aforementioned experimental literature. In that study, the two correlation modes showed different sensitivities to extractant ligand structure and solvation environment, as illustrated in Figure 5.11. The degree of closely packed uranyl dimer formation in organic solutions was found to be mediated by three primary factors: ligand–ligand packing, metal–ligand assembly and ligand–solvent solubility. The first two factors are strongly dependent on the extracting ligand structure interlocking to form electrostatically stabilized assemblies with the apical uranyl

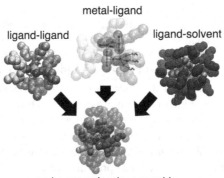

metal-ligand

ligand-ligand   ligand-solvent

uranyl supramolecular assembly

FIGURE 5.11 Short-range ordered dimer assemblies are observed in the organic phase for uranyl nitrate extracted by solvating organophosphorus extractants. Their formation is mediated by the combination of interactions between the solvent, extracting ligand, and uranyl nitrate itself. (Reprinted with permission from Servis et al. (2018). Copyright © 2018, Royal Society of Chemistry.)

oxygen and equatorially coordinating nitrate. The latter depends on how effectively the solvent solubilizes or screens the assembled metal–ligand dimers.

Caution should be taken when applying colloidal-scale techniques like modeling X-ray or neutron scattering to solvent extraction systems. Both Guilbaud et al.[260] and Servis et al.[258] reported good agreement with experimental X-ray scattering data despite reporting significantly different structures. This highlights the insensitivity of scattering data to widely variable solution-phase structures. The combined experimental and MD study conducted by Baldwin et al. found fitted parameters of the adhesive hard-sphere scattering model were strongly, and unintuitively, dependent on extractant volume fraction and the fitted "stickiness" parameters corresponded to the percolated region of the phase diagram for the modeled particles even at low uranium concentrations.[347] Diffusion measurements provide a robust, model-independent description of solution structure in colloidal systems and the scattering results were not consistent with NMR diffusometry measurements also reported in that study. These results demonstrate that assumptions inherent to applying a colloidal description required by scattering models are not necessarily valid in these types of systems. Instead, considering the extractant–solute behavior as a molecular solution, while more challenging to model, is more accurate. This emphasizes the utility of molecular simulation in describing the complex ensemble of species that form in organic-phase solvent extraction solutions.

5.3.4 The Aqueous–Organic Interface

The interface acts as the barrier to solute transport between phases. There has been a significant increase in work studying the structure, dynamics, and physicochemical properties of the interface. Recent studies, as discussed in Sections 5.3.4.1–5.3.4.3 below, have begun to examine the heterogeneity of the interface and the perturbations induced by surfactants. Herein we describe the mathematical formalism associated with analyzing interfacial characteristics, alongside new developments of our understanding of their sensitivity to solution conditions and the impact this may have upon transport mechanisms.

5.3.4.1 Macroscopic Interfacial Properties and Interfacial Roughness

Surface tensions of liquid–vapor interfaces, and interfacial tensions of liquid–liquid interfaces, are properties which relate microscopic structures and interactions to a readily measurable macroscopic quantity. There are direct and phenomenological methods for calculating surface and interfacial tensions from a simulation. A direct method using diagonal components of the pressure tensor is commonly implemented in the simulation literature because those pressure values are easily computed from interatomic forces already needed to propagate the simulation dynamics. As this method computes the interfacial tension directly from the pair potentials, its accuracy is subject to the accuracy of those potentials and any approximations made in their computation including, for example, Lennard–Jones (L–J) or Coulombic interaction cut-offs. An alternative phenomenological approach to computing surface or interfacial tensions is through fitting the capillary-wave fluctuations at the surface of a liquid. As the interfacial tension relates directly to the free-energy cost of creating an additional interfacial area, the degree to which capillary fluctuations broaden the density profile of a liquid at the interface can be related to the interfacial tension.[348] The density profile for a water–vapor surface can be fit to error function of the form:

$$\rho(z) = \frac{1}{2}\rho - \frac{1}{2}\rho \, \text{erf}\left(\frac{z - z_0}{\sqrt{2}w_c}\right) \tag{5.24}$$

where $\rho(z)$ is the density of water as a function of z position, ρ is the bulk density of water, z_0 is the Gibbs dividing surface, located at the z position corresponding to a water density equal to half its bulk value, and w_c is defined as the capillary width. The w_c parameter is derived from the interfacial tension by:

$$w_c^2 = \frac{k_B T}{2\pi\gamma} \ln \frac{L}{l_b} \tag{5.25}$$

Here, k_B is the Boltzmann constant, T is the temperature, L is the simulation box length parallel to the interfacial normal and l_b is the bulk correlation length of the fitted phase. The bulk correlation length is not known a priori, but γ can be obtained from the slope of the linear relationship between w^2 and $\ln\left(\frac{L}{l_b}\right)$ from different simulations with varying L.[349] This method assumes that the broadening of the density profile at the interface results solely from capillary-wave fluctuations and is, therefore, not applicable to systems where the density is affected by effects like surfactant adsorption.

The interfacial tension can be related to the molecular structure at the interface. Lower interfacial tensions indicate a lower free energy penalty for the creation of interfacial area. Therefore, with lower interfacial tension, the molecular roughness at the interface is increased. With the addition of a surfactant to the interface, the interfacial tension necessarily decreases. Conversely, solutes that preferentially distribute to the bulk rather than the interface, such as ions which are repelled from the water–vapor surface by their image charges, increase the interfacial tension. The concentration of the surfactant can be related to the interfacial tension through the Gibbs adsorption isotherm. From the slope of that isotherm, before the critical aggregation concentration, the distribution of surfactant between the bulk and interface can be calculated. For an interfacial tension of zero, the two phases become miscible.

Interfacial roughness, its dependence on the addition of aqueous and organic solutes in solvent extraction systems, and the interfacial structure of immiscible solvents and their solutes is an active topic of investigation in the simulation literature. Microscopic interfacial structure plays a significant role in the formation kinetics of extractable complexes. Where experimental measurements at interfaces are challenging to conduct and interpret, simulation has emerged as a tool to develop a molecular-level understanding of interfaces. Chang and Dang provide a review of the impact of aqueous ions on interfacial properties for commonly measured simulation quantities including density profiles, phase transfer PMFs, and ion hydration structures.[350] The influence of surface-active extractants upon interfacial roughness is described in further detail below.

5.3.4.2 Organization of Interfacial Water

It is a long-standing challenge to relate molecular structure at a liquid–liquid interface and its consequences on the structural, dynamical, and electrostatic properties of that interface.[351–353] Unlike bulk liquids, the asymmetry inherent to the intermolecular forces at the interface yields unique molecular dynamics and structures. This nature of the interface impacts response properties with respect to the interface. The polar nature of water in contact with nonpolar media alters water orientation and strongly modify electrostatic interactions in the interfacial systems. Water orientation near the interface can be measured by generating an orientational profile based upon the alignment of the dipole vector of H_2O to the surface normal of a "planar" interface defined by the Gibbs dividing surface. A full description of the orientational statistics of water in specific molecular conformations, two independent vectors, namely the normal vector of the molecular plane and the molecular dipole vector, are used to generate a bivariate joint distribution with respect to the local frame of the interface. For example, the intrinsic sampling method has been developed to study the intrinsic structure of water–alkane interfaces,[354,355] whereas the so-called identification of the truly interfacial molecules (ITIM) algorithm has been applied to investigate the molecular orientations of water at the vapor and organic interfaces.[356,357] In this way, the collective interfacial water molecular alignment and orientation can be detected.

This effect of water orientation upon the electrostatic interactions is quantified by means of the static dielectric permittivity tensor ε. While the dielectric response of homogeneous bulk liquids is well understood, where ε is diagonal and spatially constant, experiment, theory, and simulation all suggest that the dielectric response of bulk liquids will be strongly modified in the vicinity of an interface.[358–365] The dielectric response of the surface water layer extracted from solid–liquid interfacial simulations is represented by an oscillating function of the distance to the rigid solid surface.[363,364] The inhomogeneous nature of the liquid–liquid medium makes the

computation of the local component of dielectric permittivity even more complicated, which needs to account for the anisotropy appropriately. Liu et al. calculated
the static dielectric permittivity profile reference to the intrinsic liquid–liquid interface and presented an enhanced dielectric response at the liquid–liquid interface,
comparing to the featureless dielectric response with respect to the commonly used
Gibbs dividing interface.[365]

Concomitant to the orientation and electrostatic response of water is the change to
the hydrogen-bond network within the interfacial layers that result in spatially inhomogeneous interactions. In pure liquid, water tends toward a disordered tetrahedral
hydrogen-bonding network. While close to an interface, geometric constraints lead
it impractical to satisfy all available hydrogen bonds. Thus, the interfacial hydrogen-
bonding network is anisotropic and distorted relative to that of the bulk liquid.[366]

5.3.4.3   Interfacial Extractants, Complexation, and Transport

Surface-active extractant molecules, and the configurations they assume at the aqueous–organic interface, impact the configurations of the immiscible solvents within
the interfacial region. As such, they can have an even stronger impact on interfacial properties than aqueous metal ions.[367,368] Perhaps the most research has been
done with surface-active TBP–water–organic ternary solutions, where Wipff and
coworkers pioneered the simulation of such systems.[369] Through much study of TBP
and more recent works on broader classes of surfactants, it is now appreciated that
surfactants often enhance existing interfacial roughness, disrupt the connectivity of
interfacial water, and, depending on the surfactant, can lead to water protrusions that
eventually lead to water–extractant adducts that migrate into the organic phase.[370,371]
Servis and Clark quantified the structural changes to the hydrogen bonding of interfacial water in response to surfactant adsorption. Those structural changes led to
protrusions emerging from the extremes of induced capillary roughness and, potentially, the subsequent disengagement and extraction of water–TBP adducts from the
aqueous interface, as illustrated in Figure 5.12.[371] The relationship, or lack thereof,
between these water–surfactant protrusions and surfactant–ion complexation and ion
transport are yet to be understood.

The role of water protrusions in bare-ion transfer has been observed for quite
some time,[372–374] where PMF simulations have indicated barrier-free transport for
many ions can occur by a mechanism wherein the ion is trailed by water molecules
extending from its solvation shell as the ion leaves the aqueous interface. The details
of that mechanism and the role of the water "finger" on the transfer pathways have
been further explored in the literature.[375,376] This free-energy surface shows a barrier
not present in the one-dimensional PMF corresponding to breaking the water finger from the aqueous surface. Karnes and Benjamin complimented that analysis by
incorporating coordinates for water or organic solvent solvation of the ion to describe
the changes in free energy during interfacial deformation and "finger" formation.[377]
Further refinement of reaction coordinates for transport of bare ions and development of coordinates for ion–extractant complexes may elucidate the role of interfacial structure and solute speciation on diffusion-limited transport mechanisms.

Simulations that link interfacial structure to extractant complexation have also
been extended to aqueous metal ions. Wipff and coworkers found surface roughness

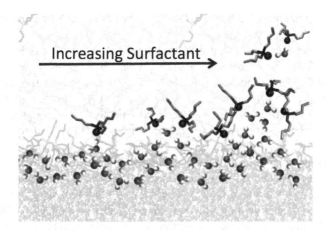

FIGURE 5.12 Increasing TBP concentration leads to induced interfacial roughness and water protrusion formation, through which water is transferred from the aqueous interface to the bulk organic phase. (Reprinted from Servis and Clark (2018) with permission of the Royal Society of Chemistry.)

enhanced upon ionophore adsorption and complexation with ions.[378,379] When investigating how extractant–solute complexes form at the interface, the simulation literature has demonstrated that, for numerous extraction systems, microscopic roughness of the aqueous–organic interface impacts the kinetic pathways separating aqueous and extracted organic-phase species. Indeed a myriad of pathways may emerge at surfactant-laden surfaces, as demonstrated in recent work by Qiao et al., who studied a combination of surfactant molecules with different morphologies, where more surface-active hexadecyltrimethylammonium (CTA^+) out-competed the extractant di(2-ethylhexyl) phosphoric acid at the water–oil interface. This led to a hypothesized extraction mechanism by which HDEHP complexes with aqueous solutes through water channels that emerge from the surface of a CTA^+-coated water-in-oil micelle. The relationship between extractant alkyl tail structures and their resulting sterics while adsorbed to a water–oil interface are discussed by Liang et al.[380] They propose those sterics affect microscopic interfacial curvature of the domain of water surrounding interfacial ions during the formation of interfacial extractant–metal complexes.

To explore different pathways, so-called mixing de-mixing studies may be pursued, and therein meta-stable interfacial configurations can emerge that do not have well-defined separation of aqueous–organic solvents. Complexation of uranyl nitrate with TBP has been observed to occur in those highly mixed interfacial regions, which contrasts with the planar interfaces typically featured in PMF studies of ion–extractant phase transfer.[381] That being said, significant work has been dedicated toward using PMFs to map out transport energetics, albeit with well-defined reaction pathways. In one such example (out of many), Jayasinghe and Beck conducted such PMF calculations for uranyl nitrate–TBP complexes at the interface to study the role of increasing interfacial roughness on that phase-transfer reaction coordinate. They found that the orientation of the complex, which is dependent on its position relative to the interface, became less limited, and the phase-transfer free-energy barrier decreased with increased adsorbed TBP concentration and concomitant surface roughness.[382]

In general, the reaction pathways through which aqueous and organic solutes complex and form extractable species are not generally known. Understanding the relative solubility of species using PMFs can inform where, relative to the interface, certain species form during the extraction process and therefore what assembly or rearrangement processes are likely to be kinetically limiting. Lauterbach et al. showed that ionophore complexes with combinations of cations and anions affected the surface activity of those complexes.[383] Ghadar et al. varied the hydrophilicity of an extractant, showing a transition in the solubility of the extractants between phases.[384] Jayasinghe and Beck demonstrated the free-energy minimum of a uranyl nitrate–TBP complex at the interface, in addition to a flat free-energy surface of the uranyl nitrate approaching the extractant-laden interface, implying assembly of that complex at the interface followed by barrier-free diffusion into the organic phase.[385] Benay and Wipff calculated PMFs of metal–ligand complex transfer between phases for a range of metal and ligand types. They demonstrated that the interfacial preference of the uranyl nitrate–TBP complex is dependent on the manner in which Coulombic charges of the constituent molecules are modeled with the most appropriate charge set shows the complex readily diffuses from the interface upon formation.[224] Similar studies of monoamide,[386] 18-crown-6,[387] bistrianzinylpyridine,[388] and bistrianzinylphenantroline,[389] metal–ligand complexes showed generally similar behavior featuring stability at the interface and bulk organic phase with changes in relative solubility depending on properties such as aqueous acidity.[386,389] Karnes and Benjamin computed PMFs for transfer of 1-bromooctance into water via occupation as a guest molecule in a β-cyclodextrin host.[390] Their findings indicated the configuration of the guest within the host molecule impacts mass transfer of the guest between phases. Although not exhaustive, these works provide some basis for understanding the complexity of solute–extractant–solvent interactions and correlating behavior with interfacial organization and fluctuations.

5.4 SUMMARY AND PROSPECTUS

As illustrated within this chapter, solvent extraction exhibits a myriad of fascinating chemistries that challenge and inspire those of us seeking to understand, predict, and control such processes. Computational chemists have at their fingertips a wide range of tools that can capture the behavior of electrons that underlie selectivity, the ensemble of intermolecular interactions that describe solvation, and the multiscale cooperative behavior that emerges at liquid–liquid interfaces. That being said, there remains both the opportunity to learn deeply about the fundamental behavior of solvent extraction and also to use this unique application as a launching point for theoretical development. To date, much emphasis has been placed upon molecular-scale approaches to understand solvent extraction. Within this domain, it is a significant challenge to understand different regimes of kinetic stability; on the one hand, there may exist rare events for which we have not yet begun to understand their impact upon extraction, and on the other hand, there may exist meta-stable long-lived configurations that significantly impact our understanding of the true thermodynamic equilibrium of these complex multicomponent, multiphase solutions. At the next scale of length and times, there exist coarse-grained non-atomistic techniques and fluid-dynamics approaches that have not yet been significantly leveraged for solvent extraction—largely because of our lack of knowledge of what physics contributions we can omit successfully and still yield reasonable results. Indeed, there is much to be learned from understanding the transfer of information across length and times-cale to perform simulations in solutions at conditions that truly resemble industrial processes. For example, the distribution of mechanical energy imparted by the stan-dard agitation of the binary solution to create a liquid–liquid dispersion is poorly understood at the level of the interface in solvent extraction systems, and how this impacts transfer rates is not understood. This provides an excellent opportunity to bridge chemistry and chemical engineering communities to create new paradigms for the simulation of solvent extraction systems. We hope that the content of this chapter illustrates not only the current capabilities of computational chemists, but also motivates such new approaches and brings new talented scientists into the field, as this topic is rich in both fundamental and applied scientific problems.

ACKNOWLEDGMENTS

AEC acknowledges the generous support of the U.S. Department of Energy, Basic Energy Sciences Separations program, grant DESC0001815. JA is grateful for the support of his actinide research by the U.S. Department of Energy, Office of Basic Energy Sciences, Heavy Element Chemistry program, grant DE-SC0001136. JS, ERB, and PY gratefully acknowledge support for this work from the U.S. Department of Energy, Office of Basic Energy Sciences, Heavy Element Chemistry Program under contract DE-AC52-06NA25396. DAD acknowledges the support of the U.S. Department of Energy, Office of Basic Energy Sciences, Separations program grant, DE-SC0018033, and Heavy Element Chemistry program, grant DE-SC0018921 as well as the Robert Ramsay Chair fund at The University of Alabama.

ABBREVIATIONS

AFM:	adaptive force matching
AIMD:	ab initio molecular dynamics
An:	actinide
AO:	atomic orbital
AOC:	average of configurations
BMHTF:	Born–Mayer–Huggins–Tosi–Fumi
CAS:	complete active space
CASPT2:	complete active space self-consistent field with second-order Moller–Plesset theory
CASSCF:	complete active space self-consistent field
CC:	coupled-cluster
CI:	configuration interaction
COSMO:	conductor-like screening model
CPMD:	Carr-Parrinello molecular dynamics

D:	distribution ratio
DFT:	density-functional theory (Kohn-Sham and generalized Kohn-Sham approximations)
DKH:	Douglas–Kroll–Hess
DMRG:	density-matrix renormalization group
EDA:	electric-dipole approximation
EOM-CC:	equation of motion coupled-cluster
ES-TDDFT:	**energy-specific time-dependent density-functional theory**
EXAFS:	extended X-ray absorption fine structure
FPD:	Feller-Peterson-Dixon
GGA:	generalized gradient approximation
HEXS:	high-energy X-ray scattering
HF:	Hartree–Fock
IEF-PCM:	integral equation formalism polarizable-continuum model
IR:	infrared
KS:	Kohn–Sham
LDA:	local-density approximation
LJ:	Lennard–Jones
Ln:	lanthanide
MBAR:	multistate Bennett acceptance ratio
MC:	multi-configuration
MC:	Monte Carlo
MD:	molecular dynamics
MR:	multi-reference
MR-CI:	multi-reference configuration interaction
NBO:	natural bond order
NEXAFS:	near-edge X-ray absorption fine structure
NMR:	nuclear magnetic resonance
NPA:	natural population analysis
PCM:	polarizable-continuum model
PES:	potential-energy surface
PMF:	potential of mean force
pNMR:	paramagnetic NMR
PSPI:	pseudo-supercritical path integration
PT:	perturbation theory
QM/MM:	quantum-mechanical/molecular-mechanical
RDF:	radial distribution function
RPP:	relativistic pseudopotential
SAXS:	small-angle X-ray scattering
SCIPCM:	self-consistent isodensity polarizable-continuum model
SCRF:	self-consistent reaction field
SF:	separation factor
SMD:	solvation model based on density
SO:	spin-orbit
SR:	scalar relativistic
TCF:	time correlation function

TDDFT:	time-dependent density-functional theory
TM:	transition metal
VSFG:	vibrational sum-frequency generation
WFT:	wave-function theory
X2C:	exact 2-component
XANES:	X-ray adsorption near-edge structure
XAS:	X-ray adsorption spectroscopy
ZORA:	zeroth-order regular approximation

REFERENCES

1. C. J. Cramer, *Essentials of Computational Chemistry*. John Wiley & Sons, Chichester, 2002.
2. T. Helgaker, P. Jørgensen, and J. Olsen, *Molecular Electronic–Structure Theory*. Wiley, New York, 2000.
3. J. Autschbach, "Orbitals: Some fiction and some facts," *J. Chem. Educ.*, vol. 89, no. 8, pp. 1032–1040, 2012.
4. R. G. Parr, and Y. Weitao, *Density Functional Theory of Atoms and Molecules*. Oxford University Press, New York, 1989.
5. E. R. Davidson, and D. Feller, "Basis set selection for molecular calculations," *Chem. Rev.*, vol. 86, no. 4, pp. 681–696, 1986.
6. K. A. Peterson, and J. G. Hill, *On the Development of Accurate Gaussian Basis Sets for f-Block Elements*, vol. 14. Elsevier, Amsterdam, 2018.
7. K. A. Peterson, *Gaussian Basis Sets Exhibiting Systematic Convergence to the Complete Basis Set Limit*, vol. 3. Elsevier, Amsterdam, 2007.
8. T. Kato, "On the eigenfunctions of many-particle systems in quantum mechanics," *Comm. Pure Appl. Math.*, vol. 10, no. 2, pp. 151–177, 1957.
9. J. P. Perdew, A. Ruzsinszky, L. A. Constantin, J. Sun, and G. I. Csonka, "Some fundamental issues in ground-state density functional theory: A guide for the perplexed," *J. Chem. Theory Comput.*, vol. 5, no. 4, pp. 902–908, 2009.
10. R. J. Bartlett, and M. Musiał, "Coupled–cluster theory in quantum chemistry," *Rev. Mod. Phys.*, vol. 79, no. 1, pp. 291–352, 2007.
11. C. Riplinger, and F. Neese, "An efficient and near linear scaling pair natural orbital based local coupled cluster method," *J. Chem. Phys.*, vol. 138, no. 3, p. 034106, 2013.
12. O. Christiansen, H. Koch, and P. Jørgensen, "The second-order approximate coupled cluster singles and doubles model CC2," *Chem. Phys. Lett.*, vol. 243, no. 5–6, pp. 409–418, 1995.
13. P. Pyykkö, "Relativistic effects in chemistry: More common than you thought," *Annu. Rev. Phys. Chem.*, vol. 63, pp. 45–64, 2012.
14. M. Reiher, and A. Wolf, *Relativistic Quantum Chemistry: The Fundamental Theory of Molecular Science*. John Wiley & Sons, Weinheim, 2014.
15. K. G. Dyall, and K. Fægri, Jr., *Relativistic Quantum Chemistry*. Oxford University Press, New York, 2007.
16. J. Autschbach, "Perspective: Relativistic effects," *J. Chem. Phys.*, vol. 136, no. 15, p. 150902, 2012.
17. P. Pyykkö, "Relativistic effects in structural chemistry," *Chem. Rev.*, vol. 88, no. 3, pp. 563–594, 1988.
18. J. Autschbach, S. Sikierski, M. Seth, P. Schwerdtfeger, and W. H. Schwarz, "Dependence of relativistic effects on the electronic configurations in the atoms of the d- and f-block elements," *J. Comput. Chem.*, vol. 23, pp. 804–813, 2002.

19. W. Liu, "Ideas of relativistic quantum chemistry," *Mol. Phys.*, vol. 108, no. 13, pp. 1679–1706, 2010.

20. T. Saue, "Relativistic Hamiltonians for chemistry: A primer," *Chemphyschem*, vol. 12, no. 17, pp. 3077–3094, 2011.

21. E. van Lenthe, E. J. Baerends, and J. G. Snijders, "Relativistic total energy using regular approximations," *J. Chem. Phys.*, vol. 101, no. 11, pp. 9783–9792, 1994.

22. C. Chang, M. Pelissier, and P. Durand, "Regular two-component Pauli-like effective Hamiltonians in Dirac theory," *Phys. Scr.*, vol. 34, no. 5, pp. 394–404, 1986.

23. B. A. Hess, "Relativistic electronic-structure calculations employing a two-component no-pair formalism with external-field projection operators," *Phys. Rev. A*, vol. 33, no. 6, pp. 3742–3748, 1986.

24. B. A. Hess, "Relativistic theory and applications," in *Encyclopedia of Computational Chemistry* (P. von Ragué Schleyer, ed.), pp. 2499–2508. John Wiley & Sons, Chichester, 1998.

25. M. Dolg, and X. Cao, "Relativistic pseudopotentials: Their development and scope of applications," *Chem. Rev.*, vol. 112, no. 1, pp. 403–480, 2012.

26. B. O. Roos, P. R. Taylor, and P. E. M. Si≐gbahn, "A complete active space SCF method (CASSCF) using a density matrix formulated super-CI approach," *Chem. Phys.*, vol. 48, no. 2, pp. 157–173, 1980.

27. G. K. L. Chan, and S. Sharma, "The density matrix renormalization group in quantum chemistry," *Annu. Rev. Phys. Chem.*, vol. 62, pp. 465–481, 2011.

28. S. Wouters, and D. van Neck, "The density matrix renormalization group for ab initio quantum chemistry," *Eur. Phys. J. D*, vol. 68, no. 9, pp. 272–291, 2014.

29. S. Knecht, E. D. Hedegård, S. Keller, A. Kovyrshin, Y. Ma, A. Muolo, C. J. Stein, and M. Reiher, "New approaches for ab-initio calculations of molecules with strong electron correlation," *Chimia*, vol. 70, no. 4, pp. 244–251, 2016.

30. D. Ganyushin, and F. Neese, "A fully variational spin-orbit coupled complete active space self-consistent field approach: Application to electron paramagnetic resonance g-tensors," *J. Chem. Phys.*, vol. 138, no. 10, p. 104113, 2013.

31. S. Knecht, Ö. Legeza, and M. Reiher, "Communication: Four-component density matrix renormalization group," *J. Chem. Phys.*, vol. 140, no. 4, p. 041101, 2014.

32. P. Å. Malmqvist, B. O. Roos, and B. Schimmelpfennig, "The restricted active space (RAS) state interaction approach with spin-orbit coupling," *Chem. Phys. Lett.*, vol. 357, no. 3–4, pp. 230–240, 2002.

33. S. Knecht, S. Keller, J. Autschbach, and M. Reiher, "A nonorthogonal state-interaction approach for matrix product state wave functions," *J. Chem. Theory Comput.*, vol. 12, no. 12, pp. 5881–5894, 2016.

34. K. Andersson, P. A. Malmqvist, B. O. Roos, A. J. Sadlej, and K. Wolinski, "Second-order perturbation theory with a CASSCF reference function," *J. Phys. Chem.*, vol. 94, no. 14, pp. 5483–5488, 1990.

35. C. Angeli, R. Cimiraglia, S. Evangelisti, T. Leininger, and J. -P. Malrieu, "Introduction of n-electron valence states for multireference perturbation theory," *J. Chem. Phys.*, vol. 114, no. 23, pp. 10252–10264, 2001.

36. R. J. Buenker, S. D. Peyerimhoff, and W. Butscher, "Applicability of the multi-reference double-excitation CI (MRD-CI) method to the calculation of electronic wavefunctions and comparison with related techniques," *Mol. Phys.*, vol. 35, no. 3, pp. 771–791, 1978.

37. A. J. Garza, A. G. Sousa Alencar, and G. E. Scuseria, "Actinide chemistry using singlet-paired coupled cluster and its combinations with density functionals," *J. Chem. Phys.*, vol. 143, no. 24, p. 244106, 2015.

38. J. A. Gomez, T. M. Henderson, and G. E. Scuseria, "Singlet-paired coupled cluster theory for open shells," *J. Chem. Phys.*, vol. 144, no. 24, p. 244117, 2016.

39. G. Li Manni, R. K. Carlson, S. Luo, D. Ma, J. Olsen, D. G. Truhlar, and L. Gagliardi, "Multiconfiguration pair-density functional theory," *J. Chem. Theory Comput.*, vol. 10, no. 9, pp. 3669–3680, 2014.

40. D. I. Lyakh, M. Musiał, V. F. Lotrich, and R. J. Bartlett, "Multireference nature of chemistry: The coupled-cluster view," *Chem. Rev.*, vol. 112, no. 1, pp. 182–243, 2012.

41. J. Tomasi, B. Mennucci, and R. Cammi, "Quantum mechanical continuum solvation models," *Chem. Rev.*, vol. 105, no. 8, pp. 2999–3093, 2005.

42. D. Rinaldi, and J.-L. Rivail, "Molecular polarizability and dielectric effect of medium in liquid-phase-theoretical study of water molecule and its dimers," *Theor. Chim. Acta*, vol. 32, no. 1, pp. 57–70, 1973.

43. J.-L. Rivail, and D. Rinaldi, "A quantum chemical approach to dielectric solvent effects in molecular liquids," *Chem. Phys.*, vol. 18, no. 1–2, pp. 233–242, 1976.

44. M. Born, "Volumen und hydratationswärme der ionen," *Z. Physik*, vol. 1, no. 1, pp. 45–48, 1920.

45. J. G. Kirkwood, "Theory of solutions of molecules containing widely separated charges with special application to zwitterions," *J. Chem. Phys.*, vol. 2, no. 7, pp. 351–361, 1934.

46. L. Onsager, "Electric moments of molecules in liquids," *J. Am. Chem. Soc.*, vol. 58, no. 8, pp. 1486–1493, 1936.

47. S. Miertuš, E. Scrocco, and J. Tomasi, "Electrostatic interaction of a solute with a continuum. A direct utilizaion of ab initio molecular potentials for the prevision of solvent effects," *Chem. Phys.*, vol. 55, no. 1, pp. 117–129, 1981.

48. A. Klamt, and G. Schüürmann, "Cosmo: A new approach to dielectric screening in solvents with explicit expressions for the screening energy and its gradient," *J. Chem. Soc. Perkin Trans.* vol. 2, no. 5, pp. 799–805, 1993.

49. J.-L. Rivail, and B. Terryn, "Énergie libre d'une distribution de charges électriques séparée d'un milieu diélectrique infini par une cavité ellipsoïdale quelconque, application à l'étude de la solvation des molécules," *J. Chim. Phys.*, vol. 79, pp. 1–6, 1982.

50. J. L. Pascual-Ahuir, E. Silla, J. Tomasi, and R. Bonaccorsi, "Electrostatic interaction of a solute with a continuum. Improved description of the cavity and of the surface cavity bound charge distribution," *J. Comput. Chem.*, vol. 8, no. 6, pp. 778–787, 1987.

51. E. Silla, F. Villar, O. Nilsson, J. L. Pascual-Ahuir, and O. Tapia, "Molecular volumes and surfaces of biomacromolecules via GEPOL : A fast and efficient algorithm," *J. Mol. Graph.*, vol. 8, no. 3, pp. 168–172, 1990.

52. E. Silla, I. Tuñón, and J. L. Pascual-Ahuir, "GEPOL: An improved description of molecular surfaces ii. Computing the molecular area and volume," *J. Comput. Chem.*, vol. 12, no. 9, pp. 1077–1088, 1991.

53. J. L. Pascual-ahuir, E. Silla, and I. Tuñon, "GEPOL: An improved description of molecular surfaces. iii. A new algorithm for the computation of a solvent-excluding surface," *J. Comput. Chem.*, vol. 15, no. 10, pp. 1127–1138, 1994.

54. I. Soteras, C. Curutchet, A. Bidon-Chanal, M. Orozco, and F. J. Luque, "Extension of the mst model to the ief formalism: Hf and b3lyp parametrizations," *J. Mol. Struct. THEOCHEM*, vol. 727, no. 1–3, pp. 29–40, 2005.

55. A. V. Marenich, C. J. Cramer, and D. G. Truhlar, "Sorting out the relative contributions of electrostatic polarization, dispersion, and hydrogen bonding to solvatochromic shifts on vertical electronic excitation energies," *J. Chem. Theory Comput.*, vol. 6, no. 9, pp. 2829–2844, 2010.

56. F. Floris, and J. Tomasi, "Evaluation of the dispersion contribution to the solvation energy. A simple computational model in the continuum approximation," *J. Comput. Chem.*, vol. 10, no. 5, pp. 616–627, 1989.

57. F. M. Floris, J. Tomasi, and J. L. P. Ahuir, "Dispersion and repulsion contributions to the solvation energy: Refinements to a simple computational model in the continuum approximation," *J. Comput. Chem.*, vol. 12, no. 7, pp. 784–791, 1991.

58. F. M. Floris, A. Tani, and J. Tomasi, "Evaluation of dispersion – repulsion contributions to the solvation energy. Calibration of the uniform approximation with the aid of rism calculations," *Chem. Phys.*, vol. 169, no. 1, pp. 11–20, 1993.

59. E. Cancès, B. Mennucci, and J. Tomasi, "A new integral equation formalism for the polarizable continuum model: Theoretical background and applications to isotropic and anisotropic dielectrics," *J. Chem. Phys.*, vol. 107, no. 8, pp. 3032–3041, 1997.

60. B. Mennucci, E. Cancès, and J. Tomasi, "Evaluation of solvent effects in isotropic and anisotropic dielectrics and in ionic solutions with a unified integral equation method: Theoretical bases, computational implementation, and numerical applications," *J. Phys. Chem. B*, vol. 101, no. 49, pp. 10506–10517, 1997.

61. E. Cancès, and B. Mennucci, "New applications of integral equations methods for solvation continuum models: Ionic solutions and liquid crystals," *J. Math. Chem.*, vol. 23, no. 3–4, pp. 309–326, 1998.

62. C. Amovilli, and B. Mennucci, "Self-consistent-field calculation of pauli repulsion and dispersion contributions to the solvation free energy in the polarizable continuum model," *J. Phys. Chem. B*, vol. 101, no. 6, pp. 1051–1057, 1997.

63. A. Pomogaeva, and D. M. Chipman, "New implicit solvation models for dispersion and exchange energies," *J. Phys. Chem. A*, vol. 117, no. 28, pp. 5812–5820, 2013.

64. A. C. Chamberlin, C. J. Cramer, and D. G. Truhlar, "Extension of a temperature-dependent aqueous solvation model to compounds containing nitrogen, fluorine, chlorine, bromine, and sulfur," *J. Phys. Chem. B*, vol. 112, no. 10, pp. 3024–3039, 2008.

65. A. Klamt, *COSMO-RS: From Quantum Chemistry to Fluid Phase Thermodynamics and Drug Design*. Elsevier, 2005.

66. J. M. Chalmers, *Spectroscopy in Process Analysis*, vol. 4. Taylor & Francis, U.S., 2000.

67. T. Vo-Dinh, and G. Gauglitz, *Handbook of Spectroscopy*. Wiley–VCH, 2003.

68. A. Maitra, and S. Bagchi, "UV-Visible spectroscopic study of solvation in ternary solvent mixtures: Ketocyanine dye in methanol+acetone+water and methanol+acetone+benzene," *J. Phys. Chem. B*, vol. 112, no. 7, pp. 2056–2062, 2008.

69. N. Nunes, R. Elvas-Leitão, and F. Martins, "UV-Vis spectroscopic study of preferential solvation and intermolecular interactions in methanol/1-propanol/acetonitrile by means of solvatochromic probes," *Spectrochim. Acta A Mol. Biomol. Spectrosc.*, vol. 124, pp. 470–479, 2014.

70. X. Li, S. M. Smith, A. N. Markevitch, D. A. Romanov, R. J. Levis, and H. B. Schlegel, "A time-dependent Hartree-Fock approach for studying the electronic optical response of molecules in intense fields," *Phys. Chem. Chem. Phys.*, vol. 7, no. 2, pp. 233–239, 2005.

71. J. J. Goings, P. J. Lestrange, and X. Li, "Real-time time-dependent electronic structure theory," *WIREs Comput. Mol. Sci.*, e1341, 2018.

72. A. Dreuw, and M. Head-Gordon, "Single-reference ab initio methods for the calculation of excited states of large molecules," *Chem. Rev.*, vol. 105, no. 11, pp. 4009–4037, 2005.

73. S. Ren, J. Harms, and M. Caricato, "An EOM-CCSD-PCM benchmark for electronic excitation energies of solvated molecules," *J. Chem. Theory Comput.*, vol. 13, no. 1, pp. 117–124, 2017.

74. T. Helgaker, S. Coriani, P. Jørgensen, K. Kristensen, J. Olsen, and K. Ruud, "Recent advances in wave function-based methods of molecular-property calculations," *Chem. Rev.*, vol. 112, no. 1, pp. 543–631, 2012.

75. M. Caricato, G. W. Trucks, M. J. Frisch, and K. B. Wiberg, "Oscillator strength: How does TDDFT compare to EOM-CCSD?," *J. Chem. Theory Comput.*, vol. 7, no. 2, pp. 456–466, 2011.

76. C. Curutchet, A. Muñoz-Losa, S. Monti, J. Kongsted, G. D. Scholes, and B. Mennucci, "Electronic energy transfer in condensed phase studied by a polarizable QM/MM model," *J. Chem. Theory Comput.*, vol. 5, no. 7, pp. 1838–1848, 2009.

77. S. Caprasecca, S. Jurinovich, L. Lagardère, B. Stamm, and F. Lipparini, "Achieving linear scaling in computational cost for a fully polarizable MM/continuum embedding," *J. Chem. Theory Comput.*, vol. 11, no. 2, pp. 694–704, 2015.

78. M. De Vetta, M. F. S. J. Menger, J. J. Nogueira, and L. González, "Solvent effects on electronically excited states: QM/continuum versus QM/explicit models," *J. Phys. Chem. B*, vol. 122, no. 11, pp. 2975–2984, 2018.

79. R. Cammi, and B. Mennucci, "Linear response theory for the polarizable continuum model," *J. Chem. Phys.*, vol. 110, no. 20, pp. 9877–9886, 1999.

80. J. Tomasi, "Thirty years of continuum solvation chemistry: A review, and prospects for the near future," *Theor. Chem. Acc.*, vol. 112, no. 4, pp. 184–203, 2004.

81. J. Tomasi, R. Cammi, B. Mennucci, C. Cappelli, and S. Corni, "Molecular properties in solution described with a continuum solvation model," *Phys. Chem. Chem. Phys.*, vol. 4, no. 23, pp. 5697–5712, 2002.

82. R. Heydová, E. Gindensperger, R. Romano, J. Sýkora, A. Vlček Jr., S. Záliš, and C. Daniel, "Spin-orbit treatment of UV-Vis absorption spectra and photophysics of rhenium (I) carbonyl-bipyridine complexes: MS-CASPT2 and TD-DFT analysis," *J. Phys. Chem. A*, vol. 116, no. 46, pp. 11319–11329, 2012.

83. C. Daniel, "Photochemistry and photophysics of transition metal complexes: Quantum chemistry," *Coord. Chem. Rev.*, vol. 282–283, pp. 19–32, 2015.

84. X. Li, B. Minaev, H., Ågren, and H. Tian, "Theoretical study of phosphorescence of iridium complexes with fluorine-substituted phenylpyridine ligands," *Eur. J. Inorg. Chem.*, vol. 2011, no. 16, pp. 2517–2524, 2011.

85. H. Brahim, C. Daniel, and A. Rahmouni, "Spin-orbit absorption spectroscopy of transition metal hydrides: A TD-DFT and MS-CASPT2 study of $HM(CO)_5$ (M = Mn, Re)," *Int. J. Quantum Chem.*, vol. 112, no. 9, pp. 2085–2097, 2012.

86. H. Brahim, and C. Daniel, "Structural and spectroscopic properties of Ir(III) complexes with phenylpyridine ligands: Absorption spectra without and with spin-orbit-coupling," *Comput. Theor. Chem.*, vol. 1040–1041, pp. 219–229, 2014.

87. S. Záliš, Y. C. Lam, H. B. Gray, and A. Vlček, "Spin-orbit TDDFT electronic structure of diplatinum (II, II) complexes," *Inorg. Chem.*, vol. 54, no. 7, pp. 3491–3500, 2015.

88. R. W. Schoenlein, S. Chattopadhyay, H. H. Chong, T. E. Glover, P. A. Heimann, C. V. Shank, A. A. Zholents, and M. S. Zolotorev, "Generation of femtosecond pulses of synchrotron radiation," *Science*, vol. 287, no. 5461, pp. 2237–2240, 2000.

89. C. Rose-Petruck, R. Jimenez, T. Guo, A. Cavalleri, C. W. Siders, F. Rksi, J. A. Squier, B. C. Walker, K. R. Wilson, and C. P. J. Barty, "Picosecond–milliångström lattice dynamics measured by ultrafast X-ray diffraction," *Nature*, vol. 398, no. 6725, p. 310–312, 1999.

90. G. T. Seidler, D. R. Mortensen, A. J. Remesnik, J. I. Pacold, N. A. Ball, N. Barry, M. Styczinski, and O. R. Hoidn, "A laboratory-based hard X-ray monochromator for high-resolution X-ray emission spectroscopy and X-ray absorption near edge structure measurements," *Rev. Sci. Instrum.*, vol. 85, no. 11, p. 113906, 2014.

91. M. Kavčič, M. Budnar, A. Mühleisen, F. Gasser, M. Žitnik, K. Bučar, and R. Bohinc, "Design and performance of a versatile curved-crystal spectrometer for high-resolution spectroscopy in the tender X-ray range," *Rev. Sci. Instrum.*, vol. 83, no. 3, p. 033113, 2012.

92. I. Llorens, E. Lahera, W. Delnet, O. Proux, A. Braillard, J. L. Hazemann, A. Prat, D. Testemale, Q. Dermigny, F. Gelebart, et al., "High energy resolution five-crystal spectrometer for high quality fluorescence and absorption measurements on an x-ray absorption spectroscopy beamline," *Rev. Sci. Instrum.*, vol. 83, no. 6, p. 063104, 2012.

93. J. Stöhr, *NEXAFS Spectroscopy*, vol. 25. Springer Science & Business Media, 2013.

94. D. C. Sergentu, T. J. Duignan, and J. Autschbach, "Ab initio study of covalency in the ground versus core-excited states and x-ray absorption spectra of actinide complexes," *J. Phys. Chem. Lett.*, vol. 9, no. 18, pp. 5583–5591, 2018.

95. E. R. Davidson, "The iterative calculation of a few of the lowest eigenvalues and corresponding eigenvectors of large real-symmetric matrices," *J. Comput. Phys.*, vol. 17, no. 1, pp. 87–94, 1975.

96. E. Vecharynski, C. Yang, and F. Xue, "Generalized preconditioned locally harmonic residual method for non-hermitian eigenproblems," *SIAM J. Sci. Comput.*, vol. 38, no. 1, pp. A500–A527, 2016.

97. W. Liang, S. A. Fischer, M. J. Frisch, and X. Li, "Energy-specific linear response TDHF/TDDFT for calculating high-energy excited states," *J. Chem. Theory Comput.*, vol. 7, no. 11, pp. 3540–3547, 2011.

98. P. J. Lestrange, F. Egidi, and X. Li, "The consequences of improperly describing oscillator strengths beyond the electric dipole approximation," *J. Chem. Phys.*, vol. 143, no. 23, p. 234103, 2015.

99. S. Bernadotte, A. J. Atkins, and C. R. Jacob, "Origin-independent calculation of quadrupole intensities in X-ray spectroscopy," *J. Chem. Phys.*, vol. 137, no. 20, p. 204106, 2012.

100. M. Bühl, "DFT computations of transition–metal chemical shifts," *Ann. Rep. NMR Spectrosc.*, vol. 64, pp. 77–125, 2008.

101. J. Autschbach, and S. Zheng, "Relativistic computations of NMR parameters from first principles: Theory and applications," *Annu. Rep. NMR Spectrosc.*, vol. 67, pp. 1–95, 2009.

102. M. Bühl, and M. Parrinello, "Medium effects on ^{51}V NMR chemical shifts: A density functional study," *Chemistry*, vol. 7, no. 20, pp. 4487–4494, 2001.

103. M. Bühl, and F. T. Mauschick, "Thermal and solvent effects on ^{57}Fe NMR chemical shifts," *Phys. Chem. Chem. Phys.*, vol. 4, no. 22, pp. 5508–5514, 2002.

104. K. Sutter, L. A. Truflandier, and J. Autschbach, "NMR J–coupling constants in cisplatin derivatives studied by molecular dynamics and relativistic density functional theory," *Chemphyschem*, vol. 12, no. 8, pp. 1448–1455, 2011.

105. L. A. Truflandier, and J. Autschbach, "Probing the solvent shell with 195Pt chemical shifts: Density functional theory molecular dynamics study of Pt(II) and Pt(IV) anionic complexes in aqueous solution," *J. Am. Chem. Soc.*, vol. 132, no. 10, pp. 3472–3483, 2010.

106. L. C. Ducati, A. Marchenko, and J. Autschbach, "NMR J-coupling constants of Tl–Pt bonded metal complexes in aqueous solution: Ab-initio molecular dynamics and localized orbital analysis," *Inorg. Chem.*, vol. 55, no. 22, pp. 12011–12023, 2016.

107. H. W. Spiess, "Rotation of molecules and nuclear spin relaxation," in *NMR Basic Principles and Progress* (R. K. P. Diehl, and E. Fluck, eds.), vol. 15, pp. 55–214. Springer, Berlin, 1978.

108. J. Mareš, M. Hanni, P. Lantto, J. Lounila, and J. Vaara, "Curie-type paramagnetic NMR relaxation in the aqueous solution of Ni(ii)," *Phys. Chem. Chem. Phys.*, vol. 16, no. 15, pp. 6916–6924, 2014.

109. J. Schnmidt, J. Vandevondele, I. F. W. Kuo, D. Sebastiani, J. I. Siepmann, J. Hutter, and C. J. Mundy, "Isobaric-isothermal molecular dynamics simulations utilizing density functional theory: An assessment of the structure and density of water at near-ambient conditions," *J. Phys. Chem. B*, vol. 113, no. 35, pp. 11959–11964, 2009.

110. A. Philips, A. Marchenko, L. A. Truflandier, and J. Autschbach, "Quadrupolar NMR relaxation from ab-initio molecular dynamics: Improved sampling and cluster models vs. periodic calculations," *J. Chem. Theory Comput.*, vol. 13, no. 9, pp. 4397–4409, 2017.

111. L. A. Truflandier, E. Brendler, J. Wagler, and J. Autschbach, "^{29}Si DFT/NMR observation of spin-orbit effect in metallasilatrane sheds some light on the strength of the metal → Si interaction," *Angew. Chem. Int. Ed.*, vol. 50, no. 1, pp. 255–259, 2011.

112. P. Hrobárik, V. Hrobáriková, A. H. Greif, and M. Kaupp, "Giant spin-orbit effects on NMR shifts in diamagnetic actinide complexes: Guiding the search of uranium(VI)

hydride complexes in the correct spectral range," *Angew. Chem. Int. Ed. Engl.*, vol. 51, no. 43, pp. 10884–10888, 2012.

113. A. H. Greif, P. Hrobárik, J. Autschbach, and M. Kaupp, "Giant spin-orbit effects on 1H and 13C NMR shifts for uranium(VI) complexes revisited: Role of the exchange-correlation response kernel, bonding analyses, and new predictions," *Phys. Chem. Chem. Phys.*, vol. 18, no. 44, pp. 30462–30474, 2016.

114. A. Bagno, and G. Saielli, "Relativistic DFT calculations of the NMR properties and reactivity of transition metal methane σ-complexes: Insights on C-H bond activation," *Phys. Chem. Chem. Phys.*, vol. 13, no. 10, pp. 4285–4291, 2011.

115. J. Vaara, O. L. Malkina, H. Stoll, V. G. Malkin, and M. Kaupp, "Study of relativistic effects on nuclear shieldings using density-functional theory and spin-orbit pseudopotentials," *J. Chem. Phys.*, vol. 114, no. 1, pp. 61–71, 2001.

116. D. G. Gusev, "Effect of weak interactions on the H...H distance in stretched dihydrogen complexes," *J. Am. Chem. Soc.*, vol. 126, no. 43, pp. 14249–14257, 2004.

117. B. Le Guennic, S. Patchkovskii, and J. Autschbach, "Density functional study of H–D coupling constants in heavy metal dihydrogen and dihydride complexes: The role of geometry, spin-orbit coupling, and gradient corrections in the exchange-correlation kernel," *J. Chem. Theory Comput.*, vol. 1, no. 4, pp. 601–611, 2005.

118. B. C. Mort, and J. Autschbach, "Zero-point corrections and temperature dependence of HD spin-spin coupling constants of heavy metal hydride and dihydrogen complexes calculated by vibrational averaging," *J. Am. Chem. Soc.*, vol. 128, no. 31, pp. 10060–10072, 2006.

119. J. Autschbach, "Two–component relativistic hybrid density functional computations of nuclear spin–spin coupling tensors using slater–type basis sets and density–fitting techniques," *J. Chem. Phys.*, vol. 129, no. 9, p. 094105, 2008.

120. D. L. Bryce, R. E. Wasylishen, J. Autschbach, and T. Ziegler, "Periodic trends in indirect nuclear spin-spin coupling tensors: Relativistic density functional calculations for interhalogen diatomics," *J. Am. Chem. Soc.*, vol. 124, no. 17, pp. 4894–4900, 2002.

121. M. Kaupp, and F. H. Köhler, "Combining NMR spectroscopy and quantum chemistry as tools to quantify spin density distributions in molecular magnetic compounds," *Coord. Chem. Rev.*, vol. 253, no. 19–20, pp. 2376–2386, 2009.

122. J. Vaara, "Chemical shift in paramagnetic systems," in *Science and Technology of Atomic, Molecular, Condensed Matter and Biological Systems* (R. H. Contreras, ed.), vol. 3, pp. 41–67. Elsevier, Amsterdam, 2013.

123. J. Autschbach, "NMR calculations for paramagnetic molecules and metal complexes," in *Annual Reports in Computational Chemistry* (D. A. Dixon, ed.), vol. 11, pp. 3–36. Elsevier, Amsterdam, 2015.

124. P. Hrobárik, R. Reviakine, A. V. Arbuznikov, O. L. Malkina, V. G. Malkin, F. H. Köhler, and M. Kaupp, "Density functional calculations of NMR shielding tensors for paramagnetic systems with arbitrary spin multiplicity: Validation on 3d metallocenes," *J. Chem. Phys.*, vol. 126, no. 2, p. 024107, 2007.

125. F. Aquino, N. Govind, and J. Autschbach, "Scalar relativistic computations of nuclear magnetic shielding and g-shifts with the zeroth-order regular approximation and range-separated hybrid density functionals," *J. Chem. Theory Comput.*, vol. 7, no. 10, pp. 3278–3292, 2011.

126. F. Rastrelli, and A. Bagno, "Predicting the ¹h and ¹³C NMR spectra of paramagnetic Ru(III) complexes by DFT," *Magn. Reson. Chem.*, vol. 48, pp. S132–S141, 2010.

127. J. Autschbach, S. Patchkovskii, and B. Pritchard, "Calculation of hyperfine tensors and paramagnetic NMR shifts using the relativistic zeroth-order regular approximation and density functional theory," *J. Chem. Theory Comput.*, vol. 7, no. 7, pp. 2175–2188, 2011.

128. S. Komorovsky, M. Repisky, K. Ruud, O. L. Malkina, and V. G. Malkin, "Four-component relativistic density functional theory calculations of NMR shielding tensors for paramagnetic systems," *J. Phys. Chem. A*, vol. 117, no. 51, pp. 14209–14219, 2013.

129. F. Rastrelli, and A. Bagno, "Predicting the NMR spectra of paramagnetic molecules by DFT: Application to organic free radicals and transition-metal complexes," *Chemistry*, vol. 15, no. 32, pp. 7990–8004, 2009.

130. A. Borgogno, F. Rastrelli, and A. Bagno, "Predicting the spin state of paramagnetic iron complexes by DFT calculation of proton NMR spectra," *Dalton Trans.*, vol. 43, no. 25, pp. 9486–9496, 2014.

131. B. Pritchard, and J. Autschbach, "Theoretical investigation of paramagnetic NMR shifts in transition metal acetylacetonato complexes: Analysis of signs, magnitudes, and the role of the covalency of ligand-metal bonding," *Inorg. Chem.*, vol. 51, no. 15, pp. 8340–8351, 2012.

132. S. Moon, and S. Patchkovskii, "First-principles calculations of paramagnetic NMR shifts," in *Calculation of NMR and EPR Parameters. Theory and Applications* (M. Kaupp, M. Bu¨hl, and V. G. Malkin, eds.), pp. 325–338. Wiley–VCH, Weinheim, 2004.

133. W. Van den Heuvel, and A. Soncini, "NMR chemical shift as analytical derivative of the Helmholtz free energy," *J. Chem. Phys.*, vol. 138, no. 5, p. 054113, 2013.

134. F. Gendron, and J. Autschbach, "Ligand NMR chemical shift calculations for paramagnetic metal complexes: 5f^1 vs. 5f^2 actinides," *J. Chem. Theory Comput.*, vol. 12, no. 11, pp. 5309–5321, 2016.

135. F. Gendron, K. Sharkas, and J. Autschbach, "Calculating NMR chemical shifts for paramagnetic metal complexes from first-principles," *J. Phys. Chem. Lett.*, vol. 6, no. 12, pp. 2183–2188, 2015.

136. J. Vaara, S. A. Rouf, and J. Mareš, "Magnetic couplings in the chemical shift of paramagnetic NMR," *J. Chem. Theory Comput.*, vol. 11, no. 10, pp. 4840–4849, 2015.

137. B. Bleaney, "Nuclear magnetic resonance shifts in solution due to lanthanide ions," *J. Mag. Res.*, vol. 8, no. 1, pp. 91–100, 1972.

138. R. M. Golding, and P. Pyykkö, "Theory of pseudocontact NMR shifts due to lanthanide complexes," *Mol. Phys.*, vol. 26, pp. 1389–1396, 1973.

139. B. W. Reilley, C. N. Good, and J. F. Desreux, "Structure-independent method for dissecting contact and dipolar NMR shifts in lanthanide complexes and its use in structure determination," *Anal. Chem.*, vol. 13, pp. 2110–2116, 2017.

140. L. Fusaro, G. Casella, and A. Bagno, "Direct detection of ^{17}O in [gd(dota)]$^-$ by NMR spectroscopy," *Chem. Eur. J.*, vol. 21, no. 5, pp. 1955–1960, 2015.

141. D. A. Dixon, D. Feller, and K. A. Peterson, *Annual Reports in Computational Chemistry*. Elsevier, Amsterdam, 2012.

142. D. Feller, K. A. Peterson, and D. A. Dixon, "Further benchmarks of a composite, convergent, statistically calibrated coupled cluster-based approach for thermochemical and spectroscopic studies," *Mol. Phys.*, vol. 110, no. 19–20, pp. 2381–2399, 2012.

143. K. A. Peterson, D. Feller, and D. A. Dixon, "Chemical accuracy in ab initio thermochemistry and spectroscopy: Current strategies and future challenges," *Theor. Chem. Acc.*, vol. 131, no. 1, p. 1079, 2012.

144. D. Feller, K. A. Peterson, and D. A. Dixon, *Annual Reports in Computational Chemistry*. Elsevier, Amsterdam, 2016.

145. T. H. Dunning, "Gaussian basis sets for use in correlated molecular calculations. I. The atoms boron through neon and hydrogen," *J. Chem. Phys.*, vol. 90, no. 2, pp. 1007–1023, 1989.

146. R. A. Kendall, T. H. Dunning, and R. J. Harrison, "Electron affinities of the first-row atoms revisited. Systematic basis sets and wave functions," *J. Chem. Phys.*, vol. 96, no. 9, pp. 6796–6806, 1992.

147. K. A. Peterson, "Correlation consistent basis sets for actinides. I. The Th and U atoms," *J. Chem. Phys.*, vol. 142, no. 7, p. 074105, 2015.

148. R. Feng, and K. A. Peterson, "Correlation consistent basis sets for actinides. II. The atoms Ac and Np–Lr," *J. Chem. Phys.*, vol. 147, no. 8, p. 084108, 2017.

149. M. Douglas, and N. M. Kroll, "Quantum electrodynamical corrections to the fine structure of helium," *Ann. Phys.*, vol. 82, no. 1, pp. 89–155, 1974.

150. B. A. Hess, "Applicability of the no-pair equation with free-particle projection operators to atomic and molecular structure calculations," *Phys. Rev. A Gen. Phys.*, vol. 32, no. 2, pp. 756–763, 1985.

151. W. A. de Jong, R. J. Harrison, and D. A. Dixon, "Parallel Douglas-Kroll energy and gradients in NWChem: Estimating scalar relativistic effects using Douglas-Kroll contracted basis sets," *J. Chem. Phys.*, vol. 114, no. 1, pp. 48–53, 2001.

152. G. D. Purvis III and R. J. Bartlett, "A full coupled-cluster singles and doubles model: The inclusion of disconnected triples," *J. Chem. Phys.*, vol. 76, pp. 1910–1918, 1982.

153. K. Raghavachari, G. W. Trucks, J. A. Pople, and M. Head-Gordon, "A fifth-order perturbation comparison of electron correlation theories," *Chem. Phys. Lett.*, vol. 157, no. 6, pp. 479–483, 479–483, 1989.

154. J. D. Watts, J. Gauss, and R. J. Bartlett, "Coupled-cluster methods with noniterative triple excitations for restricted open-shell Hartree-Fock and other general single-determinant reference functions. Energies and analytical gradients," *J. Chem. Phys.*, vol. 98, no. 11, pp. 8718–8733, 1993.

155. R. J. Bartlett, and M. Musiał, "Coupled-cluster theory in quantum chemistry," *Rev. Mod. Phys.*, vol. 79, no. 1, pp. 291–352, 2007.

156. M. J. O. Deegan, and P. J. Knowles, "Perturbative corrections to account for triple excitations in closed and open shell coupled cluster theories," *Chem. Phys. Lett.*, vol. 227, no. 3, pp. 321–326, 1994.

157. M. Rittby, and R. J. Bartlett, "An open-shell spin-restricted coupled cluster method: Application to ionization potentials in N_2," *J. Phys. Chem.*, vol. 92, no. 11, pp. 3033–3036, 1988.

158. P. J. Knowles, C. Hampel, and H.-J. Werner, "Coupled-cluster theory for high spin, open shell reference wave functions," *J. Chem. Phys.*, vol. 99, no. 7, pp. 5219–5227, 1993.

159. M. Vasiliu, K. A. Peterson, J. K. Gibson, and D. A. Dixon, "Reliable potential energy surfaces for the reactions of H_2O with ThO_2, PaO^+, UO^{2+}, and UO^+," *J. Phys. Chem. A*, vol. 119, no. 46, pp. 11422–11431, 2015.

160. Z. Fang, Z. Lee, K. A. Peterson, and D. A. Dixon, "Use of improved orbitals for CCSD(T) calculations for predicting heats of formation of group IV and group VI metal oxide monomers and dimers and UCl6," *J. Chem. Theory Comput.*, vol. 12, no. 8, pp. 3583–3592, 2016.

161. B. O. Roos, *Advances in Chemical Physics*. John Wiley & Sons, Inc, Hoboken, 1987.

162. K. Andersson, P. A. Malmqvist, B. O. Roos, A. J. Sadlej, and K. Wolinski, "Second-order perturbation theory with a CASSCF reference function," *J. Phys. Chem.*, vol. 94, no. 14, pp. 5483–5488, 1990.

163. K. Andersson, P. A. Malmqvist, and B. O. Roos, "Second-order perturbation theory with a complete active space self-consistent field reference function," *J. Phys. Chem.*, vol. 96, no. 2, pp. 1218–1226, 1992.

164. E. D. Glendening, J. K. Badenhoop, A. E. Reed, J. E. Carpenter, J. A. Bohmann, C. M. Morales, C. R. Landis, and F. Weinhold, *Natural Bond Order 6.0*. University of Wisconsin Press, Madison, 2013.

165. E. D. Glendening, C. R. Landis, and F. Weinhold, "NBO 6.0: Natural bond orbital analysis program," *J. Comput. Chem.*, vol. 34, no. 16, pp. 1429–1437, 2013.

166. A. E. Reed, L. A. Curtiss, and F. Weinhold, "Intermolecular interactions from a natural bond orbital, donor-acceptor viewpoint," *Chem. Rev.*, vol. 88, no. 6, pp. 899–926, 1988.

167. A. E. Reed, L. A. Curtiss, and F. Weinhold, "Natural population analysis," *J. Chem. Phys.*, vol. 83, pp. 735–746, 1985.

168. F. Weinhold, and C. R. Landis, *Valency and Bonding. A Natural Bond Orbital Donor–Acceptor Perspective*. University Press, Cambridge, 2005.

169. A. E. Clark, J. L. Sonnenberg, P. J. Hay, and R. L. Martin, "Density and wave function analysis of actinide complexes: What can fuzzy atom, atoms-in-molecules, Mulliken, Lowdin, and natural population analysis tell us?," *J. Chem. Phys.*, vol. 121, no. 6, pp. 2563–2570, 2004.

170. H.-J. Werner, P. J. Knowles, G. Knizia, F. R. Manby, and M. Schütz, "Molpro: A general purpose quantum chemistry program package," *Wiley Interdiscip. Rev. Comput. Mol. Sci.*, vol. 2, no. 2, pp. 242–253, 2012.

171. M. Valiev, E. J. Bylaska, N. Govind, K. Kowalski, T. P. Straatsma, H. J. J. van Dam, D. Wang, J. Nieplocha, E. Aprà, T. L. Windus, and W. A. de Jong, "NwChem: A comprehensive and scalable open-source solution for large scale molecular simulations," *Comput. Phys. Commun.*, vol. 181, no. 9, pp. 1477–1489, 2010.

172. T. P. Straatsma, E. J. Bylaska, H. J. J. van Dam, N. Govind, W. A. de Jong, K. Kowalski, and M. Valiev, *Annual Reports in Computational Chemistry*. Elsevier, Amsterdam, 2011.

173. R. A. Kendall, E. Aprà, D. E. Bernholdt, E. J. Bylaska, M. Dupuis, G. I. Fann, R. J. Harrison, J. Ju, J. A. Nichols, J. Nieplocha, T. P. Straatsma, T. L. Windus, and A. T. Wong, "High performance computational chemistry: an overview of NWChem a distributed parallel application," *Comput. Phys. Commun.*, vol. 128, no. 1–2, pp. 260–283, 2000.

174. J. Thyssen, *Development and Applications of Methods for Correlated Relativistic Calculations of Molecular Properties*. University of Southern Denmark, Odense, 2001.

175. R. Bast, K. Dyall, U. Ekström, E. Eliav, T. Enevoldsen, T. Fleig, A. Gomes, J. Henriksson, M. Ilias, C. Jacob, et al., "DIRAC, a relativistic ab initio electronic structure program, release DIRAC11," 2012.

176. P. Nichols, N. Govind, E. J. Bylaska, and W. A. de Jong, "Gaussian basis set and planewave relativistic spin-orbit methods in NWCHEM," *J. Chem. Theory Comput.*, vol. 5, no. 3, pp. 491–499, 2009.

177. G.T. Te Velde, F. M. Bickelhaupt, E. J. Baerends, C. Fonseca Guerra, S. J. A. van Gisbergen, J. G. Snijders, and T. Ziegler, "Chemistry with ADF," *J. Comput. Chem.*, vol. 22, no. 9, pp. 931–967, 2001.

178. E. J. Baerends, T. Ziegler, A. J. Atkins, J. Autschbach, D. Bashford, O. Baseggio, A. Bérces, F. M. Bickelhaupt, C. Bo, P. M. Boerritger, et al., *ADF2017, SCM, Theoretical Chemistry*. Vrije Universiteit, Amsterdam, The Netherlands.

179. E. van Lenthe, A. E. Ehlers, and E. J. Baerends, "Geometry optimizations in the zero order regular approximation for relativistic effects," *J. Chem. Phys.*, vol. 110, no. 18, pp. 8943–8953, 1999.

180. E. van Lenthe, E. J. Baerends, and J. G. Snijders, "Relativistic regular two-component Hamiltonians," *J. Chem. Phys.*, vol. 99, pp. 4597–4610, 1993.

181. E. van Lenthe, E. J. Baerends, and J. G. Snijders, "Relativistic total energy using regular approximations," *J. Chem. Phys.*, vol. 101, no. 11, pp. 9783–9792, 1994.

182. E. van Lenthe, J. G. Snijders, and E. J. Baerends, "The zero-order regular approximation for relativistic effects: The effect of spin-orbit coupling in closed shell molecules," *J. Chem. Phys.*, vol. 105, no. 15, pp. 6505–6516, 1996.

183. E. Van Lenthe, R. Van Leeuwen, E. J. Baerends, and J. G. Snijders, "Relativistic regular two-component Hamiltonians," *Int. J. Quantum Chem.*, vol. 57, no. 3, pp. 281–293, 1996.

184. A. D. Becke, "Density-functional exchange-energy approximation with correct asymptotic behavior," *Phys. Rev.*, vol. 38, pp. 785–789, 1988.

185. C. Lee, W. Yang, and R. G. Parr, "Development of the Colle-Salvetti correlation–energy formula into a functional of the electron density," *Phys. Rev. B*, vol. 37, no. 2, pp. 785–789, 1988.

186. P. D. Dau, M. Vasiliu, K. A. Peterson, D. A. Dixon, and J. K. Gibson, "Remarkably high stability of late actinide dioxide cations: Extending chemistry to pentavalent berkelium and californium," *Chemistry*, vol. 23, no. 68, pp. 17369–17378, 2017.

187. S. Knecht, H. J. A. Jensen, and T. Fleig, "Large-scale parallel configuration interaction. II. Two-and four-component double-group general active space implementation with application to BiH," *J. Chem. Phys.*, vol. 132, p. 014108, 2010.

188. L. Visscher, E. Eliav, and U. Kaldor, "Formulation and implementation of the relativistic Fock-space coupled cluster method for molecules," *J. Chem. Phys.*, vol. 115, no. 21, pp. 9720–9726, 2001.

189. L. Visscher, T. J. Lee, and K. G. Dyall, "Formulation and implementation of a relativistic unrestricted coupled-cluster method including noniterative connected triples," *J. Chem. Phys.*, vol. 105, no. 19, pp. 8769–8776, 1996.

190. K. A. Peterson, and K. G. Dyall, *Computational Methods in Lanthanide and Actinide Chemistry*. John Wiley & Sons, Ltd., Chichester, 2015.

191. K. G. Dyall, "Relativistic double-zeta, triple-zeta, and quadruple-zeta basis sets for the actinides ac–lr," *Theor. Chem. Acc.*, vol. 117, no. 4, pp. 491–500, 2007.

192. K. G. Dyall, "Core correlating basis functions for elements 31–118," *Theor. Chem. Acc.*, vol. 131, no. 5, p. 1217, 2012.

193. S. K. Cary, M. Vasiliu, R. E. Baumbach, J. T. Stritzinger, T. D. Green, K. Diefenbach, J. N. Cross, K. L. Knappenberger, G. Liu, M. A. Silver, et al., "Emergence of californium as the second transitional element in the actinide series," *Nat. Commun.*, vol. 6, p. 6827, 2015.

194. M. A. Silver, S. K. Cary, J. A. Johnson, R. E. Baumbach, A. A. Arico, M. Luckey, M. Urban, J. C. Wang, M. J. Polinski, A. Chemey, et al., "Characterization of berkelium (III) dipicolinate and borate compounds in solution and the solid state," *Science*, vol. 353, no. 6302, p. aaf3762, 2016.

195. J. N. Cross, J. Su, E. R. Batista, S. K. Cary, W. J. Evans, S. A. Kozimor, V. Mocko, B. L. Scott, B. W. Stein, C. J. Windorff, and P. Yang, "Covalency in americium(III) hexachloride," *J. Am. Chem. Soc.*, vol. 139, no. 25, pp. 8667–8677, 2017.

196. R. M. Diamond, J. K. Street, and G. T. Seaborg, "An ion-exchange study of possible hybridized 5f bonding in the actinides," *J. Am. Chem. Soc.*, vol. 76, pp. 1461–1469, 1954.

197. J. Tomasi, B. Mennucci, and R. Cammi, "Quantum mechanical continuum solvation models," *Chem. Rev.*, vol. 105, no. 8, pp. 2999–3093, 2005.

198. C.-G. Zhan, and D. A. Dixon, "Absolute hydration free energy of the proton from first-principles electronic structure calculations," *J. Phys. Chem. A*, vol. 105, no. 51, pp. 11534–11540, 2001.

199. C.-G. Zhan, and D. A. Dixon, "First-principles determination of the absolute hydration free energy of the hydroxide ion†," *J. Phys. Chem. A*, vol. 106, no. 42, pp. 9737–9744, 2002.

200. C.-G. Zhan, and D. A. Dixon, "The nature and absolute hydration free energy of the solvated electron in water," *J. Phys. Chem. B*, vol. 107, no. 18, pp. 4403–4417, 2003.

201. C.-G. Zhan, and D. A. Dixon, "Hydration of the fluoride anion: Structures and absolute hydration free energy from first-principles electronic structure calculations," *J. Phys. Chem. A*, vol. 108, no. 11, pp. 2020–2029, 2004.

202. Y. Alexeev, T. L. Windus, C.-G. Zhan, and D. A. Dixon, "Accurate heats of formation and acidities for H_3PO_4, H_2SO_4, and H_2CO_3 from ab initio electronic structure calculations," *Int. J. Quantum Chem.*, vol. 102, no. 5, pp. 775–784, 2005.

203. D. A. Dixon, D. Feller, C.-G. Zhan, and J. S. Francisco, "The gas and solution phase acidities of HNO, HOOONO, HONO, and $HONO_2$," *Int. J. Mass Spectrom.*, vol. 227, no. 3, pp. 421–438, 2003.

204. K. E. Gutowski, and D. A. Dixon, "Ab initio prediction of the gas- and solution-phase acidities of strong Brønsted acids: The calculation of pk_a values less than −10," *J. Phys. Chem. A*, vol. 110, no. 43, pp. 12044–12054, 2006.

205. V. E. Jackson, A. R. Felmy, and D. A. Dixon, "Prediction of the pKa's of aqueous metal ion +2 complexes," *J. Phys. Chem. A*, vol. 119, no. 12, pp. 2926–2939, 2015.

206. A. Klamt, *Quantum Chemistry to Fluid Phase Thermodynamics and Drug Design*. Elsevier, Amsterdam, 2005.

207. A. Klamt, and G. Schüürmann, "Cosmo: A new approach to dielectric screening in solvents with explicit expressions for the screening energy and its gradient," *J. Chem. Soc. Perkin Trans.* vol. 2, no. 5, pp. 799–805, 1993.

208. M. Frisch, G. Trucks, H. Schlegel, G. Scuseria, M. Robb, J. Cheeseman, G. Scalmani, V. Barone, G. Petersson, H. Nakatsuji, et al., "Gaussian 16 rev. b. 01; Gaussian," 2016.

209. C. J. Cramer, and D. G. Truhlar, *Trends and Perspectives in Modern Computational Science*. Brill/VSP, Leiden, 2006.

210. A. V. Marenich, C. J. Cramer, and D. G. Truhlar, "Universal solvation model based on solute electron density and on a continuum model of the solvent defined by the bulk dielectric constant and atomic surface tensions," *J. Phys. Chem. B*, vol. 113, no. 18, pp. 6378–6396, 2009.

211. D. G. Truhlar, and J. R. Pliego Jr., *Continuum Solvation Models in Chemical Physics: Theory and Application*. Wiley, Chichester, 2007.

212. J. Neuefeind, L. Soderholm, and S. Skanthakumar, "Experimental coordination environment of uranyl(VI) in aqueous solution," *J. Phys. Chem. A*, vol. 108, no. 14, pp. 2733–2739, 2004.

213. L. Soderholm, S. Skanthakumar, and J. Neuefeind, "Determination of actinide speciation in solution using high-energy X-ray scattering," *Anal. Bioanal. Chem.*, vol. 383, no. 1, pp. 48–55, 2005.

214. K. E. Gutowski, and D. A. Dixon, "Predicting the energy of the water exchange reaction and free energy of solvation for the uranyl ion in aqueous solution," *J. Phys. Chem. A*, vol. 110, no. 28, pp. 8840–8856, 2006.

215. M. W. Feyereisen, D. Feller, and D. A. Dixon, "Hydrogen bond energy of the water dimer," *J. Phys. Chem.*, vol. 100, no. 8, pp. 2993–2997, 1996.

216. P. Parmar, A. Samuels, and A. E. Clark, "Applications of polarizable continuum models to determine accurate solution-phase thermochemical values across a broad range of cation charge – the case of U(III–VI)," *J. Chem. Theory Comput.*, vol. 11, no. 1, pp. 55–63, 2015.

217. W. L. Jorgensen, D. S. Maxwell, and J. Tirado-Rives, "Development and testing of the OPLS all-atom force field on conformational energetics and properties of organic liquids," *J. Am. Chem. Soc.*, vol. 118, no. 45, pp. 11225–11236, 1996.

218. D. A. Case, T. E. Cheatham, T. Darden, H. Gohlke, R. Luo, K. M. Merz, A. Onufriev, C. Simmerling, B. Wang, and R. J. Woods, "The amber biomolecular simulation programs," *J. Comput. Chem.*, vol. 26, no. 16, pp. 1668–1688, 2005.

219. J. Wang, R. M. Wolf, J. W. Caldwell, P. A. Kollman, and D. A. Case, "Development and testing of a general amber force field," *J. Comput. Chem.*, vol. 25, no. 9, pp. 1157–1174, 2004.

220. K. Vanommeslaeghe, E. Hatcher, C. Acharya, S. Kundu, S. Zhong, J. Shim, E. Darian, O. Guvench, P. Lopes, I. Vorobyov, and A. D. Mackerell, "Charmm general force field: A force field for drug-like molecules compatible with the CHARMM all-atom additive biological force fields," *J. Comput. Chem.*, vol. 31, no. 4, pp. 671–690, 2010.

221. D. Frenkel, and B. Smit, *Understanding Molecular Simulation: From Algorithms to Applications*, vol. 1. Elsevier, 2001.

222. D. C. Rapaport, and D. C. R. Rapaport, *The Art of Molecular Dynamics Simulation*. Cambridge University Press, 2004.

223. P. Guilbaud, and G. Wipff, "Hydration of uranyl (UO22+) cation and its nitrate ion and 18-crown-6 adducts studied by molecular dynamics simulations," *J. Phys. Chem.*, vol. 97, no. 21, pp. 5685–5692, 1993.

224. G. Benay, and G. Wipff, "Liquid–liquid extraction of uranyl by TBP: The TBP and ions models and related interfacial features revisited by MD and PMF simulations," *J. Phys. Chem. B*, vol. 118, no. 11, pp. 3133–3149, 2014.

225. E. R. Irish, and W. H. Reas, *The Purex Process: A Solvent Extraction Reprocessing Method for Irradiated Uranium*. Hanford Atomic Products Operation, Richland, WA, 1957.

226. M. Baaden, R. Schurhammer, and G. Wipff, "Molecular dynamics study of the uranyl extraction by tri-n-butylphosphate (TBP): Demixing of water/'oil'/TBP solutions with a comparison of supercritical CO_2 and chloroform," *J. Phys. Chem. B*, vol. 106, no. 2, pp. 434–441, 2002.

227. D. C. Stepinski, B. A. Young, M. P. Jensen, P. G. Rickert, J. A. Dzielawa, A. A. Dilger, D. J. Rausch, and M. L. Dietz, "Application of ionic liquids in actinide and fission product separations: Progress and prospects," in *Separations for the Nuclear Fuel Cycle in the 21st Century, ACS Symposium Series*, American Chemical Society, Washington DC, pp. 233–247, 2006.

228. Z. Liu, J. Timmermann, K. Reuter, and C. Scheurer, "Benchmarks and dielectric constants for reparametrized OPLS and polarizable force field models of chlorinated hydrocarbons," *J. Phys. Chem. B*, vol. 122, no. 2, pp. 770–779, 2018.

229. D. Hagberg, G. Karlström, B. O. Roos, and L. Gagliardi, "The coordination of uranyl in water: A combined quantum chemical and molecular simulation study," *J. Am. Chem. Soc.*, vol. 127, no. 41, pp. 14250–14256, 2005.

230. D. Hagberg, E. Bednarz, N. M. Edelstein, and L. Gagliardi, "A quantum chemical and molecular dynamics study of the coordination of Cm(III) in water," *J. Am. Chem. Soc.*, vol. 129, no. 46, pp. 14136–14137, 2007.

231. F. Réal, M. Trumm, V. Vallet, B. Schimmelpfennig, M. Masella, and J. P. Flament, "Quantum chemical and molecular dynamics study of the coordination of Th(IV) in aqueous solvent," *J. Phys. Chem. B*, vol. 114, no. 48, pp. 15913–15924, 2010.

232. C. Beuchat, D. Hagberg, R. Spezia, and L. Gagliardi, "Hydration of lanthanide chloride salts: A quantum chemical and classical molecular dynamics simulation study," *J. Phys. Chem. B*, vol. 114, no. 47, pp. 15590–15597, 2010.

233. N. Rai, S. P. Tiwari, and E. J. Maginn, "Force field development for actinyl ions via quantum mechanical calculations: An approach to account for many body solvation effects," *J. Phys. Chem. B*, vol. 116, no. 35, pp. 10885–10897, 2012.

234. V. Pomogaev, S. P. Tiwari, N. Rai, G. S. Goff, W. Runde, W. F. Schneider, and E. J. Maginn, "Development and application of effective pairwise potentials for UO^{n+}, NpO^{n+}, PuO^{n+}, and AmO^{n+} (n = 1, 2) ions with water," *Phys. Chem. Chem. Phys.*, vol. 15, pp. 15954–15963, 2013.

235. J. Li, and F. Wang, "Pairwise-additive force fields for selected aqueous monovalent ions from adaptive force matching," *J. Chem. Phys.*, vol. 143, no. 19, p. 194505, 2015.

236. F. Ercolessi, and J. B. Adams, "Interatomic potentials from first-principles calculations: The force-matching method," *Europhys. Lett.*, vol. 26, no. 8, pp. 583–588, 1994.

237. S. Izvekov, M. Parrinello, C. J. Burnham, and G. A. Voth, "Effective force fields for condensed phase systems from ab initio molecular dynamics simulation: A new method for force-matching," *J. Chem. Phys.*, vol. 120, no. 23, p. 10896–10913, 2004.

238. L. Koziol, L. E. Fried, and N. Goldman, "Using force matching to determine reactive force fields for water under extreme thermodynamic conditions," *J. Chem. Theory Comput.*, vol. 13, no. 1, pp. 135–146, 2017.

239. J. R. Maple, U. Dinur, and A. T. Hagler, "Derivation of force fields for molecular mechanics and dynamics from ab initio energy surfaces," *Proc. Natl. Acad. Sci. U.S.A.*, vol. 85, no. 15, pp. 5350–5354, 1988.

240. M. Doemer, P. Maurer, P. Campomanes, I. Tavernelli, and U. Rothlisberger, "Generalized QM/MM force matching approach applied to the 11-cis protonated Schiff base chromophore of rhodopsin," *J. Chem. Theory Comput.*, vol. 10, no. 1, pp. 412–422, 2014.

241. O. Akin-Ojo, Y. Song, and F. Wang, "Developing ab initio quality force fields from condensed phase quantum-mechanics/molecular-mechanics calculations through the adaptive force matching method," *J. Chem. Phys.*, vol. 129, no. 6, p. 064108, 2008.

242. O. Akin-Ojo, and F. Wang, "Improving the point-charge description of hydrogen bonds by adaptive force matching," *J. Phys. Chem. B*, vol. 113, no. 5, pp. 1237–1240, 2009.

243. O. Akin-Ojo, and F. Wang, "The quest for the best nonpolarizable water model from the adaptive force matching method," *J. Comput. Chem.*, vol. 32, no. 3, pp. 453–462, 2011.

244. I. S. Huang, and M. K. Tsai, "Interplay between polarizability and hydrogen bond network of water: Reparametrizing the flexible single-point-charge water model by the nonlinear adaptive force matching approach," *J. Phys. Chem. A*, vol. 122, no. 19, pp. 4654–4662, 2018.

245. C. M. Baker, "Polarizable force fields for molecular dynamics simulations of biomolecules," *Wiley Interdiscip. Rev. Comput. Mol. Sci.*, vol. 5, no. 2, pp. 241–254, 2015.

246. R. Spezia, V. Migliorati, and P. D'Angelo, "On the development of polarizable and Lennard-Jones force fields to study hydration structure and dynamics of actinide(III) ions based on effective ionic radii," *J. Chem. Phys.*, vol. 147, no. 16, p. 161707, 2017.

247. S. D. Fried, L. P. Wang, S. G. Boxer, P. Ren, and V. S. Pande, "Calculations of the electric fields in liquid solutions," *J. Phys. Chem. B*, vol. 117, no. 50, pp. 16236–16248, 2013.

248. A. Grossfield, P. Ren, and J. W. Ponder, "Ion solvation thermodynamics from simulation with a polarizable force field," *J. Am. Chem. Soc.*, vol. 125, no. 50, pp. 15671–15682, 2003.

249. A. Marjolin, C. Gourlaouen, C. Clavaguéra, P. Y. Ren, J. P. Piquemal, and J. P. Dognon, "Hydration Gibbs free energies of open and closed shell trivalent lanthanide and actinide cations from polarizable molecular dynamics," *J. Mol. Model.*, vol. 20, no. 10, p. 2471, 2014.

250. R. Atta-Fynn, D. F. Johnson, E. J. Bylaska, E. S. Ilton, G. K. Schenter, and W. A. de Jong, "Structure and hydrolysis of the U(IV), U(V), and U(VI) aqua ions from ab initio molecular simulations," *Inorg. Chem.*, vol. 51, no. 5, pp. 3016–3024, 2012.

251. R. Atta-Fynn, E. J. Bylaska, G. K. Schenter, and W. A. de Jong, "Hydration shell structure and dynamics of curium(III) in aqueous solution: First principles and empirical studies," *J. Phys. Chem. A*, vol. 115, no. 18, pp. 4665–4677, 2011.

252. R. Atta-Fynn, E. J. Bylaska, and W. A. de Jong, "Importance of counteranions on the hydration structure of the curium ion," *J. Phys. Chem. Lett.*, vol. 4, no. 13, pp. 2166–2170, 2013.

253. R. Atta-Fynn, E. J. Bylaska, and W. A. de Jong, "Strengthening of the coordination shell by counter ions in aqueous Th^{4+} solutions," *J. Phys. Chem. A*, vol. 120, no. 51, pp. 10216–10222, 2016.

254. M. Pouvreau, M. Dembowski, S. B. Clark, J. G. Reynolds, K. M. Rosso, G. K. Schenter, C. I. Pearce, and A. E. Clark, "Ab initio molecular dynamics reveal spectroscopic siblings and ion pairing as new challenges for elucidating prenucleation aluminum speciation," *J. Phys. Chem. B*, vol. 122, no. 29, pp. 7394–7402, 2018.

255. J. E. Davies, N. L. Doltsinis, A. J. Kirby, C. D. Roussev, and M. Sprik, "Estimating pK_a values for pentaoxyphosphoranes," *J. Am. Chem. Soc.*, vol. 124, no. 23, pp. 6594–6599, 2002.

256. M. Sulpizi, and M. Sprik, "Acidity constants from DFT-based molecular dynamics simulations," *J. Phys. Condens. Matter*, vol. 22, no. 28, p. 284116, 2010.

257. J. Mu, R. Motokawa, K. Akutsu, S. Nishitsuji, and A. J. Masters, "A novel microemulsion phase transition: Toward the elucidation of third-phase formation in spent nuclear fuel reprocessing," *J. Phys. Chem. B*, vol. 122, no. 4, pp. 1439–1452, 2018.

258. M. J. Servis, D. T. Wu, J. C. Shafer, and A. E. Clark, "Square supramolecular assemblies of uranyl complexes in organic solvents," *Chem. Commun.*, vol. 54, no. 72, pp. 10064– 10067, 2018.

259. B. Qiao, K. C. Littrell, and R. J. Ellis, "Liquid worm-like and proto-micelles: Water solubilization in amphiphile–oil solutions," *Phys. Chem. Chem. Phys.*, vol. 20, no. 18, pp. 12908–12915, 2018.

260. P. Guilbaud, L. Berthon, W. Louisfrema, O. Diat, and N. Zorz, "Determination of the structures of uranyl–tri-n-butyl-phosphate aggregates by coupling experimental results with molecular dynamic simulations," *Chemistry*, vol. 23, no. 65, pp. 16660–16670, 2017.

261. L. X. Dang, Q. N. Vo, M. Nilsson, and H. D. Nguyen, "Rate theory on water exchange in aqueous uranyl ion," *Chem. Phys. Lett.*, vol. 671, pp. 58–62, 2017.

262. S. Kerisit, E. J. Bylaska, M. S. Massey, M. E. McBriarty, and E. S. Ilton, "Ab initio molecular dynamics of uranium incorporated in goethite (α-FeOOH): Interpretation of X-ray absorption spectroscopy of trace polyvalent metals," *Inorg. Chem.*, vol. 55, no. 22, pp. 11736–11746, 2016.

263. I. Farkas, I. Bányai, Z. Szabó, U. Wahlgren, and I. Grenthe, "Rates and mechanisms of water exchange of UO^{2+}(aq) and UO_2(oxalate) $F(H_2O)^-$: A variable-temperature ^{17}O and ^{19}F NMR study," *Inorg. Chem.*, vol. 39, no. 4, pp. 799–805, 2000.

264. Y. Nagata, C. S. Hsieh, T. Hasegawa, J. Voll, E. H. Backus, and M. Bonn, "Water bending mode at the water-vapor interface probed by sum-frequency generation spectroscopy: A combined molecular dynamics simulation and experimental study," *J. Phys. Chem. Lett.*, vol. 4, no. 11, pp. 1872–1877, 2013.

265. T. R. Graham, M. Dembowski, E. Martinez-Baez, X. Zhang, N. R. Jaegers, J. Hu, M. S. Gruszkiewicz, H. W. Wang, A. G. Stack, M. E. Bowden, et al., "In situ ^{27}Al NMR spectroscopy of aluminate in sodium hydroxide solutions above and below saturation with respect to gibbsite," *Inorg. Chem.*, vol. 57, no. 19, 11864–11873, 2018.

266. A. Wildman, E. Martinez-Baez, J. Fulton, G. Schenter, C. Pearce, A. E. Clark, and X. Li, "Anticorrelated contributions to pre-edge features of aluminate near-edge X-ray absorption spectroscopy in concentrated electrolytes," *J. Phys. Chem. Lett.*, vol. 9, no. 10, pp. 2444–2449, 2018.

267. B. J. Palmer, D. M. Pfund, and J. L. Fulton, "Direct modeling of EXAFS spectra from molecular dynamics simulations," *J. Phys. Chem.*, vol. 100, no. 32, pp. 13393–13398, 1996.

268. S. Pérez-Conesa, F. Torrico, J. M. Martínez, R. R. Pappalardo, and E. Sánchez Marcos, "A hydrated ion model of $[UO2]^{2+}$ in water: structure, dynamics, and spectroscopy from classical molecular dynamics," *J. Chem. Phys.*, vol. 145, no. 22, p. 224502, 2016.

269. S. Pérez-Conesa, J. M. Martínez, R. R. Pappalardo, and E. Sánchez Marcos, "Extracting the americyl hydration from an americium cationic mixture in solution: A combined X-ray absorption spectroscopy and molecular dynamics study," *Inorg. Chem.*, vol. 57, no. 14, pp. 8089–8097, 2018.

270. D. S. Walker, D. K. Hore, and G. L. Richmond, "Understanding the population, coordination, and orientation of water species contributing to the nonlinear optical spectroscopy of the vapor-water interface through molecular dynamics simulations," *J. Phys. Chem. B*, vol. 110, no. 41, pp. 20451–20459, 2006.

271. W. Rock, B. Qiao, T. Zhou, A. E. Clark, and A. Uysal, "Heavy anionic complex creates a unique water structure at a soft charged interface," 2018.

272. R. W. Zwanzig, "High-temperature equation of state by a perturbation method. I. Nonpolar gasses," *J. Chem. Phys.*, vol. 22, pp. 1420–1426, 1954.

273. M. R. Shirts, and J. D. Chodera, "Statistically optimal analysis of samples from multiple equilibrium states," *J. Chem. Phys.*, vol. 129, no. 12, p. 124105, 2008.

274. N. Bhatnagar, G. Kamath, I. Chelst, and J. J. Potoff, "Direct calculation of 1-octanol-water partition coefficients from adaptive biasing force molecular dynamics simulations," *J. Chem. Phys.*, vol. 137, no. 1, p. 014502, 2012.

275. C. C. Bannan, G. Calabró, D. Y. Kyu, and D. L. Mobley, "Calculating partition coefficients of small molecules in octanol/water and cyclohexane/water," *J. Chem. Theory Comput.*, vol. 12, no. 8, pp. 4015–4024, 2016.

276. J. L. Aragones, E. Sanz, and C. Vega, "Solubility of NaCl in water by molecular simulation revisited," *J. Chem. Phys.*, vol. 136, no. 24, p. 244508, 2012.

277. I. S. Joung, and T. E. Cheatham, "Molecular dynamics simulations of the dynamic and energetic properties of alkali and halide ions using water-model-specific ion parameters," *J. Phys. Chem. B*, vol. 113, no. 40, pp. 13279–13290, 2009.

278. A. S. Paluch, S. Jayaraman, J. K. Shah, and E. J. Maginn, "A method for computing the solubility limit of solids: Application to sodium chloride in water and alcohols," *J. Chem. Phys.*, vol. 133, no. 12, p. 124504, 2010.

279. G. M. Torrie, and J. P. Valleau, "Nonphysical sampling distributions in Monte Carlo free-energy estimation: Umbrella sampling," *J. Comp. Phys.*, vol. 23, no. 2, pp. 187–199, 1977.

280. B. Roux, "The calculation of the potential of mean force using computer simulations," *Comput. Phys. Commun.*, vol. 91, no. 1–3, pp. 275–282, 1995.

281. K. J. Schweighofer, and I. Benjamin, "Transfer of small ions across the water/1,2-dichloroethane interface," *J. Phys. Chem.*, vol. 99, no. 24, pp. 9974–9985, 1995.

282. K. J. Schweighofer, and I. Benjamin, "Transfer of a tetramethylammonium ion across the water–nitrobenzene interface: Potential of mean force and nonequilibrium dynamics," *J. Phys. Chem. A*, vol. 103, no. 49, pp. 10274–10279, 1999.

283. L. X. Dang, "A mechanism for ion transport across the water/dichloromethane interface: A molecular dynamics study using polarizable potential models," *J. Phys. Chem. B*, vol. 105, no. 4, pp. 804–809, 2001.

284. S. Bonhommeau, N. Ottosson, W. Pokapanich, S. Svensson, W. Eberhardt, O. Björneholm, and E. F. Aziz, "Solvent effect of alcohols at the L-edge of iron in solution: X-ray absorption and multiplet calculations," *J. Phys. Chem. B*, vol. 112, no. 40, pp. 12571–12574, 2008.

285. Y. Marcus, and G. Hefter, "Ion pairing," *Chem. Rev.*, vol. 106, no. 11, pp. 4585–4621, 2006.

286. Y. Marcus, "Preferential solvation of ions in mixed solvents. Part 2.–The solvent composition near the ion," *J. Chem. Soc., Faraday Trans. I*, vol. 84, no. 5, pp. 1465–1473, 1988.

287. B. O. Strasser, M. Shamsipur, and A. I. Popov, "Kinetics of complexation of the cesium ion with large crown ethers in acetone and in methanol solutions," *J. Phys. Chem.*, vol. 89, no. 22, pp. 4822–4824, 1985.

288. M. P. Kelley, P. Yang, S. B. Clark, and A. E. Clark, "Competitive interactions within Cm(III) solvation in binary water/methanol solutions," *Inorg. Chem.*, vol. 57, no. 16, pp. 10050–10058, 2018.

289. M. P. Kelley, P. Yang, S. B. Clark, and A. E. Clark, "Structural and thermodynamic properties of the Cm(II) ion solvated by water and methanol," *Inorg. Chem.*, vol. 55, no. 10, pp. 4992–4999, 2016.

290. M. Kelley, A. Donley, S. Clark, and A. Clark, "Structure and dynamics of NaCl ion pairing in solutions of water and methanol," *J. Phys. Chem. B*, vol. 119, no. 51, pp. 15652–15661, 2015.

291. V. E. Jackson, K. E. Gutowski, and D. A. Dixon, "Density functional theory study of the complexation of the uranyl dication with anionic phosphate ligands with and without water molecules," *J. Phys. Chem. A*, vol. 117, no. 36, pp. 8939–8957, 2013.

292. J. R. Rustad, D. A. Dixon, K. M. Rosso, and A. R. Felmy, "Trivalent ion hydrolysis reactions: A linear free-energy relationship based on density functional electronic structure calculations," *J. Am. Chem. Soc.*, vol. 121, no. 13, pp. 3234–3235, 1999.

293. V. Neck, and J. I. Kim, "Solubility and hydrolysis of tetravalent actinides," *Radiochim. Acta*, vol. 89, no. 1, pp. 1–16, 2001.

294. I. Grenthe, J. Fuger, R. J. Konings, R. J. Lemire, A. B. Muller, C. Nguyen-Trung, and H. Wanner, *Chemical Thermodynamics of Uranium*, vol. 1. Elsevier, North-Holland, Amsterdam, 1992.

295. R. Guillaumont, F. J. Mompean, et al., "Update on the chemical thermodynamics of uranium, neptunium, plutonium, americium and technetium," vol. 5. Elsevier, Amsterdam, 2003.

296. M. P. Kelley, J. Su, M. Urban, M. Luckey, E. R. Batista, P. Yang, and J. C. Shafer, "On the origin of covalent bonding in heavy actinides," *J. Am. Chem. Soc.*, vol. 139, no. 29, pp. 9901–9908, 2017.

297. M. P. Kelley, G. J.-P. Deblonde, J. Su, C. H. Booth, R. J. Abergel, E. R. Batista, and P. Yang, "Bond covalency and oxidation state of actinide ions complexed with therapeutic chelating agent 3,4,3-LI(1,2-HOPO)," *Inorg. Chem.*, vol. 57, no. 9, p. 5352–5363, 2018.

298. G. J. -P. Deblonde, M. P. Kelley, J. Su, E. R. Batista, P. Yang, C. H. Booth, and R. J. Abergel, "Spectroscopic and computational characterization of diethylenetriamine-pentaacetic acid/transplutonium chelates: Evidencing heterogeneity in the heavy actinide(III) series," *Angew. Chem. Int. Ed.*, vol. 57, no. 17, pp. 4521–4526, 2018.

299. S. K. Cary, J. Su, S. S. Galley, T. E. Albrecht-Schmitt, E. R. Batista, M. G. Ferrier, S. A. Kozimor, V. Mocko, B. L. Scott, C. E. Van Alstine, F. D. White, and P. Yang, "A series of dithiocarbamates for americium, curium, and californium," *Dalton Trans.*, vol. 47, no. 41, pp. 14452–14461, 2018.

300. S. D. Reilly, J. Su, J. M. Keith, P. Yang, E. R. Batista, A. J. Gaunt, L. M. Harwood, M. J. Hudson, F. W. Lewis, B. L. Scott, C. A. Sharrad, and D. M. Whittaker, "Plutonium coordination and redox chemistry with the CyMe$_4$-BTPhen polydentate N-donor extractant ligand," *Chem. Comm.*, vol. 54, no. 89, pp. 12582–12585, 2018.

301. R. V. Davies, J. Kennedy, R. W. McIlroy, R. Spence, and K. M. Hill, "Extraction of uranium from sea water," *Nature*, vol. 203, pp. 1110–1115, 1964.

302. G. R. Choppin, "Soluble rare earth and actinide species in seawater," *Mar. Chem.*, vol. 28, no. 1–3, pp. 19–26, 1989.

303. J. Kim, C. Tsouris, R. T. Mayes, T. Oyola, T. Saito, C. J. Janke, S. Dai, E. Schneider, and D. Sachde, "Recovery of uranium from seawater: A review of current status and future research needs," *Sep. Sci. Technol.*, vol. 48, no. 3, pp. 367–387, 2013.

304. S. Vukovic, L. A. Watson, S. O. Kang, R. Custelcean, and B. P. Hay, "How amidoximate binds the uranyl cation," *Inorg. Chem.*, vol. 51, no. 6, pp. 3855–3859, 2012.

305. N. Mehio, M. A. Lashely, J. W. Nugent, L. Tucker, B. Correia, C. L. Do-Thanh, S. Dai, R. D. Hancock, and V. S. Bryantsev, "Acidity of the amidoxime functional group in aqueous solution: A combined experimental and computational study," *J. Phys. Chem. B*, vol. 119, no. 8, pp. 3567–3576, 2015.

306. S. Vukovic, B. P. Hay, and V. S. Bryantsev, "Predicting stability constants for uranyl complexes using density functional theory," *Inorg. Chem.*, vol. 54, no. 8, pp. 3995–4001, 2015.

307. A. P. Ladshaw, A. S. Ivanov, S. Das, V. S. Bryantsev, C. Tsouris, and S. Yiacoumi, "First-principles integrated adsorption modeling for selective capture of uranium from seawater by polyamidoxime sorbent materials," *ACS Appl. Mater. Interfaces*, vol. 10, no. 15, pp. 12580–12593, 2018.

308. J. M. Keith, and E. R. Batista, "Theoretical examination of the thermodynamic factors in the selective extraction of Am^{3+} from Eu^{3+} by dithiophosphinic acids," *Inorg. Chem.*, vol. 51, no. 1, pp. 13–15, 2012.

309. L. E. Roy, N. J. Bridges, and L. R. Martin, "Theoretical insights into covalency driven f element separations," *Dalton Trans.*, vol. 42, pp. 2636–2642, 2012.

310. S. M. Ali, S. Pahan, A. Bhattacharyya, and P. K. Mohapatra, "Complexation thermodynamics of diglycolamide with f-elements: Solvent extraction and density functional theory analysis," *Phys. Chem. Chem. Phys.*, vol. 18, no. 14, pp. 9816–9828, 2016.

311. S. M. Ali, "Enhanced free energy of extraction of Eu^{3+} and Am^{3+} ions towards diglycolamide appended calix[4]arene: Insights from DFT-D3 and COSMO-RS solvation models," *Dalton Trans.*, vol. 46, no. 33, pp. 10886–10898, 2017.

312. M. Kaneko, S. Miyashita, and S. Nakashima, "Bonding study on the chemical separation of Am(III) from Eu(III) by S-, N-, and O-donor ligands by means of all-electron ZORA-DFT calculation," *Inorg. Chem.*, vol. 54, no. 14, pp. 7103–7109, 2015.

313. Y. Yang, J. Liu, L. Yang, K. Li, H. Zhang, S. Luo, and L. Rao, "Probing the difference in covalence by enthalpy measurements: A new heterocyclic n-donor ligand for actinide/lanthanide separation," *Dalton Trans.*, vol. 44, no. 19, pp. 8959–8970, 2015.

314. H. Wu, Q. Y. Wu, C. Z. Wang, J. H. Lan, Z. R. Liu, Z. F. Chai, and W. Q. Shi, "Theoretical insights into the separation of Am(III) over Eu(III) with PhenBHPPA," *Dalton Trans.*, vol. 44, no. 38, pp. 16737–16745, 2015.

315. H. Wu, Q. Y. Wu, C. Z. Wang, J. H. Lan, Z. R. Liu, Z. F. Chai, and W. Q. Shi, "New insights into the selectivity of four 1,10-phenanthroline-derived ligands toward the separation of trivalent actinides and lanthanides: A DFT based comparison study," *Dalton Trans.*, vol. 45, no. 19, pp. 8107–8117, 2016.

316. V. S. Bryantsev, and B. P. Hay, "Theoretical prediction of Am(III)/Eu(III) selectivity to aid the design of actinide-lanthanide separation agents," *Dalton Trans.*, vol. 44, pp. 26969–26979, 2017.

317. Q. Y. Wu, Y. T. Song, L. Ji, C. Z. Wang, Z. F. Chai, and W. Q. Shi, "Theoretically unraveling the separation of Am(III)/Eu(III): Insights from mixed N,O-donor ligands with variations of central heterocyclic moieties," *Phys. Chem. Chem. Phys.*, vol. 19, no. 39, pp. 26969–26979, 2017.

318. I. Lehman-Andino, J. Su, K. E. Papathanasiou, T. M. Eaton, J. Jian, D. Dan, T. E. Albrecht-Schmitt, C. J. Dares, E. R. Batista, P. Yang, J. K. Gibson, and K. Kavallieratos, "Soft-donor dipicolinamide derivatives for selective actinide(III)/lanthanide(III) separation: A solution, gas phase, and theoretical study," *Chem. Comm.*, vol. 55, no. 17, pp. 2441–2444, 2019.

319. F. W. Lewis, M. J. Hudson, and L. M. Harwood, "Development of highly selective ligands for separations of actinides from lanthanides in the nuclear fuel cycle," *Synlett*, vol. 18, pp. 2609–2632, 2011.

320. H. V. Lavrov, N. A. Ustynyuk, P. I. Matveev, I. P. Gloriozov, S. S. Zhokhov, M. Y. Alyapyshev, L. I. Tkachenko, I. G. Voronaev, V. A. Babain, S. N. Kalmykov, and Y. A. Ustynyuk, "A novel highly selective ligand for separation of actinides and lanthanides in the nuclear fuel cycle. Experimental verification of the theoretical prediction," *Dalton Trans.*, vol. 46, no. 33, pp. 10926–10934, 2017.

321. C. L. Xiao, C. Z. Wang, L. Y. Yuan, B. Li, H. He, S. Wang, Y. L. Zhao, Z. F. Chai, and W. Q. Shi, "Excellent selectivity for actinides with a tetradentate 2,9-diamide-1,10-phenanthroline ligand in highly acidic solution: A hard–soft donor combined strategy," *Inorg. Chem.*, vol. 53, no. 3, pp. 1712–1720, 2014.

322. J. Bisson, J. Dehaudt, M. C. Charbonnel, D. Guillaneux, M. Miguirditchian, C. Marie, N. Boubals, G. Dutech, M. Pipelier, V. Blot, and D. Dubreuil, "1,10-phenanthroline and non-symmetrical 1,3,5-triazine dipicolinamide-based ligands for group actinide extraction," *Chem. Eur. J.*, vol. 20, no. 25, pp. 7819–7829, 2014.

323. D. Aravena, M. Atanasov, and F. Neese, "Periodic trends in lanthanide compounds through the eyes of multireference ab initio theory," *Inorg. Chem.*, vol. 55, no. 9, pp. 4457–4469, 2016.

324. J. Jung, M. Atanasov, and F. Neese, "Ab initio ligand-field theory analysis and covalency trends in actinide and lanthanide free ions and octahedral complexes," *Inorg. Chem.*, vol. 56, no. 15, pp. 8802–8816, 2017.

325. M. Dolg, X. Cao, and J. Ciupka, "Misleading evidence for covalent bonding from EuIIIX and AmIIIX density functional theory bond lengths," *J. Electron. Spectrosc. Relat. Phenom.*, vol. 194, pp. 8–13, 2014.

326. J. R. Klaehn, D. R. Peterman, M. K. Harrup, R. D. Tillotson, T. A. Luther, J. D. Law, and L. M. Daniels, "Synthesis of symmetric dithiophosphinic acids for 'minor actinide' extraction," *Inorg. Chim. Acta*, vol. 361, no. 8, pp. 2522–2532, 2008.

327. W. Wang, and C. Y. Cheng, "Separation and purification of scandium by solvent extraction and related technologies: A review," *J. Chem. Technol. Biotechnol.*, vol. 86, pp. 1237–1246, 2011.

328. P. R. V. Rao, and Z. Kolarik, "A review of third phase formation in extraction of actinides by neutral organophosphorus extractants," *Solvent Extr. Ion Exch.*, vol. 14, no. 6, pp. 955–993, 1996.

329. S. Cui, V. F. de Almeida, B. P. Hay, X. Ye, and B. Khomami, "Molecular dynamics simulation of tri-n-butyl-phosphate liquid: A force field comparative study," *J. Phys. Chem. B*, vol. 116, no. 1, pp. 305–313, 2012.

330. S. Cui, V. F. de Almeida, and B. Khomami, "Molecular dynamics simulations of tri-n-butyl-phosphate/n-dodecane mixture: Thermophysical properties and molecular structure," *J. Phys. Chem. B*, vol. 118, no. 36, pp. 10750–10760, 2014.

331. Q. N. Vo, C. A. Hawkins, L. X. Dang, M. Nilsson, and H. D. Nguyen, "Computational study of molecular structure and self-association of tri-n-butyl phosphates in n-dodecane," *J. Phys. Chem. B*, vol. 119, no. 4, pp. 1588–1597, 2015.

332. M. J. Servis, C. A. Tormey, D. T. Wu, and J. C. Braley, "A molecular dynamics study of tributyl phosphate and diamyl amyl phosphonate self-aggregation in dodecane and octane," *J. Phys. Chem. B*, vol. 120, no. 10, pp. 2796–2806, 2016.

333. J. Mu, R. Motokawa, C. D. Williams, K. Akutsu, S. Nishitsuji, and A. J. Masters, "Comparative molecular dynamics study on tri-n-butyl phosphate in organic and aqueous environments and its relevance to nuclear extraction processes," *J. Phys. Chem. B*, vol. 120, no. 23, pp. 5183–5193, 2016.

334. Q. N. Vo, L. X. Dang, M. Nilsson, and H. D. Nguyen, "Quantifying dimer and trimer formation by tri-n-butyl phosphates in n-dodecane: Molecular dynamics simulations," *J. Phys. Chem. B*, vol. 120, no. 28, pp. 6985–6994, 2016.

335. M. J. Servis, D. T. Wu, and J. C. Shafer, "The role of solvent and neutral organophosphorus extractant structure in their organization and association," *J. Mol. Liq.*, vol. 253, pp. 314–325, 2018.

336. X. Ye, S. Cui, V. F. de Almeida, and B. Khomami, "Molecular simulation of water extraction into a tri-n-butylphosphate/n-dodecane solution," *J. Phys. Chem. B*, vol. 117, no. 47, pp. 14835–14841, 2013.

337. M. J. Servis, D. T. Wu, and J. C. Braley, "Network analysis and percolation transition in hydrogen bonded clusters: Nitric acid and water extracted by tributyl phosphate," *Phys. Chem. Chem. Phys.*, vol. 19, no. 18, pp. 11326–11339, 2017.

338. R. Chiarizia, A. Briand, M. Jensen, and P. Thiyagarajan, "SANS study of reverse micelles formed upon the extraction of inorganic acids by TBP in n-octane," *Solvent Extr. Ion Exch.*, vol. 26, pp. 333–359, 2008.

339. P. Ivanov, J. Mu, L. Leay, S.-Y. Chang, C. Sharrad, A. Masters, and S. Schroeder, "Organic and third phase in HNO_3/TBP/n-dodecane system: No reverse micelles," *Solvent Extr. Ion Exch.*, vol. 35, no. 4, pp. 251–265, 2017.

340. B. Qiao, G. Ferru, and R. J. Ellis, "Complexation enhancement drives water-to-oil ion transport: A simulation study," *Chemistry*, vol. 23, no. 2, pp. 427–436, 2017.

341. R. J. Ellis, D. J. Brigham, L. Delmau, A. S. Ivanov, N. J. Williams, M. N. Vo, B. Reinhart, B. A. Moyer, and V. S. Bryantsev, "'Straining' to separate the rare earths: How the lanthanide contraction impacts chelation by diglycolamide ligands," *Inorg. Chem.*, vol. 56, pp. 1152–1160, 2017.

342. D. M. Brigham, A. S. Ivanov, B. A. Moyer, L. H. Delmau, V. S. Bryantsev, and R. J. Ellis, "Trefoil-shaped outer-sphere ion clusters mediate lanthanide(III) ion transport with diglycolamide ligands," *J. Am. Chem. Soc.*, vol. 139, no. 48, pp. 17350–17358, 2017.

343. A. G. Baldwin, A. S. Ivanov, N. J. Williams, R. J. Ellis, B. A. Moyer, V. S. Bryantsev, and J. C. Shafer, "Outer-sphere water clusters tune the lanthanide selectivity of diglycolamides," *ACS Cent. Sci.*, vol. 4, no. 6, pp. 739–747, 2018.

344. B. Qiao, T. Demars, M. Olvera de la Cruz, and R. J. Ellis, "How hydrogen bonds affect the growth of reverse micelles around coordinating metal ions," *J. Phys. Chem. Lett.*, vol. 5, no. 8, pp. 1440–1444, 2014.

345. R. Chiarizia, M. Jensen, B. M., J. Ferraro, P. Thiyagarajan, and K. Littrell, "Third phase formation revisited: The U(VI), HNO_3-TBP, n-Dodecane system," *Solvent Extr. Ion Exch.*, vol. 21, no. 1, pp. 1–27, 2003.

346. R. Chiarizia, K. Nash, M. Jensen, P. Thiyagarajan, and K. Littrell, "Application of the Baxter model for hard spheres with surface adhesion to SANS data for the U(VI)-HNO_3, TBP-n-dodecane system," *Langmuir*, vol. 19, pp. 9592–9599, 2003.

347. A. G. Baldwin, M. J. Servis, Y. Yang, N. J. Bridges, D. T. Wu, and J. C. Shafer, "The structure of tributyl phosphate solutions: Nitric acid, uranium (VI), and zirconium (IV)," *J. Mol. Liq.*, vol. 246, pp. 225–235, 2017.

348. A. E. Ismail, G. S. Grest, and M. J. Stevens, "Capillary waves at the liquid–vapor interface and the surface tension of water," *J. Chem. Phys.*, vol. 125, no. 1, p. 014702, 2006.

349. S. Senapati, and M. Berkowitz, "Computer simulation study of the interface width of the liquid/liquid interface," *Phys. Rev. Lett.*, vol. 87, no. 17, pp. 20–23, 2001.

350. T. M. Chang, and L. X. Dang, "Recent advances in molecular simulations of ion solvation at liquid interfaces," *Chem. Rev.*, vol. 106, no. 4, pp. 1305–1322, 2006.

351. D. Chandler, "Interfaces and the driving force of hydrophobic assembly," *Nature*, vol. 437, no. 7059, p. 640–647, 2005.

352. F. Sedlmeier, J. Janecek, C. Sendner, L. Bocquet, R. R. Netz, and D. Horinek, "Water at polar and nonpolar solid walls," *Biointerphases*, vol. 3, no. 3, pp. FC23–FC39, 2008.

353. D. Laage, G. Stirnemann, and J. T. Hynes, "Why water reorientation slows without iceberg formation around hydrophobic solutes," *J. Phys. Chem. B*, vol. 113, no. 8, pp. 2428–2435, 2009.

354. F. Bresme, E. Chacón, and P. Tarazona, "Molecular dynamics investigation of the intrinsic structure of water–fluid interfaces via the intrinsic sampling method," *Phys. Chem. Chem. Phys.*, vol. 10, no. 32, pp. 4704–4715, 2008.

355. F. Bresme, E. Chacón, P. Tarazona, and K. Tay, "Intrinsic structure of hydrophobic surfaces: The oil-water interface," *Phys. Rev. Lett.*, vol. 101, no. 5, p. 056102, 2008.

356. L. B. Pártay, G. Hantal, P. Jedlovszky, A. Vincze, and G. Horvai, "A new method for determining the interfacial molecules and characterizing the surface roughness in computer simulations. Application to the liquid–vapor interface of water," *J. Comput. Chem.*, vol. 29, no. 6, pp. 945–956, 2008.

357. G. Hantal, M. Darvas, L. B. Pártay, G. Horvai, and P. Jedlovszky, "Molecular-level properties of the free water surface and different organic liquid/water interfaces, as seen from ITIM analysis of computer simulation results," *J. Phys. Condens. Matter*, vol. 22, no. 28, p. 284112, 2010.

358. K. J. Tielrooij, D. Paparo, L. Piatkowski, H. J. Bakker, and M. Bonn, "Dielectric relaxation dynamics of water in model membranes probed by terahertz spectroscopy," *Biophys. J.*, vol. 97, no. 9, pp. 2484–2492, 2009.

359. H. A. Stern, and S. E. Feller, "Calculation of the dielectric permittivity profile for a nonuniform system: Application to a lipid bilayer simulation," *J. Chem. Phys.*, vol. 118, no. 7, pp. 3401–3412, 2003.

360. V. Ballenegger, and J.-P. Hansen, "Local dielectric permittivity near an interface," *Europhys. Lett.*, vol. 63, no. 3, p. 381, 2003.

361. V. Ballenegger, and J. P. Hansen, "Dielectric permittivity profiles of confined polar fluids," *J. Chem. Phys.*, vol. 122, no. 11, p. 114711, 2005.

362. A. Ghoufi, A. Szymczyk, R. Renou, and M. Ding, "Calculation of local dielectric permittivity of confined liquids from spatial dipolar correlations," *Europhys. Lett.*, vol. 99, no. 3, p. 37008, 2012.

363. D. J. Bonthuis, S. Gekle, and R. R. Netz, "Profile of the static permittivity tensor of water at interfaces: Consequences for capacitance, hydration interaction and ion adsorption," *Langmuir*, vol. 28, no. 20, pp. 7679–7694, 2012.

364. D. J. Bonthuis, and R. R. Netz, "Beyond the continuum: How molecular solvent structure affects electrostatics and hydrodynamics at solid–electrolyte interfaces," *J. Phys. Chem. B*, vol. 117, no. 39, pp. 11397–11413, 2013.

365. Z. Liu, T. Stecher, H. Oberhofer, K. Reuter, and C. Scheurer, "Response properties at the dynamic water/dichloroethane liquid–liquid interface," *Mol. Phys.*, vol. 116, no. 21–22, pp. 3409–3416, 2018.

366. P. L. Geissler, "Water interfaces, solvation, and spectroscopy," *Annu. Rev. Phys. Chem.*, vol. 64, pp. 317–337, 2013.

367. P. Beudart, V. Lamare, J.-F. Dozol, L. Troxler, and G. Wipff, "Theoretical studies on tri-n-butyl phosphate: Md simulations in vacuo, in water, in chloroform, and at a water/chloroform interface," *Solvent Extr. Ion Exch.*, vol. 16, no. 2, pp. 597–618, 1998.

368. Y. Ghadar, P. Parmar, A. C. Samuels, and A. E. Clark, "Solutes at the liquid: Liquid phase boundary–solubility and solvent conformational response alter interfacial microsolvation," *J. Chem. Phys.*, vol. 142, no. 10, p. 104707, 2015.

369. M. Baaden, M. Burgard, and G. Wipff, "Tbp at the water–oil interface: The effect of TBP concentration and water acidity investigated by molecular dynamics simulations," *J. Phys. Chem. B*, vol. 105, pp. 11131–11141, 2001.

370. X. Ye, S. Cui, V. F. de Almeida, and B. Khomami, "Molecular simulation of water extraction into a tri-n-butylphosphate/n-dodecane solution," *J. Phys. Chem. B*, vol. 117, no. 47, pp. 14835–14841, 2013.

371. M. J. Servis, and A. E. Clark, "Interfacial heterogeneity is essential to water extraction into organic solvents," *Phys. Chem. Chem. Phys.*, vol. 21, pp. 2866–2874, 2019.

372. P. A. Fernandes, M. N. D. Cordeiro, and J. A. Gomes, "Molecular dynamics study of the transfer of iodide across two liquid/liquid interfaces," *J. Phys. Chem. B*, vol. 103, no. 42, pp. 8930–8939, 1999.

373. P. A. Fernandes, M. N. D. S. Cordeiro, and J. A. N. F. Gomes, "Influence of ion size and charge in ion transfer processes across a liquid–liquid interface," *J. Phys. Chem. B*, vol. 104, no. 10, pp. 2278–2286, 2000.

374. M. Darvas, M. Jorge, M. N. Cordeiro, and P. Jedlovszky, "Calculation of the intrinsic solvation free energy profile of methane across a liquid/liquid interface in computer simulations," *J. Mol. Liq.*, vol. 189, pp. 39–43, 2014.

375. N. Kikkawa, L. Wang, and A. Morita, "Microscopic barrier mechanism of ion transport through liquid–liquid interface," *J. Am. Chem. Soc.*, vol. 137, no. 25, pp. 8022–8025, 2015.

376. N. Kikkawa, L. Wang, and A. Morita, "Computational study of effect of water finger on ion transport through water–oil interface," *J. Chem. Phys.*, vol. 145, no. 1, p. 014702, 2016.

377. J. J. Karnes, and I. Benjamin, "Geometric and energetic considerations of surface fluctuations during ion transfer across the water-immiscible organic liquid interface," *J. Chem. Phys.*, vol. 145, no. 1, p. 014701, 2016.

378. A. Varnek, L. Troxler, and G. Wipff, "Adsorption of ionophores and of their cation complexes at the water/chloroform interface: A molecular dynamics study of a [2.2.2] cryptand and of phosphoryl-containing podands," *Chem. Eur. J.*, vol. 3, pp. 552–560, 1997.

379. M. Lauterbach, E. Engler, N. Muzet, L. Troxler, and G. Wipff, "Migration of ionophores and salts through a water–chloroform liquid–liquid interface: Molecular dynamics potential of mean force investigations," *J. Phys. Chem. B*, vol. 102, pp. 245–256, 1998.

380. Z. Liang, W. Bu, K. J. Schweighofer, D. J. Walwark, J. S. Harvey, G. R. Hanlon, D. Amoanu, C. Erol, I. Benjamin, and M. L. Schlossman, "Nanoscale view of assisted ion transport across the liquid–liquid interface," *Proc. Nat. Acad. Sci. U. S. A.*, pii: 201701389, 2018.

381. G. Benay, and G. Wipff, "Liquid–liquid extraction of uranyl by TBP: The TBP and ions models and related interfacial features revisited by MD and PMF simulations," *J. Phys. Chem. B*, vol. 118, no. 11, pp. 3133–3149, 2014.

382. M. Jayasinghe, and T. L. Beck, "Molecular dynamics simulations of the structure and thermodynamics of carrier-assisted uranyl ion extraction," *J. Phys. Chem. B*, vol. 113, no. 34, pp. 11662–11671, 2009.

383. M. Lauterbach, E. Engler, N. Muzet, L. Troxler, and G. Wipff, "Migration of iono-phores and salts through a water–chloroform liquid–liquid interface: Molecular dynamics potential of mean force investigations," *J. Phys. Chem. B*, vol. 5647, no. 97, pp. 245–256, 1998.

384. Y. Ghadar, P. Parmar, A. C. Samuels, and A. E. Clark, "Solutes at the liquid:liquid phase boundary – Solubility and solvent conformational response alter interfacial microsolvation," *J. Chem. Phys.*, vol. 142, no. 10, p. 104707, 2015.

385. M. Jayasinghe, and T. L. Beck, "Molecular dynamics simulations of the structure and thermodynamics of carrier-assisted uranyl ion extraction," *J. Phys. Chem. B*, vol. 113, no. 34, pp. 11662–11671, 2009.

386. G. Benay, and G. Wipff, "Liquid–liquid extraction of uranyl by an amide ligand: Interfacial features studied by MD and PMF simulations," *J. Phys. Chem. B*, vol. 117, no. 24, pp. 7399–7415, 2013.

387. G. Benay, and G. Wipff, "Liquid–liquid extraction of alkali cations by 18-crown-6: Complexation and interface crossing studied by MD and PMF simulations," *New J. Chem.*, vol. 40, no. 3, pp. 2102–2114, 2016.

388. G. Benay, R. Schurhammer, and G. Wipff, "BTP-based ligands and their complexes with Eu^{3+} at 'oil'/water interfaces. A molecular dynamics study," *Phys. Chem. Chem. Phys.*, vol. 12, no. 36, pp. 11089–11102, 2010.

389. G. Benay, and G. Wipff, "Oil-soluble and water-soluble BTPhens and their europium complexes in octanol/water solutions: Interface crossing studied by MD and PMF simulations," *J. Phys. Chem. B*, vol. 117, no. 4, pp. 1110–1122, 2013.

390. J. J. Karnes, and I. Benjamin, "Structure and dynamics of host/guest complexation at the liquid/liquid interface: Implications for inverse phase transfer catalysis," *J. Phys. Chem. C*, vol. 121, no. 9, pp. 4999–5011, 2017.

6 A Review of Mass-Transfer and Reaction-Kinetics Studies in Microfluidic Solvent Extraction Processes

Fang Zhao, Kai Wang, and Guangsheng Luo

CONTENTS

6.1 INTRODUCTION

Microfluidics have been investigated and researched for more than 20 years. Since the initial appearance of microchips, the regular and controllable multiphase flows in microspace keep attracting researchers to test for the various applications, which include chemical synthesis (Britton and Raston 2017), detection and analysis (Shui et al. 2014), highly efficient separation (Jaritsch et al. 2014), emulsifications (Morais et al. 2010), nanomaterial preparation (Niu et al. 2015), and bio-engineering processes (Mehta and Linderman 2006). The long-lasting vitality of microfluidic studies has been evidenced in many new microchip- or microtube-based technologies. In possibly the most complicated example, a refrigerator-sized micro plant was implemented by Adamo et al. (2016) at the Massachusetts Institute of Technology; in one of the simplest, microchannel junctions were widely used as effective setups in preparing uniform emulsions (Zhu and Wang 2016). Although most of the flowing patterns are laminar flows in microfluidic processes, the much smaller mixing scale and larger specific areas of droplets and bubbles largely intensify the mass- and heat-transport rates (Wang et al. 2008; Xu et al. 2008), and, therefore, microfluidic technologies are popular approaches for process intensifications (Charpentier 2012).

In the chemical engineering industry, process intensification for separation and purification is very important. Microstructured extraction, adsorption, and distillation equipment with high efficiency and low energy consumption (Abolhasani et al. 2012; Ko et al. 2016; Weeranoppanant et al. 2017) have been demonstrated, and, therefore, they are very important components in continuous-flow chemistry systems (Porta et al. 2016). However, for applying those devices in industrial processes, a lot of research still needs to proceed for understanding the basic laws of micro-device scale-up (Kuijpers et al. 2017). Some devices, like the Corning reactor (Woitalka et al. 2014) and IMM reactor (Kirschneck and Tekautz 2007), have been proven to give good performance in treating substances lower than hundreds of liters per hour, but those amounts are still far away from large-scale chemical engineering applications.

Besides using microfluidic devices in industrial applications, using small-scale microfluidic setups as effective platforms for laboratory research is another significant approach for exhibiting the advantages of microfluidic technology. Because mixing, mass transfer, and heat transfer are much faster at the microscale, the equipment with a micromixer, microchip, and microtube are very good in transport-limited processes, especially for kinetic measurement (Castor et al. 2015; McMullen and Jensen 2011). Other excellent properties further make kinetic measurement much easier in microfluidic equipment, such as the easy access to *in-situ* observation (Seemann et al. 2012), narrow residence time distribution (RTD), and accurate residence time control (Gobert et al. 2017), effective phase separation (Phillips et al. 2015), as well as excellent temperature control (Mielke et al. 2017). Considering the advantages of the microfluidic process for determining the kinetics, we, therefore, prepared this review chapter to show the recent developments in mass-transfer and reaction-kinetic studies, mainly concerning the application of microfluidic platforms in studying solvent extraction processes. The fundamental knowledge for liquid–liquid biphasic flows in microspaces is introduced first. Then, some important discoveries in understanding mass transfer and reaction kinetics are summarized. Finally, the conclusions of the recent developments in microflow-based kinetic studies are proposed, and a general outlook provides suggestions for further research.

6.2 LIQUID–LIQUID BIPHASIC FLOW IN MICROSPACES

6.2.1 LIQUID–LIQUID BIPHASIC FLOW PATTERNS

The inside of a microchannel or microtube is a narrow and confined space. According to the basic knowledge of fluid dynamics, the Reynolds number is small in the microfluidic process; therefore, the liquids are mostly under the laminar flow condition. Although turbulence flow is more often used in chemical engineering processes, the laminar flow is also useful, especially for stabilizing the regular flow patterns for the liquid–liquid biphasic systems. Figure 6.1 gives three classical flow patterns in liquid–liquid microfluidic processes (Vural Gürsel et al. 2017), namely parallel flow (Figure 6.1a), segmented flow (Figure 6.1b), and tiny swarming droplet flow (Figure 6.1c). Because the tiny droplet flow has the largest specific area between the two phases, it has the best mass- and heat-transfer performance (Wang et al. 2008). However, this flow pattern is mostly used for mass-transfer and reaction-enhancement applications (Liu et al. 2013; Nieves-Remacha et al. 2012), especially for the scale-up of microfluidic systems (Wang et al. 2014). Since it is difficult to quickly trace the composition variation in different phases in this flow pattern, it is less used in the kinetic studies.

In the kinetic study of mass transfer and chemical reactions, the liquid–liquid segmented flow and parallel flow are commonly used. Although the droplet length in some segmented flows is larger than 1 mm, the mass transfer in this flow pattern is still enhanced by the circulation flow of fluids (Taha and Cui 2006) as shown in Figure 6.1b. These circulations are the typical characteristics of multiphase

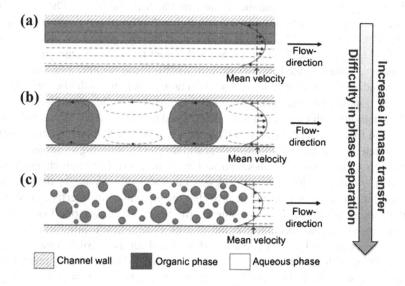

FIGURE 6.1 Classical flow patterns of liquid–liquid biphasic system in microspace. (a) Liquid–liquid parallel flow, (b) liquid–liquid segmented flow, and (c) tiny swarming droplet flow. (Adapted from Vural Gürsel et al. (2017) with the permission of Elsevier.)

microfluidic processes, and we will give an in-depth introduction to the flow details of this flow pattern in the following paragraphs. In the liquid–liquid parallel flow, the mass transfer between the two phases is mainly dependent on molecular diffusion. Although the mass transfer is relatively slow, the mass-transfer rate can be characterized by classical mathematical models, and it is easy to realize phase separation in a microchip (Guillot et al. 2008; Zhao et al. 2001); therefore, the parallel flow is favored by researchers, especially for determining some key parameters for the mass-transfer and reaction processes (Foroozan Jahromi et al. 2018; Priest et al. 2012).

6.2.2 Liquid–Liquid Segmented Flow

Considering the fluidic dynamics in microspace is fundamental to kinetic studies on mass transfer and reaction in microfluidic systems; we will give some more introductions to the characteristics of liquid–liquid segmented flow and parallel flow in this section. A typical liquid–liquid segmented flow is shown in Figure 6.2 (Yin et al. 2018b), which was generated from a microchannel junction in Figure 6.2a. This flow pattern can also be created from commercial T- or Y-type connectors in round cross-section microtube devices (Ghaini et al. 2010). As shown in the microscope photo in Figure 6.2, the two phases occupy different parts of the microchannel. However, liquid films, whose thicknesses are labeled by h, actually exist between the droplets and the channel wall, as exhibited by Figure 6.2c. This is because one of the two phases preferentially wets the microchannel wall (Shui et al. 2009). Owing to the connection of the liquid films, the wall wetting phase is called the continuous phase in the microfluidic process, and the droplets are the dispersed phase. The liquid films work as lubricating layers around the droplets, but the continuous phase flows slowly in the films. Therefore, the average velocities of the two phases are different, and the liquid holdup of the dispersed phase is different from the volume proportion of the dispersed phase in feeding (Baroud et al. 2010). These parameters are crucial for understanding the mass-transfer and reaction processes in microfluidic systems, and it is very important to understand the control mechanism of the liquid film thickness.

The liquid film thickness is mainly determined by the viscosity of the continuous phase and the interfacial tension under a certain flow rate (Christopher et al. 2008), and we can observe these films clearly around droplets when a high viscous polymer solution is used as the continuous phase (Fu et al. 2011). For microchannels with rectangular cross-sections, we still lack general models for calculating the thickness of the film, but some literature has provided the equations for interface curvatures in the corners of channel cross-section (Glawdel et al. 2012; Ransohoff et al. 1987). Besides the microchannels in microchip devices, commercial capillaries made from glass (Wei et al. 2011) or fluoroplastics (Snead and Jamison 2015) have also been commonly used for their excellent chemical stabilities. Those tubular devices have round cross-sections and the liquid film thickness can be represented from the previous studies on the liquid–liquid or gas–liquid annular flow in small tubes (Bico and Quéré 2000; Kashid et al. 2005). Some equations for calculating film thickness are summarized in Table 6.1, which have similar calculated results at low capillary numbers (Ghaini et al. 2010). Note that sometimes the liquid film is not stable:

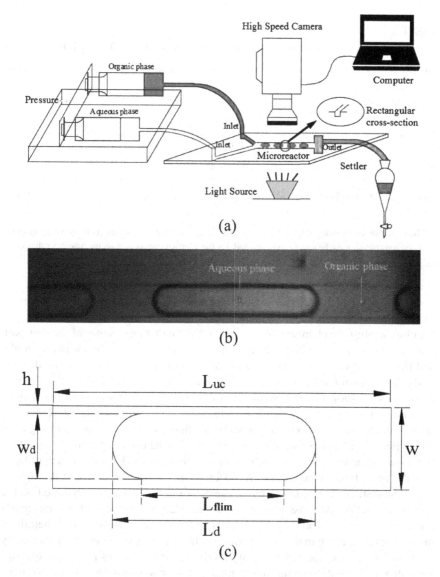

(a)

(b)

(c)

FIGURE 6.2 Liquid–liquid segmented flow in a microchannel. (a) The generation of liquid–liquid segmented flow; (b) microscope picture with colored dispersed phase; (c) geometric construction of a flow unit, where h is the film thickness; w_d is the droplet width; w is the microchannel width; L_{uc} is the flow unit length; L_{film} is the liquid film length; L_d is the droplet length. (Adapted from Yin et al. (2018b) with the permission of Elsevier.)

Günther et al. have shown the film break-up phenomenon in polydimethylsiloxane (PDMS) microchannels (Günther et al. 2005).

Using the liquid film thickness, the liquid holdup of the dispersed phase can be calculated by the equation given in our previous review chapter (Wang and Luo 2017). In most microfluidic processes, the droplet capillary number Ca_d is much

TABLE 6.1

Summary of Models for the Liquid Film Thickness Around Droplets

References	Equations[a]	Numbers
Bretherton (1961)	$\dfrac{h}{d_{in}} = 0.67Ca_d^{2/3}$, at $Ca_d < 3 \times 10^{-3}$	(6.1)
Marchessault and Mason (1960)	$\dfrac{h}{d_{in}} = 0.5\sqrt{\dfrac{\mu_c}{\gamma}}\left(-0.05 + 0.89\sqrt{u_d}\right)$, at $Ca_d < 2 \times 10^{-4}$	(6.2)
Irandoust and Andersson (1989)	$\dfrac{h}{d_{in}} = 0.18\left(1 - \exp\left(-3.08Ca_d^{0.54}\right)\right)$, at $Ca_d < 2$	(6.3)
Aussillous and Quéré (2000)	$\dfrac{h}{d_{in}} = \dfrac{0.67Ca_d^{2/3}}{1 + 3.35Ca_d^{2/3}}$, at $Ca_d < 1.4$	(6.4)

[a] The definition of capillary number Ca_d in the equations is $Ca_d = \mu_c u_d / \gamma$, where μ_c is the viscosity of the continuous phase, u_d is the droplet velocity, and γ is the interfacial tension. d_{in} is the tube inner diameter.

smaller than 1, and, therefore, the film thickness is just around several micrometers. The liquids in these films are almost immobilized; therefore, edges of the droplet are close against the channel wall. However, the fluids in the center of the flow path moves faster as shown in Figure 6.1b, and the fluids exhibit circulations in the droplet and the segmented continuous phase, if our sight moves with the running droplet. Early in the microfluidic study, both circulations in the dispersed and the continuous phases are shown by two asymmetrical eddies in the midsection of the channel (Kashid et al. 2005). Recent work with improved simulation resolution shows that six asymmetrical eddies exist, as shown by the flow stream in Figure 6.3 (Yang et al. 2017). Four small eddies run in the regions close to the caps of the droplet, and thus the surface-renewing effect is much stronger in these areas. This phenomenon shows that the mass-transfer behaviors are different in the droplet caps and the droplet edges (Susanti et al. 2016). The mass transfer from the cap is mainly based on the convection, while the mass transfer from the edge mainly depends on molecular diffusion (Tsaoulidis and Angeli 2015). We will continue to show more details of mass-transfer research in the next section. Although the two phases are separated by the liquid–liquid interface, the velocity profile in the continuous phase still exhibits a parabolic distribution along the channel cross-section, as shown in Figure 6.3c. Therefore, the strength of circulation flow can be evaluated by the droplet average velocity, and the centers of the vortexes are at the zero velocity position.

When using microfluidic devices to determine the mass-transfer and reaction kinetics, accurate time recording is very important. The RTD shows the time dispersion of different fluid elements, which has a narrow profile in the liquid–liquid segmented flow (Song et al. 2003). Regulated by the closed interface, the droplets in microchannels or microtubes work as flowing batch reactors (Song et al. 2006), and therefore the RTD for the dispersed phase is close to the Dirichlet function if the tracer is not soluble in the continuous phase. For the continuous phase which wets the channel wall and is connected by the liquid films around the droplets, the tracer

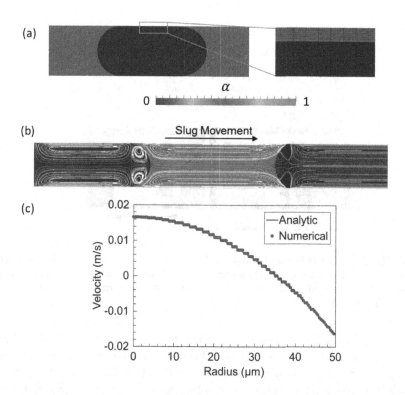

FIGURE 6.3 Simulation of the circulations in the microfluidic liquid–liquid segmented flow. (a) The volume distribution of the continuous phase; (b) streamlines to show the circulations in both phases; (c) velocity profile along the flow path cross-section with the comparison of analytic result from the literature of Thulasidas et al. (1997). (Adapted from Yang et al. (2017) with the permission of Elsevier.)

can disperse into the neighboring segmentations, and thus the flow path is extended; therefore, RTD of the continuous phase will become wider. However, because the liquid film is very thin at low capillary numbers, the RTD of the continuous phase is still very narrow in this situation (Kuhn et al. 2011). If the distributions of the residence time for both phases is narrow, turning the wetting property of the channel wall to partially wet both phases will lead both phases to work as the batch reactors (Kazemi Oskooei and Sinton 2010); however, both two phases are dispersed phases in this situation and mass transfer through the liquid film around the droplet does not exist.

6.2.3 Liquid–Liquid Parallel Flow

In contrast to the liquid–liquid segmented flow, liquid–liquid parallel flow is relatively simpler. However, if the 3D distribution of the two phases is considered, we will find it is not simple at all. Figure 6.4 shows two typical distributions of the two-phase fluids (Guillot et al. 2006). When the phase ratio of the fluids is not extremely high or extremely low, the two phases occupy each half part of the channel cross-section

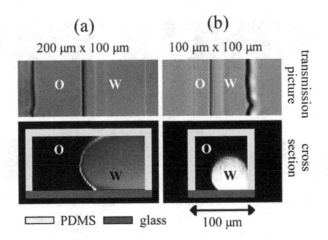

(a) 200 μm x 100 μm

(b) 100 μm x 100 μm

transmission picture

cross section

O   W   O   W

O   O

W   W

PDMS   glass   100 μm

FIGURE 6.4 Different liquid–liquid parallel flows in the microchannels. (a) The two phases occupy each end of the channel cross-section; (b) one phase only occupies a corner of the cross-section. (Adapted from Guillot et al. (2006) with the permission of the American Chemical Society.)

respectively. A cutoff line in the middle of the channel will be observed, but a curved interface is actually shown in the channel (Figure 6.4a). The curvature of the interface determines the mass-transfer area between the two phases, which depends on the fluid contact angle and the pressure difference between phases (Xu et al. 2012). Since the interface curvature is stable along the microchannel (Guillot et al. 2006), a constant pressure difference exists between two phases:

$$p_w - p_o = \gamma/R \tag{6.5}$$

where, p_w and p_o are the pressures in the water and oil phases respectively, and R is the interface radius. By contrast, as one-phase flow rate is much lower than the other, this low-flow-rate phase cannot occupy half of the channel cross-section as shown in Figure 6.4b. In this situation, the low-flow-rate phase will be localized in one of the corners of the channel cross-section to reduce the surface energy, and it is very likely to wet the local part of the channel, which has a good wetting property for that phase. In this situation, it is very easy to inaccurately calculate the interface area if only 2D observation is implemented. Confocal microscopy will be useful for solving this problem.

The position of two phases in the liquid–liquid parallel flow is balanced by the shearing stress between them. At the interface, the velocities of two phases are equal ($u_w = u_o$) and the continuity of the shear stress at the fluid–fluid interface leads to:

$$\mu_w \frac{\partial u_w}{\partial n} = \mu_o \frac{\partial u_o}{\partial n} + \eta_s \frac{\partial^2 u_o}{\partial^2 t_n} \tag{6.6}$$

where, μ_w and μ_o are the viscosities of two phases, respectively; n and t_n, stand for the normal and tangent vectors to the interface, respectively; and η_s is the interface

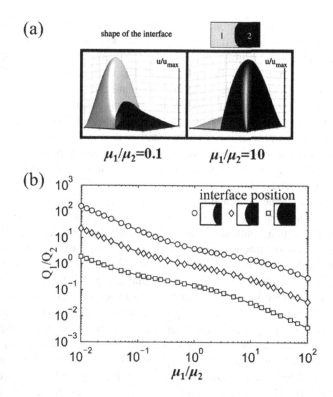

FIGURE 6.5 Velocity profiles and the interface position in the microfluidic liquid–liquid parallel flow. (a) The 3D velocity profiles in a microchannel cross-section. (Adapted from Guillot et al. (2008) with the permission of Springer.) (b) The relationship for the flow-rate ratio and viscosity ratio at different interface positions. (Adapted from Guillot et al. (2006) with the permission of the American Chemical Society.)

shear viscosity. As analyzed by (Guillot et al. 2006), the typical value of η_s ranges from 10^{-9} to 10^{-7} Pa·s·m for the surfactant monolayers at the oil–water interface. In the case of a 100×100 μm^2 channel, the viscosity surface term is negligible for fluid viscosities $>10^{-3}$ Pa·s. Using the relationship in Eq. 6.6, the velocity profiles for the liquid–liquid parallel flow can be calculated by Navier–Stokes equations (Galambos and Forster 1998; Malengier et al. 2012). Figure 6.5a shows typical velocity profiles at different viscosity ratios, which shows that the highest velocity is at the center of the low-viscosity fluid. From the velocity profiles, the position of the interface can be further confirmed by the inflection point of the velocity gradient, and the results are represented by the lines in Figure 6.5b.

6.3 MICROFLUIDIC SETUP FOR KINETIC STUDY OF LIQUID–LIQUID EXTRACTION

Kinetic information, especially the intrinsic reaction kinetics (kinetics without mass-transfer limitations), is essential for the design and operation of chemical engineering processes. Traditionally, kinetic investigations for extraction process are performed in a constant interface cell (or Lewis cell, Wasewar et al. 2002; Weigl et al. 2006) or a highly stirred flask (Lee 2004; Li et al. 2013). The former uses low stirring rates to keep the liquid–liquid interface "tranquil," leading to slow mass transfer, and thus could not eliminate the mass-transfer limitations, especially for highly concentrated extraction systems accompanied by a fast reaction. Rapid mixing could be achieved in a vigorously stirred flask; however, device-dependent kinetic parameters are usually obtained due to the uneven dispersion and the unknown and unpredictable liquid–liquid interfacial area. Besides, heat-transfer limitations could exist in bulk reactors for strongly exothermic/endothermic extraction and reaction process.

Microfluidics enable chemical engineers to conduct kinetic experiments under well-defined conditions with rapid mass transfer and a regular interface shape. Figure 6.6a presents a continuous microfluidic platform primarily composed of a feeding unit, a microextraction unit, a temperature control unit, a phase separation unit, and an analyzing unit. Typically, liquid–liquid parallel flow (Hellé et al. 2015) and liquid–liquid segmented flow (Sattari-Najafabadi et al. 2017) are generated in a microextraction device for kinetic studies to ensure good dispersion/contact of the two liquids. These flows can be readily characterized, giving important information for kinetic analysis, namely the interfacial area between the two liquids. The inline phase separation unit, ensuring rapid phase separation and highly temporal resolution, can be homemade based on the selective wettability of the two liquids (Figure 6.6b and 6c) or the balance of capillary force (Figure 6.6d). Besides, commercial liquid–liquid separators are also available, such as Zaiput and Dolomite. An analyzing apparatus for concentration detection, such as UV-Vis spectrometry

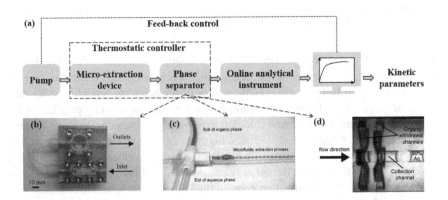

FIGURE 6.6 Microfluidic setup for kinetic analysis. (a) Schematics of a continuous and automated setup from the reagent tank to the final output of the kinetic parameters. Three examples of inline liquid–liquid-phase separators based on (b) a hydrophobic PTFE membrane (adapted from Adamo et al. (2013) with the permission of American Chemical Society), (c) a hydrophilic stainless-steel needle (adapted from Liu et al. (2017) with the permission of Elsevier), and (d) withdraw channels (adapted from Nichols et al. (2011) with the permission of the American Chemical Society.)

(Ciceri et al. 2013b), ICP-MS (Inductively coupled plasma mass spectrometry, Hellé et al. 2015), and LIF (laser-induced fluorescence microscopy, Tokimoto et al. 2003), can also be readily telescoped into the microfluidic systems thanks to the advancement of modern analytical techniques.

To acquire accurate kinetic information for the extraction and reaction process, a large number of experiments are required, which can be tedious and consume a lot of time and labor. Recently, automatic control and operation have been integrated into the above-discussed platform to enable high-throughput acquisition of kinetic data (Abolhasani and Jensen 2016). By using programmable pumps (to change the residence time) and some simple codes in a computer, or simply a microcontroller, kinetic curves under different extraction and operation conditions can be acquired in a "plug-and-forget" fashion. With a carefully designed algorithm and more complex feedback control, extraction kinetics can be identified, and kinetic parameters can be determined automatically (McMullen and Jensen 2011; Reizman and Jensen 2016). Ultimately, fast, accurate, and efficient kinetic analysis can be achieved.

6.4 MASS-TRANSFER KINETICS IN MICROFLUIDIC EXTRACTION PROCESSES

6.4.1 Mass-Transfer Performance of Microflow Systems

Mass-transfer kinetics is mostly explored through the value of the mass-transfer coefficient. One of the most attractive characters of microfluidics to chemical engineers is the acceleration of mass-transfer kinetics. For liquid–liquid microfluidic systems, mass-transfer coefficients are significantly larger than those of conventional liquid–liquid extraction processes (Wang and Luo 2017). As a typical example, a comparison of microfluidic extraction to traditional extraction for metal ions summarized by Yin et al. (2018a) is shown in Table 6.2. The table clearly exhibits a more than 10 times increase of the volumetric mass-transfer coefficient by using different kinds of microfluidic extractors. The mass-transfer enhancement comes from both the enlargement of the mass-transfer area and the enforced convection flow in the confined microspace, but the dominating effect is different in different flow patterns. For the flow with a swarm of tiny droplets, the increase of the mass-transfer area contributes more (Su et al. 2010), but for the liquid–liquid segmented flow the convective mass transfer is more important (Li et al. 2018). Although the flow patterns are very regular in microchannels and microtubes, the mass-transfer processes in microspace are still complicated as shown in Figure 6.7 (Sattari-Najafabadi et al. 2017), which exhibits the non-uniform distribution of dye concentration.

To characterize the mass-transfer coefficient with a simple expression, equations have been proposed for microfluidic processes, and most of them are semi-empirical correlations (Wang and Luo 2017). Those equations consider the most important parameter affecting the mass-transfer kinetics, and they are more preferred in engineering applications. We give a summary of the mass-transfer coefficient equations in the following sections to help with the understanding of the rules of microfluidic mass transfer in microfluidic extraction processes.

TABLE 6.2

Comparison of Microfluidic and Traditional Mass-Transfer Coefficients[a]

Type of Extractors	Working Systems[b]	$K_L a$ (1/s)	References
Y-junction serpentine microextractor	$RECl_3$–Lactic acid–1.0 M P507–sulphonated kerosene	0.007–0.57	Yin et al. 2018a
Microfluidic-based hollow droplet	$NdCl_3$–1.0 M P507–sulphonated kerosene	0.04–0.8	Chen et al. (2017)
Y–Y microchip with an integrated guide structure channel	$RECl_3$–1.0 M Cyanex 572	0.05–0.17	Kolar et al. (2016)
Membrane dispersion microextractor	$CeCl_3$–saponified P507–sulphonated kerosene	0.1–0.54	Hou et al. (2015a, 2015b)
Hollow fiber supported liquid membrane	$Nd_2(SO_4)_3$–0.5 M DNPPA–n-dodecane	0.008–0.074	Wannachod et al. (2015)
Hollow fiber contactor	$Nd(NO_3)_3$–DNPPA–TOPO	0.047–0.081	Ambare et al. (2013)
Mixer-settler	Hydroxyoxime (palladium) hydrochloric acid	0.01–0.0197	Singh et al. (2015)
Annular centrifugal extractor	Water (succinic acid) n-butanol system	0.0155–0.0465	Singh et al. (2015)

[a] Adapted from Yin et al. (2018a) with the permission of Elsevier.

[b] RE represents rare earth element.

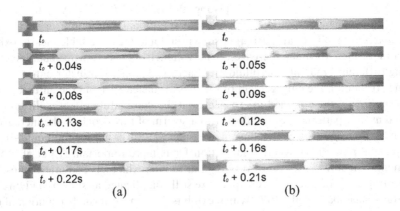

FIGURE 6.7 Concentration distribution of NaOH represented by phenol red in a mass-transfer study of hexane/acetic acid/NaOH solution. (a) Droplet generated from a cross-junction channel; (b) droplet generated from a T-junction channel. Phenol red changes the color of the aqueous phase from red to yellow at pH of 8.2. (Adapted from Sattari-Najafabadi et al. (2017) with the permission of Elsevier.)

6.4.2 MASS-TRANSFER KINETICS IN LIQUID–LIQUID SEGMENTED FLOW

Mass transfer in a liquid–liquid segmented microflow is similar to that in the dispersed droplet flow in extraction columns; therefore, the traditional formula for the mass-transfer coefficient derived from dimensionless numbers, such as the Reynolds number ($Re = du\rho/\mu$), Sherwood number ($Sh = k_L d/D$), and Schmidt number ($Sc = \mu/\rho D$), which are constantly used in the research of microflow extraction (Matsuoka et al. 2016). Additionally, considering the similarity of the gas–liquid and liquid–liquid segmented flow in microspace, some equations from gas–liquid microfluidic mass-transfer studies can also be applied in liquid–liquid systems (Kashid et al. 2011). Kashid et al. (2010) proposed a semi-empirical equation (Eq. 6.7) for the volumetric mass-transfer coefficient of two phases ($K_L a$) in capillaries with a round cross-section. This equation employs the capillary number and considers the average velocity and physical properties of two phases. Di Miceli Raimondi et al. (2008) proposed a 2D CFD (Computational Fluid Dynamics) simulation-based equation (Eq. 6.8) to predict the mass-transfer coefficient (k_D) in the dispersed phase during a liquid–liquid segmented flow. This equation uses the spherical equivalent diameter of slug-shaped droplets to normalize the length scale. Besides approaches based on the dimensionless numbers, another approach for predicting the volumetric mass-transfer coefficients comes from penetration theory, which employs the structure parameters of a liquid–liquid segmented flow. Yue et al. (2009) proposed an equation for the volumetric mass-transfer coefficient of the continuous phase ($k_C a$, Eq. 6.9) by using the velocity, the channel width, and the length of flow unit shown in Figure 6.2c. This equation is actually from a gas–liquid mass-transfer study, where the mass-transfer resistance in the dispersed phase was neglected.

According to the above analysis of the fluid-dynamic character of a liquid–liquid segmented flow, the droplet caps and the liquid film surrounding the droplets should give different contributions to the mass-transfer kinetics (Dessimoz et al. 2008; Ghaini et al. 2010). In order to give a precise determination to the mass-transfer coefficient, some equations distinguish the contributions of the caps and the film from the total mass-transfer coefficient (Kashid et al. 2011):

$$ka = k_{cap}a_{cap} + k_{film}a_{film} \tag{6.15}$$

In the above theory, Berčič and Pintar (1997) proposed an equation for the volumetric mass-transfer coefficient in circular capillaries of 1.5, 2.5, and 3.1 mm diameters by employing the liquid cell length, the holdup of the dispersed phase, and the superficial velocity of two phases, as shown by Eq. 6.10 in Table 6.3. This equation exhibits the contribution of the droplet caps during the absorption of methane in water because of the rapid saturation of the film in the experiment. The liquid-phase diffusivity D is not included for the single working system in the work of Berčič and Pintar. For the mass-transfer contribution of the liquid film, Vandu et al. (2005) used a simpler method by building a relationship between the mass-transfer coefficient and the superficial velocity of the dispersed phase and the liquid cell length (Eq. 6.11). 1-, 2- and 3-mm-diameter capillaries of circular and square cross-sections were investigated by Vandu et al. with an oxygen absorption experiment. The contribution of the

TABLE 6.3
Equations for Mass-Transfer Coefficients in Microflow Extraction

References	Equations[a]	Numbers
Kashid et al. (2010)	$K_L a \dfrac{l}{U} = \left(\dfrac{U\mu_m}{\gamma}\right)^{-0.09} \left(\dfrac{U d_{in}\rho_m}{\mu_m}\right)^{-0.09} \left(\dfrac{d_{in}}{l}\right)^{-0.1}$, at $1 < K_L a \dfrac{l}{U} < 5$	(6.7)
Di Miceli Raimondi et al. (2008)	$k_D = 2.77 \times 10^{-4} \dfrac{1}{d_d}\left(\dfrac{V_d}{V_{uc}}\right)^{0.17} (u_d w)^{0.69}\left(\dfrac{u_d}{\gamma}\right)^{-0.07}\left(\dfrac{w}{d_d}\right)^{0.75}$, at $k_D < 1.5 \times 10^{-3}$ m/s	(6.8)
Yue et al. (2009)	$k_C a = \dfrac{2}{w}\left(\dfrac{D u_d}{L_{uc}}\right)^{0.5}\left(\dfrac{L_d}{L_{uc}}\right)^{0.3}$, at $u_d = 0.4 - 2$ m/s, $\dfrac{L_d}{w} = 1.4 - 6.3$, $\dfrac{L_{uc}}{w} = 1 - 3.2$	(6.9)
Berčič and Pintar (1997)	$k_{C,cap}a_{cap} = \dfrac{0.111 U^{1.19}}{\left[(1-\varphi_d)L_{uc}\right]^{0.57}}$, at $k_{C,cap}a_{cap} = 0.005 - 0.3\,\mathrm{s}^{-1}$	(6.10)
Vandu et al. (2005)	$k_{C,film}a_{film} = \dfrac{4.5}{d_{in}}\left(\dfrac{D U_D}{L_{uc}}\right)^{0.5}$, at $k_{c,film}a_{film} = 0.09 - 0.72\,\mathrm{s}^{-1}$	(6.11)
van Baten and Krishna (2004)	$k_{C,cap}a_{cap} = \sqrt{\dfrac{2 D u_d}{\pi^2 w_d}}\dfrac{4}{L_{uc}} \approx \dfrac{8}{\pi L_{uc}}\sqrt{\dfrac{2 D u_d}{d_{in}}}$ at $L_{uc} = 0.015 - 0.05$ m; $u_d = 0.15 - 0.55$ m/s; $\varphi_D = 0.136 - 0.5$	(6.12)
	$k_{C,film}a_{film} \approx \begin{cases} \dfrac{8 L_{film}}{L_{uc}d_{in}}\sqrt{\dfrac{D L_{film}}{\pi u_d}}\dfrac{\ln(1/\Delta)}{1-\Delta}, & \text{at Fo} < 0.1 \\[2ex] \dfrac{13.6 D L_{film}}{h L_{uc}d_{in}}, & \text{at Fo} > 1 \end{cases}$ where $\Delta = 0.7857\exp(-5.121\mathrm{Fo}) + 0.1001\exp(-39.21\mathrm{Fo}) + \ldots$ at $L_{uc} = 0.015 - 0.05$ m; $u_d = 0.15 - 0.55$ m/s; $\varphi_D = 0.136 - 0.5$	(6.13)
Susanti et al. (2016)	$K_L a = 2.6\left(\dfrac{1}{\dfrac{1}{2}\sqrt{\dfrac{\pi t}{D_D}} + \dfrac{1}{2}\sqrt{\dfrac{\pi t}{D_C}}}\right)\left(\dfrac{4 l_d}{d_{in}l_{uc}}\right)$, at $K_L a = 0.005 - 0.035\,\mathrm{s}^{-1}$	(6.14)

[a] ρ_m and μ_m are the mixture density and the mixture viscosity calculated using the phase fraction, respectively; d_d is the diameter of a spherical droplet with the volume equal to that of the dispersed phase slug; l is the channel length; U is the total superficial velocity of two phases; U_D is superficial velocities of the dispersed phase; V_d is the volume of the droplet; V_{uc} is the volume of the flow unit in Figure 2c; D is the diffusion coefficient; D_C is the diffusion coefficient in the continuous phase; D_D is the diffusion coefficient in the dispersed phase; φ_D is the holdup of the dispersed phase; m is the partition coefficient.

liquid film was found important for the large mass-transfer area around the bubble. van Baten and Krishna (2004) provided another approach for estimating the volumetric mass-transfer coefficient in circular cross-section monolith reactors. The contribution of droplet caps were calculated from the penetration theory, which is shown in Eq. 6.12. This equation is based on assuming the film is much smaller than the channel diameter and the droplet diameter w_d is approximated by flow path diameter d_{in}. According to the value of the Fourier number ($Fo = Du_d/L_{film}h^2$), the mass-transfer coefficient of the liquid film part was expressed to Eq. 6.13 by van Baten and Krishna. This equation considers the different contact time of a fresh liquid element at the film interface, and the model of Eq. 6.12 and 6.13 obtained good predictions of $k_L a$ with the experiment using a 1.5, 2, and 3 mm circular capillary. Similar to the method of Berčič and Pintar, Susanti et al. (2016) reported Eq. 6.14 in Table 6.3, for a liquid–liquid extraction process containing dilute solutions in capillaries, as the Fourier numbers for both phases were $<5 \times 10^{-2}$. Thus, the overall mass-transfer coefficient from penetration theory and the partition coefficient were embedded in this equation.

Another advantage of a liquid–liquid segmented flow in a study is that it is easy to demonstrate flow and concentration details by using CFD methods. An example is shown in Figure 6.8, according to the results of Lubej et al. (2015). A full 3D mathematical model, incorporating convection and diffusion mass transfer was developed, and the finite elements method was used to solve the momentum and mass equations. The color field in Figure 6.8 clearly shows a low concentration gradient around the droplet cap, where mass transfer is enhanced by convection flow.

6.4.3 Mass-Transfer Kinetics in Liquid–Liquid Parallel Flow

For the mass-transfer kinetics in a liquid–liquid parallel flow, there are not many reported equations for the mass-transfer coefficient. Because mass transfer is mainly dependent on molecular diffusion in this flow pattern (Dessimoz et al. 2008), CFD is usually directly used in the modeling of the mass-transfer process. As shown by the schematic diagram in Figure 6.9 for a 2D simulation (Vir et al. 2014), the main parameters considered in the simulation are the velocity, viscosity, and concentrations in two phases. Controlling equations for the mass-transfer simulation are shown by Eqs. 6.16 through 6.17.

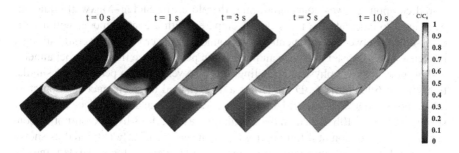

FIGURE 6.8 Concentration profiles of azobenzene inside and outside of a droplet at different residence times. C/C_0 is the dimensionless concentration. (Adapted from Lubej et al. (2015) which is an open access paper of the Royal Society of Chemistry.)

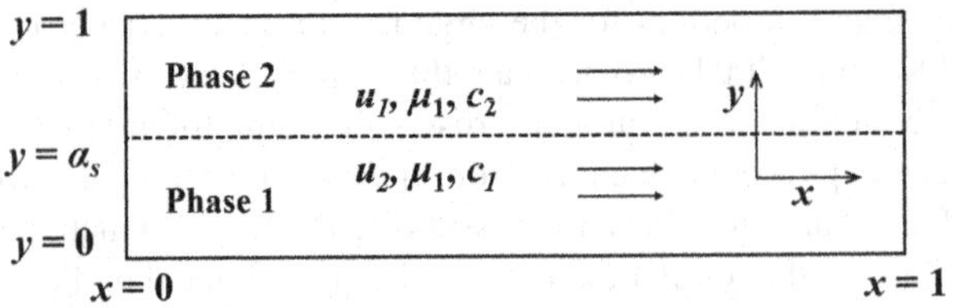

FIGURE 6.9 Schematic of mass transfer in liquid–liquid parallel flow. (Adapted from Vir et al. (2014) with the permission of the American Chemical Society.)

$$\text{phase-1}: \frac{\partial c_1}{\partial x} = \left(\frac{\varphi}{u_1}\right)\left(\frac{1}{Pe_1}\right)\frac{\partial^2 c_1}{\partial y^2}, \quad \text{at } 0 \le y \le \varphi$$

$$\text{phase-2}: \frac{\partial c_2}{\partial x} = \left(\frac{1-\varphi}{u_2}\right)\left(\frac{1}{Pe_2}\right)\frac{\partial^2 c_2}{\partial y^2}, \quad \text{at } \varphi \le y \le 1$$

(6.16)

where φ is the liquid holdup of phase 1, Pe is the Peclet number. The boundary conditions are:

$$y = 0: \frac{\partial c_1}{\partial y} = 0$$

$$y = 1: \frac{\partial c_2}{\partial y} = 0;$$

(6.17)

$$y = \varphi: c_1 = mc_2 \text{ and } D_1 \frac{\partial c_1}{\partial y} = D_2 \frac{\partial c_2}{\partial y};$$

In a more complicated process, Chasanis et al. (2010) reported a theoretical study of multicomponent mass transfer in a liquid–liquid parallel flow with water/toluene/acetone/MIPK as the working system. The diffusional interaction between the transferred components was considered, and, therefore, the Stefan–Maxwell equations were used to calculate the mass flux. Compared with a control model without the cross-effects between components, it was found that the cross-effects could strongly influence mass-transfer behavior and extraction performance in the microchannels.

For more accurate physical modeling, it is also very easy to perform 3D simulation. Figure 6.10 shows a 3D model for a Y–Y shaped microchannel (Mason et al. 2013). Finite-volume numerical simulations were used, and the lattices are clearly shown in Figure 6.10b. Based on the calculation results, the 8-hydroxyquinoline concentration distribution at different cross-sections can be clearly exhibited as shown in Figure 6.10c, and these simulation results are important for understanding the mass-transfer resistance. Similar to Mason's work, Novak et al. (2015) studied mass transfer in a three-phase parallel flow. The Ψ–Ψ microchannel assisted in creating

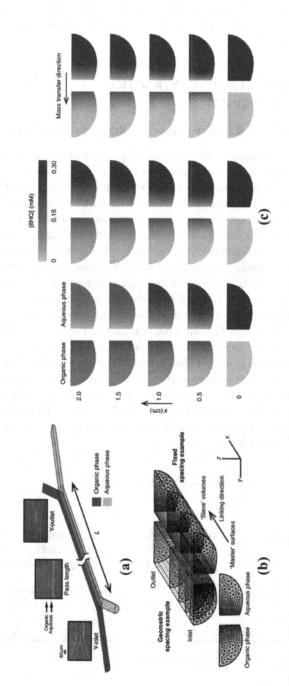

FIGURE 6.10 3D simulation model and concentration distribution at the cross-sections for the liquid–liquid parallel flow. (a) Two-phase microchannel; (b) mesh geometry; (c) cross-sectional distribution of 8-hydroxyquinoline concentration. (Adapted from Mason et al. (2013) with the permission of Springer.)

two interfacial layers for α-amylase transferring to the PEG-rich phase. The simulation results show not only the mass transfer by convection in the flow direction but also the diffusion in the spatial directions, both of which make a major contribution to the higher extraction efficiency.

6.5 MICROFLUIDIC REACTIVE EXTRACTION AND REACTION KINETICS

6.5.1 DETERMINATION OF RATE CONSTANTS FOR REACTION-CONTROLLED SYSTEMS

Herein, we move on to reaction-based liquid–liquid microextraction processes, which are widely used for separating metal ions and charged chemicals. In reaction-controlled microfluidic extraction or ion-exchange processes, mass transfer in the bulk continuous and dispersed phases is much faster than the interfacial mass transfer, or interfacial reaction. In such a kinetic regime, the rate constant for the interfacial reaction could be determined very easily, and the obtained reaction rate constant is thus device-independent. To eliminate the mass-transfer limitations, Nichols et al. (2011) employed a liquid–liquid segmented flow, where strong circulation occurs in both slugs of the dispersed phase and continuous phase, to perform the extraction of lanthanides under TALSPEAK (trivalent actinide–lanthanide separation by phosphorus reagent extraction from aqueous komplexes) conditions as shown in Figure 6.11. Capillaries with 125 μm inner diameter and flow rates under 33 μL/min were used to generate short slugs (length/width ratio 1.5–2) with a constant specific surface area (slug size did not change with flow rates when the flow rate was changed). The process was assumed to be controlled by the equilibrium shown at the bottom of Figure 6.11. The kinetics of the forward and backward reactions were both simplified to be first order in metal M (M^{3+} and $MA_3(HA)_3$ for the forward and backward reactions respectively), with a forward and backward reaction rate constant k_+ and k_- respectively. These lead to:

$$\frac{1}{a} \cdot \frac{d[M]}{dt} = -k_+[M] + k_-[\overline{M}]$$ (6.17)

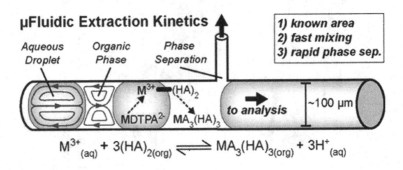

$$M^{3+}_{(aq)} + 3(HA)_{2(org)} \rightleftharpoons MA_3(HA)_{3(org)} + 3H^+_{(aq)}$$

FIGURE 6.11 TALSPEAK metal solvent extraction in liquid–liquid segmented flow. (Adapted from Nichols et al. (2011) with the permission of the American Chemical Society.)

where k_+ and k_- are both interfacial rate constants and have the unit of m/s. [M] and [$\bar{\text{M}}$] are the concentrations of metal in the aqueous and oil phases respectively and were obtained by the analysis of the aqueous sample and material-balance calculation, respectively. $\bar{\text{M}}$ represents $MA_3(HA)_3$. The equilibrium constant of the interfacial ion-exchange reaction was acquired via extraction equilibrium experiment, and at dynamic equilibrium it can be expressed as:

$$K_{eq} = k_+/k_- \tag{6.18}$$

where the subscript "eq" denotes "equilibrium." Consequently, only one unknown parameter, k_+ or k_-, was left to be determined in Eq. 6.17. Integration of Eq. 6.17 gives the relationship between [M] and t, and then the experimental data can be used to correlate the kinetic constant k_+ or k_-.

By employing the above-described microfluidic method, the authors obtained the extraction kinetics of heavy lanthanides, which were not investigated in previous studies. Interestingly, the authors found that, beyond the Dy element in the lanthanide series, the extraction rate increased as the lanthanide atomic number increased, which is different from the common belief. Moreover, according to the authors, k_+ and k_- remained unchanged even in the presence of other lanthanide cations, which means the cations did not interfere with each other when transferred through the oil–aqueous interface. Consequently, the extraction kinetics for 15 cations were determined at the same time, producing no more than 30 µL of liquid waste. Launiere and Gelis (2016) adopted a liquid–liquid segmented flow as well but with a different microfluidic setup, to investigate the variation of lanthanide extraction kinetics under different pH values and using different extractants. The values of k_+ and k_- were in good agreement with those from Nichols et al., demonstrating that the interfacial reaction rate constants determined in such microfluidic process were indeed device-dependent. In both above papers, the authors found that the values of k_+ and k_- obtained via conventional methods (e.g., constant interface cell) were smaller than those they obtained, which was partially due to the diffusion zone at the liquid–liquid interface being present in the constant interface cell.

It is important to note that the above-mentioned two papers both ignored mass transfer in the film between the droplet and the tube wall and only took the droplet cap into account when calculating the interfacial area. This is applicable at lower flow rates as in the case of these two papers. However, when the flow rate is elevated to higher values, the effect of this thin film will become significant, and the effective interfacial area will increase markedly and approach the physical (geometric) interfacial area including the droplet caps and cylindrical area (Ghaini et al. 2010). Zhao et al. (2015) managed to avoid the above-discussed issue by utilizing spherical droplets to investigate the kinetics for the ion exchange of $H_2PO_4^-$ and Cl^- between an aqueous KCl solution and H_3PO_4-loaded oil phase (trioctylamine + n-octanol):

$$\overline{TOA \cdot H_3PO_4} + KCl \rightleftharpoons KH_2PO_4 + \overline{TOA \cdot HCl} \tag{6.19}$$

where the bar above the symbol represents the oil phase, and TOA denotes trioctylamine. The forward and backward reactions were both assumed to proceed as first order in each reactant, resulting in more complexed mass-transfer equations as expressed below:

$$\frac{d[\overline{TOA \cdot HCl}]}{dt} = k_+[\overline{TOA \cdot H_3PO_4}][Cl^-] - \frac{k_+}{K_{eq}}[\overline{TOA \cdot HCl}][H_2PO_4^-] \quad (6.20)$$

$$\frac{d[\overline{TOA \cdot H_3PO_4}]}{dt} = -k_+[\overline{TOA \cdot H_3PO_4}][Cl^-] + \frac{k_+}{K_{eq}}[\overline{TOA \cdot HCl}][H_2PO_4^-] \quad (6.21)$$

where k_+, with a unit of m^3/(mol·s) is the forward reaction rate constant and the only parameter to be determined. The changes in concentrations of the two acids in the oil phase were monitored in experiments. Thus k_+ was solved by fitting these experimental data to Eq. 6.20 and 6.21 with the nonlinear least-squares method in MATLAB©, and subsequently the backward reaction rate constant k_- was calculated using Eq. 6.18. In the same setup, Zhao et al. (2016b) also acquired the back-extraction kinetics of HCl with ammonia from the extractant (TOA + n-octanol). The process was assumed to be controlled by the reversible interfacial neutralization reaction between $\overline{TOA \cdot HCl}$ and NH$_3$·H$_2$O.

For chemical extraction with an irreversible reaction, the kinetic model could be much simpler than the above-discussed models due to its simple reaction rate law, especially for the first-order reaction. Morita et al. (2010) studied the extraction of Cu(II) by potassium dioctyldithiocarbamate with parallel flow. The mass-transfer equation was given using the following plain equation:

$$\frac{d[Cu(II)]}{dt} = -k_r[Cu(II)] \quad (6.22)$$

where k_r was regarded as the apparent reaction rate constant. Ultimately after the determination of reaction rate constants, validation of elimination of mass-transfer limitations should be done, either by comparing the values of k_+ or k_- under different flow rates or by comparing the characteristic reaction time (τ_r) and mixing time (τ_m). If the k_+ or $k-$ values keep constant under different flow rates or $\tau_r \gg \tau_m$, the assumption that the process is kinetically limited holds true, and the kinetics determined are correct. Besides, Zhao et al. (2015) calculated the Hatta number (Wasewar et al. 2004), Ha, to verify that the process was not affected by mass transfer and the ion-exchange reaction took place at the interface.

6.5.2 DETERMINATION OF KINETICS CONTROLLED BY DIFFUSION AND REACTION SIMULTANEOUSLY

In many (probably more usual) cases, the extraction process combined with a chemical reaction operates in a mixed-diffusion-reaction regime, where mass-transfer resistance and kinetic resistance are both important (Jafari et al. 2016; Yang et al. 2013). The mass-transfer equations become even more complicated than Eqs. 6.17 and 6.18. To understand the kinetics in such cases, we can examine the overall mass-transfer coefficient, or build a numerical model. The extraction of solute X can be expressed by the following reaction:

$$X \underset{k_-}{\overset{k_+}{\rightleftharpoons}} \overline{X} \tag{6.23}$$

with forward and backward reactions both being first order in X. Thus, the change in the concentration of X in the aqueous phase with time is:

$$\frac{d[X]}{dt} = -k_+[X]_{int} + k_-[\overline{X}]_{int} \tag{6.24}$$

where the subscript "int" denotes "interface." The equilibrium constant of the reaction in Eq. 6.23 is defined as K_{eq} and can be obtained with Eq. 6.25 by performing extraction equilibrium tests.

$$K_{eq} = [\overline{X}]_{eq}/[X]_{eq} \tag{6.25}$$

Thus, Eq. 6.24 can be converted to:

$$\frac{d[X]}{dt} = k_+ \left(-[X]_{int} + \frac{[X]_{eq}}{[\overline{X}]_{eq}}[\overline{X}]_{int} \right) \tag{6.26}$$

Besides the reaction-kinetic model, the mass-transfer equations on the aqueous and oil sides can be expressed by Eqs. 6.27 and 6.28, respectively:

$$V\frac{d[X]}{dt} = k_w A \left([X]_{int} - [X] \right) \tag{6.27}$$

$$V\frac{d[X]}{dt} = k_o A \left([\overline{X}] - [\overline{X}]_{int} \right) \tag{6.28}$$

where k_w and k_o are the mass-transfer coefficients in the aqueous and oil phases, respectively; and A and V are the interfacial area and the reacting phase volume, respectively. Combining Eqs. 6.26 through 6.28, we have:

$$\frac{d[X]}{dt}\left(\frac{1}{k_+} + \frac{V}{k_w A} + \frac{[X]_{eq}}{[\overline{X}]_{eq}}\frac{V}{k_o A} \right) = \frac{[X]_{eq}}{[\overline{X}]_{eq}}[\overline{X}] - [X] \tag{6.29}$$

If we express the reaction extraction process to a common mass-transfer process, and define:

$$\frac{1}{K_L a} = \frac{1}{k_+} + \frac{1}{k_w a} + \frac{[X]_{eq}}{[\overline{X}]_{eq}}\frac{1}{k_o a} \tag{6.30}$$

where a is defined as A/V, then Eq. 6.29 can be converted to a simple equation:

$$\frac{d[X]}{dt} = K_L a \left(\frac{[X]_{eq}}{[\overline{X}]_{eq}} [\overline{X}] - [X] \right) \tag{6.31}$$

By integrating Eq. 6.31 over the residence time t in the microfluidic device, the volumetric overall mass-transfer coefficient can be determined:

$$K_L a = \ln \left(\frac{[X]_{eq} - [X]_0}{[X]_{eq} - [X]_t} \right) \Big/ \left(t + \frac{[X]_{eq}}{[\overline{X}]_{eq}} \frac{Q_w}{Q_o} t \right) \tag{6.32}$$

where $[X]_t$ is the concentration of X at residence time t (outlet concentration), and Q_w and Q_o are the flow rates of the aqueous and oil phases respectively. It can be seen from Eq. 6.30 that the volumetric overall mass-transfer coefficient $K_L a$ takes fully into account the kinetic resistance, the aqueous-side mass-transfer resistance, and the oil-side mass-transfer resistance. Thus, K_L is also the apparent overall mass-transfer coefficient. In addition, it is worth mentioning that $K_L a$ determined by Eq. 6.32 is an average value over residence time t in the microfluidic device. And as t is decreased, the value of $K_L a$ increases with t (Foroozan Jahromi et al. 2018). Tsaoulidis and Angeli (2015) used CFD simulation to model the extraction of uranyl UO_2^{2+} from nitric acid solutions into TBP/IL mixtures during slug flow. The simulated $K_L a$ values were in reasonable agreement with all the experimental results.

Besides the overall mass-transfer coefficient, the reaction rate constant can also be determined via numerical modeling. Ciceri et al. (2013a) established a CFD model to determine the extraction kinetics of Co(II) and Fe(III) by di-(2-ethylhexyl) phosphoric acid (DEHPA) in liquid–liquid parallel flows, respectively. The model simulated the diffusion in the aqueous and organic phases as well as the reaction at the interface for the path-length domain in Figure 6.12a. Incompressible continuity and steady-state Naiver–Stokes equations in their conservative forms were used for

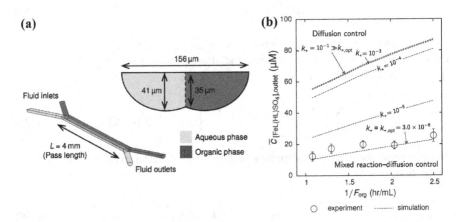

FIGURE 6.12 Determination of extraction kinetics by simulation. (a) Schematic and cross-sectional geometry of the Y–Y microfluidic device; (b) Effect of k_+ on outlet concentration of the extraction product $FeL(HL)SO_4$. (Adapted from Ciceri et al. (2014) with the permission of Elsevier.)

fluid dynamics and the advection–diffusion equation for mass transfer. The reaction rate law used, taking Fe(III) extraction as an example, was in the form of:

$$r = k_+[\overline{FeSO_4}][\overline{(HL)_2}] - k_-[\overline{Fe(HL)SO_4}][H^+] \tag{6.33}$$

for the equilibrium at the interface:

$$\overline{FeSO_4} + \overline{(HL)_2} \rightleftharpoons \overline{Fe(HL)SO_4} + H^+ \tag{6.34}$$

where HL represents the monomeric form of DEHPA and $(HL)_2$ the dimeric form. The forward interfacial reaction rate constant of the equilibrium k_+ was regarded as the undetermined parameter in the simulation, which was obtained to be $k_{+,opt} = (3.0 \pm 0.1) \times 10^{-6}$ m^4/(mol·s) via least-squares fitting to the experimental kinetic data. The reaction rate determined by this method corresponded well with that obtained in a constant interface cell. In addition, the extraction regime was judged by simulating the extraction process under diffusion-controlled conditions using a reaction rate constant $k_+ \gg k_{+,opt}$. As shown in Figure 6.12b, when the value of k_+ was $>10^{-3}$, the kinetic curve changed little, indicating the appearance of the diffusion-controlled regime. It can be obviously seen that the kinetic curve in this regime deviated greatly from the experimental data. Therefore, reaction played an important role in the extraction process here and a mixed diffusion-reaction controlled mechanism was determined.

6.5.3 MICROFLUIDIC EXTRACTION KINETICS FOR SEPARATING MULTI-COMPONENTS

It is not surprising that the separation of two different solutes/components could be more easily and rapidly realized in a microextraction process than in a traditional process. Kolar et al. (2016) conducted extraction of rare earth elements (REEs) from a leached mixed rare earth oxide mineral concentrate by Cyanex® 572 to separate the heavy REEs from the light REEs. Tests were done in a Y–Y shaped microchip and a stirred beaker, respectively. It was found that after the same contacting time 10 s, more heavy REEs (66% vs. 31%) and meanwhile less light REEs were extracted in the microchip than in the bulk extraction process, as shown in Figure 6.13a, indicating that higher selectivity of heavy REEs to light REEs could be obtained via the microextraction process. This satisfying result was exhibited more clearly by the work of Jiang et al. (2018) where the dynamic change of selectivity as a function of time was investigated for both microfluidic and conventional extraction. Extraction and separation of La(III) and Ce(III) from chloride solution was done in microfluidic segmented flow with EHEHPA (2-ethylhexyl phosphoric acid-2-ethylhexyl ester) acting as the extractant. It can be seen from Figure 6.13b that a higher La-to-Ce separation ratio could be obtained in a much shorter time in microfluidics as compared to the conventional results (selectivity of 8.5 in 10 s vs. selectivity of 5.7 in 600 s).

Intriguingly, some unexpected separation performance could also be achieved in microfluidics. Zhang et al. (2017) studied the recovery of Co(II) from Ni(II) sulfate solution by solvent extraction with Cyanex 272 (Diisooctylphosphinic acid, $C_{16}H_{35}O_2P$). A coiled flow inverter was employed to enhance the mass transfer in the

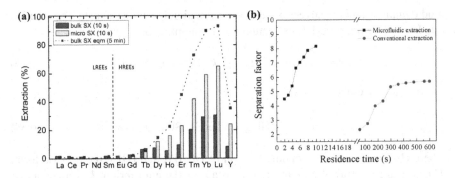

FIGURE 6.13 Comparison of separation performances between microfluidic and conventional extraction. (a) Extraction ratio of different REEs in microextraction and under bulk conditions after a contact time of 10 s. (Adapted from Kolar et al. (2016) with the permission of Elsevier). (b) Separation ratio of La-to-Ce in microfluidic and conventional extraction. (Adapted from Jiang et al. (2018) with the permission of Elsevier.)

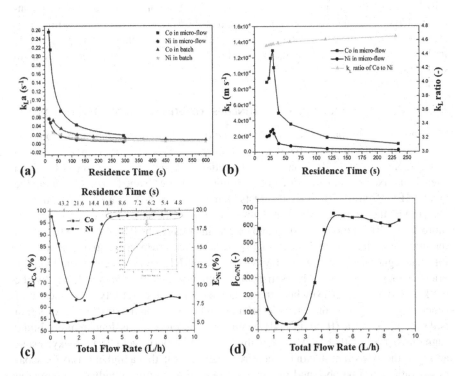

FIGURE 6.14 Dynamics for the extraction and separation of Co and Ni by Cyanex 272 in microscale and batch reactors. The plots of (a) overall volumetric mass-transfer coefficient and (b) overall mass-transfer coefficient versus residence time for Co and Ni. (Both (a) and (b) are adapted from Zhang et al. (2017) with the permission of Elsevier). The plots of (c) extraction ratio and (d) selectivity versus total flow rate for Co and Ni. (Both (c) and (d) are adapted from Zhang et al. (2018) with the permission of Elsevier.)

liquid–liquid segmented flow, and a separating funnel was used for bulk extraction. The authors examined the dynamic change of overall volumetric mass-transfer coefficient, $k_L a$, along with residence time and found that Co and Ni showed different changing trends. As displayed by Figure 6.14a, for Co extraction, $k_L a$ was increased remarkably at the microscale in a shorter residence time range (i.e., at higher flow

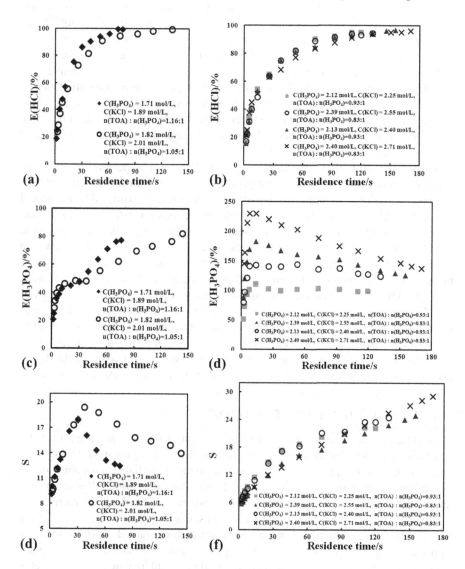

FIGURE 6.15 Dynamics for the selective extraction of HCl and H_3PO_4 by (TOA + n-octanol) under different conditions. The variation of (a) HCl extraction efficiency, (b) H_3PO_4 extraction efficiency and (c) HCl-to-H_3PO_4 selectivity versus residence time with initial TOA in excess of H_3PO_4; the variation of (d) HCl extraction efficiency, (e) H_3PO_4 extraction efficiency and (f) HCl-to-H_3PO_4 selectivity versus residence time with initial H_3PO_4 in excess of TOA. (Adapted from Zhao et al. (2016a) with the permission of Elsevier.)

rates), indicating the obvious intensification of mass transfer in Co extraction. By contrast, $k_L a$ did not differ much for Ni extraction in micro and batch extraction. This is possibly ascribed to the fact that Co is much more easily extracted by Cyanex 272 than Ni.

Further, Zhang et al. (2018) developed the microfluidic extraction process to the pilot scale and examined the dynamic change of extraction ratio and separation ratio for Co and Ni, respectively. As shown by Figure 6.14d, in a relatively low total flow-rate range from 1 to 3 L/h (herein, flow rate was inversely proportional to residence time), a low selectively of Co to Ni was found due to the very low extraction efficiency of Co. And as the total flow rate increased to be larger than 5 L/h, the selectively for Co decreased gradually from a high value due to increased extraction efficiency of Ni and saturated Co extraction in the oil phase. We can see that both extraction ratio and separation ratio exhibited even more complicated changes vs. the total flow rate thanks to the mutual effects of several parameters including flow rate, slug size, and contact time of the two phases.

Zhao et al. (2016a) eliminated the effects of flow rates by performing microextraction in a capillary tube whose length could be easily varied to obtain different residence times. The aqueous solution of ($KCl + H_3PO_4$) was extracted by the mixed solvent of ($TOA + n$-octanol), which was actually the selective extraction of HCl than H_3PO_4. The authors obtained different selectivity changing profiles under different initial conditions. As shown in Figure 6.15c, when the initial amount of TOA was in excess of H_3PO_4, the selectivity of HCl-to-H_3PO_4 increased first and then decreased because of the second-stage increase in H_3PO_4 extraction rate after the first plateau (Figure 6.15b). By contrast, the extraction efficiency of H_3PO_4 first increased swiftly and then decreased gradually (Figure 6.15e) when the initial amount of H_3PO_4 was in excess of TOA. This leads to the continuous augment of HCl-to-H_3PO_4 selectivity (Figure 6.15f). These results are exciting because they indicate that selectivity exceeding the thermodynamic equilibrium can be achieved by exploiting the different extraction kinetics of the different components in a selective extraction process.

6.6 CONCLUSIONS AND OUTLOOK

Microfluidic solvent extraction is not only a potential intensification approach for separation processes but also an ideal method for determining the mass-transfer and reaction kinetics of liquid–liquid systems. Within the controllable and stable liquid–liquid flows in microspace, the segmented droplet flow and parallel flow are preferable to researchers when implementing kinetic experiments, and typical microfluidic platforms have been shown in the above sections. Combining with CFD simulation, the mass-transfer and reaction-kinetic parameters can be obtained with high accuracy as presented by the above examples. We hope the equations and mathematic models summarized from recent or classic literature will be useful for readers in the areas of microfluidics, mass transfer, and separation engineering. As a young research area, the microfluidic extraction processes still have many unknown rules and knowledge waiting to be discovered, which should be further based on new experimental and detection methods, more accurate models, and highly efficient simulations. With the minimization of ultraviolet (UV) and infrared (IR) spectrometers, nuclear magnetic resonance (NMR), and other advanced equipment, microfluidic platforms could be more functionalized, which is beneficial for identifying and quantifying chemicals. With the help of automated control and artificial intelligence (AI) methods, the microflow extraction platform could be more efficient and intelligent. With more accurate models considering the complicated interface effects, such as Marangoni effect and interface flow during microflow extraction, more details of microflow extraction could be exhibited. Therefore, we can expect more and more effective applications of microfluidic extractions in the areas of fine chemical synthesis, pharmaceutical, biological products, and so on.

ACKNOWLEDGMENTS

We gratefully acknowledge financial support from the National Natural Science Foundation of China (91334201, 21776150, U1463208, U1302271).

NOMENCLATURE

A	interfacial area, m^2
a	specific mass-transfer area, m^2/m^3
a_{cap}	specific mass-transfer area of droplet caps, m^2/m^3
a_{film}	specific mass-transfer area of liquid films around droplets, m^2/m^3
Ca_d	droplet capillary number, $Ca_d = \mu_C u_d/\gamma$
c	concentration, mol/m^3
D	diffusion coefficient, m^2/s
D_C	diffusion coefficient in the continuous phase, m^2/s
D_D	diffusion coefficient in the dispersed phase, m^2/s
d	diameter, m
d_d	diameter of spherical droplet, m
d_{in}	microtube or capillary inner diameter, m
Fo	Fourier number, $Fo = D u_d/L_{film} h^2$
h	film thickness, m
K_{eq}	reaction equilibrium constant
K_L	total mass-transfer coefficient of liquid phases, m/s
k	mass-transfer coefficient, m/s
k_C	local mass-transfer coefficient in the continuous phase, s^{-1}
$k_{C,cap}$	local mass-transfer coefficient in the continuous phase near the droplet caps, s^{-1}
$k_{C,film}$	local mass-transfer coefficient in the continuous phase in the liquid films, s^{-1}
k_{cap}	mass-transfer coefficient near the droplet caps, s^{-1}
k_D	local mass-transfer coefficient in the dispersed phase, s^{-1}
k_{film}	mass-transfer coefficient in the liquid films, s^{-1}
k_L	mass-transfer coefficient of the liquid phase, m/s
k_o	local mass-transfer coefficient of the oil phase, m/s
k_r	reaction constant, m/s
k_w	local mass-transfer coefficient of the water phase, m/s

k_+	forward (interfacial) reaction rate constant
k_-	backward (interfacial) reaction rate constant
L_d	droplet length, m
L_{film}	liquid film length, m
L_{uc}	flow unit length, m
l	microchannel length, m
m	partition coefficient
n	normal vector to the interface
p_o	pressure of the oil phase, Pa
p_w	pressure of the water phase, Pa
Q	flow rate, m^3/s
R	interface radius, m
Re	Reynolds number, $Re = du\rho/\mu$
Sc	Schmidt number, $Sc = \mu/\rho D$
Sh	Sherwood number, $Sh = k_L d/D$
t	time, s
t_n	tangent vector to the interface
U	total superficial velocity of two phases, m/s
U_D	superficial velocities of the dispersed phase, m/s
u	velocity, m/s
u_d	droplet velocity, m/s
u_o	oil phase velocity, m/s
u_w	water phase velocity, m/s
V	reacting phase volume, m^3
V_d	droplet volume, m^3
V_{uc}	volume of a flow unit, m^3
w	microchannel width, m
w_d	droplet width, m
X	concentration of transfer substance, mol/m^3
φ	liquid holdup
φ_D	holdup of the dispersed phase
γ	interfacial tension, N/m
η_s	interface shear viscosity, Pa·s·m
μ	viscosity, Pa·s
μ_C	continuous phase viscosity, Pa·s
μ_m	mixture viscosity of two phases, Pa·s
μ_o	oil phase viscosity, Pa·s
μ_w	water phase viscosity, Pa·s
ρ	density, kg/m^3
ρ_m	mixture density of two phases, kg/m^3
τ_r	characteristic reaction time, s
τ_m	characteristic mixing time, s

REFERENCES

Abolhasani, M., and K. F. Jensen. 2016. Oscillatory multiphase flow strategy for chemistry and biology. *Lab Chip* 16:2775–84.

Abolhasani, M., M. Singh, E. Kumacheva, and A. Günther. 2012. Automated microfluidic platform for studies of carbon dioxide dissolution and solubility in physical solvents. *Lab Chip* 12:1611–8.

Adamo, A., R. L. Beingessner, M. Behnam, et al. 2016. On-demand continuous-flow production of pharmaceuticals in a compact, reconfigurable system. *Science* 352:61–7.

Adamo, A., P. L. Heider, N. Weeranoppanant, and K. F. Jensen. 2013. Membrane-based, liquid–liquid separator with integrated pressure control. *Ind. Eng. Chem. Res.* 52:10802–8.

Ambare, D. N., S. A. Ansari, M. Anitha, et al. 2013. Non-dispersive solvent extraction of neodymium using a hollow fiber contactor: Mass transfer and modeling studies. *J. Membr. Sci.* 446:106–12.

Aussillous, P., and D. Quéré. 2000. Quick deposition of a fluid on the wall of a tube. *Phys. Fluids* 12:2367–71.

Baroud, C. N., F. Gallaire, and R. Dangla. 2010. Dynamics of microfluidic droplets. *Lab Chip* 10:2032–45.

Berčič, G., and A. Pintar. 1997. The role of gas bubbles and liquid slug lengths on mass transport in the Taylor flow through capillaries. *Chem. Eng. Sci.* 52:3709–19.

Bico, J., and D. Quéré. 2000. Liquid trains in a tube. *Europhys. Lett.* 51:546–50.

Bretherton, F. P. 1961. The motion of long bubbles in tubes. *J. Fluid Mech.* 10:166–88.

Britton, J., and C. L. Raston. 2017. Multi-step continuous-flow synthesis. *Chem. Soc. Rev.* 46:1250–71.

Castor, C. A., A. Pontier, J. Durand, J. C. Pinto, and L. Prat. 2015. Real time monitoring of the quiescent suspension polymerization of methyl methacrylate in microreactors – Part 1. A kinetic study by Raman spectroscopy and evolution of droplet size. *Chem. Eng. Sci.* 131:340–52.

Charpentier, J. C. 2012. Did you say "Reactor Design & Process Intensification," or how to produce much more and better while consuming much less? Review of Editorial Material. *Chem. Eng. Technol.* 35:1118–9.

Chasanis, P., M. Brass, and E. Y. Kenig. 2010. Investigation of multicomponent mass transfer in liquid–liquid extraction systems at microscale. *Int. J. Heat Mass Transf.* 53:3758–63.

Chen, Z., W. T. Wang, F. N. Sang, J. H. Xu, G. S. Luo, and Y. D. Wang. 2017. Fast extraction and enrichment of rare earth elements from waste water via microfluidic-based hollow droplet. *Sep. Purif. Technol.* 174:352–61.

Christopher, G. F., N. N. Noharuddin, J. A. Taylor, and S. L. Anna. 2008. Experimental observations of the squeezing-to-dripping transition in T-shaped microfluidic junctions. *Phys Rev E Stat. Nonlin. Soft Matter. Phys.* 78:036317.

Ciceri, D., L. R. Mason, D. J. E. Harvie, J. M. Perera, and G. W. Stevens. 2013a. Modelling of interfacial mass transfer in microfluidic solvent extraction: Part II. Heterogeneous transport with chemical reaction. *Microfluid. Nanofluid.* 14:213–24.

Ciceri, D., K. Nishi, G. W. Stevens, and J. M. Perera. 2013b. An integrated UV-Vis microfluidic device for the study of concentrated solvent extraction reactions. *Solvent Extr. Res. Dev. Jpn* 20:197–203.

Ciceri, D., L. R. Mason, D. J. E. Harvie, J. M. Perera, and G. W. Stevens. 2014. Extraction kinetics of Fe(III) by di-(2-ethylhexyl) phosphoric acid using a Y–Y shaped microfluidic device. *Chem. Eng. Res. Des.* 92:571–80.

Dessimoz, A. L., L. Cavin, A. Renken, and L. Kiwi-Minsker. 2008. Liquid–liquid two-phase flow patterns and mass transfer characteristics in rectangular glass microreactors. *Chem. Eng. Sci.* 63:4035–44.

Di Miceli Raimondi, N., L. Prat, C. Gourdon, and P. Cognet. 2008. Direct numerical simulations of mass transfer in square microchannels for liquid–liquid slug flow. *Chem. Eng. Sci.* 63:5522–30.

Foroozan Jahromi, P. F., J. Karimi-Sabet, and Y. Amini. 2018. Ion-pair extraction-reaction of calcium using Y-shaped microfluidic junctions: An optimized separation approach. *Chem. Eng. J.* 334:2603–15.

Fu, T. T., Y. G. Ma, D. Funfschilling, and H. Z. Li. 2011. Gas-liquid flow stability and bubble formation in non-Newtonian fluids in microfluidic flow-focusing devices. *Microfluid. Nanofluid.* 10:1135–40.

Galambos, P., and F. Forster. 1998. An optical micro fluidic viscometer. *ASME Int. Mech. Eng. Cong. Exp.*, Anaheim, CA:187–91.

Ghaini, A., M. N. Kashid, and D. W. Agar. 2010. Effective interfacial area for mass transfer in the liquid–liquid slug flow capillary microreactors. *Chem. Eng. Process. Process Intensif.* 49:358–66.

Glawdel, T., C. Elbuken, and C. L. Ren. 2012. Droplet formation in microfluidic T-junction generators operating in the transitional regime. II. Modeling. *Phys Rev E Stat. Nonlin. Soft Matter. Phys.* 85:016323.

Gobert, S. R. L., S. Kuhn, L. Braeken, and L. C. J. Thomassen. 2017. Characterization of milli- and microflow reactors: Mixing efficiency and residence time distribution. *Org. Process Res. Dev.* 21:531–42.

Guillot, P., T. Moulin, R. Kötitz, et al. 2008. Towards a continuous microfluidic Rheometer. *Microfluid. Nanofluid.* 5:619–30.

Guillot, P., P. Panizza, J. B. Salmon, et al. 2006. Viscosimeter on a microfluidic chip. *Langmuir* 22:6438–45.

Günther, A., M. Jhunjhunwala, M. Thalmann, M. A. Schmidt, and K. F. Jensen. 2005. Micromixing of miscible liquids in segmented gas–liquid flow. *Langmuir* 21:1547–55.

Hellé, G., C. Mariet, and G. Cote. 2015. Liquid–liquid extraction of uranium(VI) with Aliquat® 336 from HCl media in microfluidic devices: Combination of micro-unit operations and online ICP-MS determination. *Talanta* 139:123–31.

Hou, H. L., Y. Jing, Y. Wang, Y. D. Wang, J. H. Xu, and J. Chen. 2015a. Solvent extraction performance of Ce(III) from chloride acidic solution with 2-ethylhexyl phosphoric acid-2-ethylhexyl ester (EHEHPA) by using membrane dispersion micro-extractor. *J. Rare Earths* 33:1114–21.

Hou, H. L., J. H. Xu, Y. D. Wang, and J. Chen. 2015b. Solvent extraction performance of Pr (III) from chloride acidic solution with 2-ethylhexyl phosphoric acid-2-ethylhexyl ester (EHEHPA) by using membrane dispersion micro-extractor. *Hydrometallurgy* 156:116–23.

Irandoust, S., and B. Andersson. 1989. Liquid-film in Taylor flow through a capillary. *Ind. Eng. Chem. Res.* 28:1684–88.

Jafari, O., M. Rahimi, and F. H. Kakavandi. 2016. Liquid–liquid extraction in twisted micromixers. *Chem. Eng. Process. Process Intensif.* 101:33–40.

Jaritsch, D., A. Holbach, and N. Kockmann. 2014. Counter-current extraction in microchannel flow: Current status and perspectives. *J. Fluids Eng.* 136:091211.

Jiang, F., S. Yin, C. Srinivasakannan, S. Li, and J. Peng. 2018. Separation of lanthanum and cerium from chloride medium in presence of complexing agent along with EHEHPA (P507) in a serpentine microreactor. *Chem. Eng. J.* 334:2208–14.

Kashid, M. N., I. Gerlach, S. Goetz, et al. 2005. Internal circulation within the liquid slugs of a liquid–liquid slug-flow capillary microreactor. *Ind. Eng. Chem. Res.* 44:5003–10.

Kashid, M. N., A. Gupta, A. Renken, and L. Kiwi-Minsker. 2010. Numbering-up and mass transfer studies of liquid–liquid two-phase microstructured reactors. *Chem. Eng. J.* 158:233–40.

Kashid, M. N., A. Renken, and L. Kiwi-Minsker. 2011. Gas–liquid and liquid–liquid mass transfer in microstructured reactors. *Chem. Eng. Sci.* 66:3876–97.

Kazemi Oskooei, S. A, and D. Sinton. 2010. Partial wetting gas–liquid segmented flow microreactor. *Lab Chip* 10:1732–4.

Kirschneck, D., and G. Tekautz. 2007. Integration of a microreactor in an existing production plant. *Chem. Eng. Technol.* 30:305–8.

Ko, D. H., W. Ren, J. O. Kim, et al. 2016. Superamphiphobic silicon-nanowire-embedded microsystem and in-contact flow performance of gas and liquid streams. *ACS Nano* 10:1156–62.

Kolar, E., R. P. R. Catthoor, F. H. Kriel, et al. 2016. Microfluidic solvent extraction of rare earth elements from a mixed oxide concentrate leach solution using Cyanex® 572. *Chem. Eng. Sci.* 148:212–8.

Kuhn, S., R. L. Hartman, M. Sultana, K. D. Nagy, S. Marre, and K. F. Jensen. 2011. Teflon-coated silicon microreactors: Impact on segmented liquid–liquid multiphase flows. *Langmuir* 27:6519–27.

Kuijpers, K. P. L., M. A. H. van Dijk, Q. G. Rumeur, V. Hessel, Y. Su, and T. Noël. 2017. A sensitivity analysis of a numbered-up photomicroreactor system. *React. Chem. Eng.* 2:109–15.

Launiere, C. A., and A. V. Gelis. 2016. High precision droplet-based microfluidic determination of americium(III) and lanthanide(III) solvent extraction separation kinetics. *Ind. Eng. Chem. Res.* 55:2272–6.

Lee, S. C. 2004. Kinetics of reactive extraction of penicillin G by Amberlite LA-2 in kerosene. *AIChE J.* 50:119–26.

Li, G., M. Shang, Y. Song, and Y. Su. 2018. Characterization of liquid–liquid mass transfer performance in a capillary microreactor system. *AIChE J.* 64:1106–16.

Li, Y., J. Zhu, Y. Wu, J. Liu, and J. Aalders. 2013. Reactive-extraction of 2,3-butanediol from fermentation broth by propionaldehyde – Equilibrium and kinetic study. *Korean J. Chem. Eng.* 30:73–81.

Liu, D., K. Wang, Y. Wang, Y. D. Wang, and G. S. Luo. 2017. A simple online phase separator for the microfluidic mass transfer studies. *Chem. Eng. J.* 325:342–9.

Liu, Z. D., Y. C. Lu, M. Zhang, W. M. Wan, and G. S. Luo. 2013. Controllable preparation of uniform polystyrene nanospheres with premix membrane emulsification. *J. Appl. Polym. Sci.* 129:1202–11.

Lubej, M., U. Novak, M. Liu, M. Martelanc, M. Franko, and I. Plazl. 2015. Microfluidic droplet-based liquid–liquid extraction: Online model validation. *Lab Chip* 15:2233–9.

Malengier, B., J. L. Tamalapakula, and S. Pushpavanam. 2012. Comparison of laminar and plug flow-fields on extraction performance in micro-channels. *Chem. Eng. Sci.* 83:2–11.

Marchessault, R. N., and S. G. Mason. 1960. Flow of entrapped bubbles through a capillary. *Ind. Eng. Chem.* 52:79–84.

Mason, L. R., D. Ciceri, D. J. E. Harvie, J. M. Perera, and G. W. Stevens. 2013. Modelling of interfacial mass transfer in microfluidic solvent extraction: Part I. Heterogenous transport. *Microfluid. Nanofluid.* 14:197–212.

Matsuoka, A., K. Noishiki, and K. Mae. 2016. Experimental study of the contribution of liquid film for liquid–liquid Taylor flow mass transfer in a microchannel. *Chem. Eng. Sci.* 155:306–13.

McMullen, J. P., and K. F. Jensen. 2011. Rapid determination of reaction kinetics with an automated microfluidic system. *Org. Process Res. Dev.* 15:398–407.

Mehta, K., and J. J. Linderman. 2006. Model-based analysis and design of a microchannel reactor for tissue engineering. *Biotechnol. Bioeng.* 94:596–609.

Mielke, E., P. Plouffe, N. Koushik, et al. 2017. Local and overall heat transfer of exothermic reactions in microreactor systems. *React. Chem. Eng.* 2:763–75.

Morais, J. M., O. D. H. Santos, and S. E. Friberg. 2010. Some fundamentals of the one-step formation of double emulsions. *J. Dispers. Sci. Technol.* 31:1019–26.

Morita, K., T. Hagiwara, N. Hirayama, and H. Imura. 2010. Extraction of Cu(II) with dioctyldithiocarbamate and a kinetic study of the extraction using a two-phase microflow system. *Solvent Extr. Res. Dev. Jpn* 17:209–14.

Nichols, K. P., R. R. Pompano, L. Li, A. V. Gelis, and R. F. Ismagilov. 2011. Toward mechanistic understanding of nuclear reprocessing chemistries by quantifying lanthanide solvent extraction kinetics via microfluidics with constant interfacial area and rapid mixing. *J. Am. Chem. Soc.* 133:15721–9.

Nieves-Remacha, M. J., A. A. Kulkarni, and K. F. Jensen. 2012. Hydrodynamics of liquid–liquid dispersion in an advanced-flow reactor. *Ind. Eng. Chem. Res.* 51:16251–62.

Niu, G., A. Ruditskiy, M. Vara, and Y. Xia. 2015. Toward continuous and scalable production of colloidal nanocrystals by switching from batch to droplet reactors. *Chem. Soc. Rev.* 44:5806–20.

Novak, U., M. Lakner, I. Plazl, and P. Žnidaršič-Plazl. 2015. Experimental studies and modeling of α-amylase aqueous two-phase extraction within a microfluidic device. *Microfluid. Nanofluid* 19:75–83.

Phillips, T. W., J. H. Bannock, and J. C. DeMello. 2015. Microscale extraction and phase separation using a porous capillary. *Lab Chip* 15:2960–7.

Porta, R., M. Benaglia, and A. Puglisi. 2016. Flow chemistry: Recent developments in the synthesis of pharmaceutical products. *Org. Process Res. Dev.* 20:2–25.

Priest, C., J. Zhou, S. Klink, R. Sedev, and J. Ralston. 2012. Microfluidic solvent extraction of metal ions and complexes from leach solutions containing nanoparticles. *Chem. Eng. Technol.* 35:1312–9.

Ransohoff, T. C., P. A. Gauglitz, and C. J. Radke. 1987. Snap-off of gas-bubbles in smoothly constricted noncircular capillaries. *AIChE J.* 33:753–65.

Reizman, B. J., and K. F. Jensen. 2016. Feedback in flow for accelerated reaction development. *Acc. Chem. Res.* 49:1786–96.

Sattari-Najafabadi, M., M. Nasr Esfahany, Z. Wu, and B. Sundén. 2017. Hydrodynamics and mass transfer in liquid–liquid non-circular microchannels: Comparison of two aspect ratios and three junction structures. *Chem. Eng. J.* 322:328–38.

Seemann, R., M. Brinkmann, T. Pfohl, and S. Herminghaus. 2012. Droplet based microfluidics. *Rep. Prog. Phys.* 75:016601.

Shui, L. L., R. A. Hayes, M. Jin, et al. 2014. Microfluidics for electronic paper-like displays. *Lab Chip* 14:2374–84.

Shui, L. L., A. van den Berg, and J. C. T. Eijkel. 2009. Interfacial tension controlled W/O and O/W 2-phase flows in microchannel. *Lab Chip* 9:795–801.

Singh, K. K., A. U. Renjith, and K. T. Shenoy. 2015. Liquid–liquid extraction in microchannels and conventional stage-wise extractors: A comparative study. *Chem. Eng. Process. Process Intensif.* 98:95–105.

Snead, D. R., and T. F. Jamison. 2015. A three-minute synthesis and purification of ibuprofen: Pushing the limits of continuous-flow processing. *Angew. Chem. Int. Ed.* 54:983–7.

Song, H., D. L. Chen, and R. F. Ismagilov. 2006. Reactions in droplets in microfluidic channels. *Angew. Chem. Int. Ed. Engl.* 45:7336–56.

Song, H., J. D. Tice, and R. F. Ismagilov. 2003. A microfluidic system for controlling reaction networks in time. *Angew. Chem. Int. Ed.* 42:768–72.

Su, Y. H., Y. C. Zhao, G. W. Chen, and Q. A. Yuan. 2010. Liquid–liquid two-phase flow and mass transfer characteristics in packed microchannels. *Chem. Eng. Sci.* 65:3947–56.

Susanti, J. G. M. Winkelman, B. Schuur, H. J. Heeres, and J. Yue. 2016. Lactic acid extraction and mass transfer characteristics in slug flow capillary microreactors. *Ind. Eng. Chem. Res.* 55:4691–702.

Taha, T., and Z. F. Cui. 2006. CFD modelling of slug flow inside square capillaries. *Chem. Eng. Sci.* 61:665–75.

Thulasidas, T. C., M. A. Abraham, and R. L. Cerro. 1997. Flow patterns in liquid slugs during bubble-train flow inside capillaries. *Chem. Eng. Sci.* 52:2947–62.

Tokimoto, T., S. Tsukahara, and H. Watarai. 2003. Kinetic study of fast complexation of zinc(II) with 8-quinolinol and 5-octyloxymethyl-8-quinolinol at 1-butanol/water interface by two-phase sheath flow/laser-induced fluorescence microscopy. *Bull. Chem. Soc. Jpn* 76:1569–76.

Tsaoulidis, D., and P. Angeli. 2015. Effect of channel size on mass transfer during liquid–liquid plug flow in small scale extractors. *Chem. Eng. J.* 262:785–93.

van Baten, J. M., and R. Krishna. 2004. CFD simulations of mass transfer from Taylor bubbles rising in circular capillaries. *Chem. Eng. Sci.* 59:2535–45.

Vandu, C. O., H. Liu, and R. Krishna. 2005. Mass transfer from Taylor bubbles rising in single capillaries. *Chem. Eng. Sci.* 60:6430–7.

Vir, A. B., A. S. Fabiyan, J. R. Picardo, and S. Pushpavanam. 2014. Performance comparison of liquid–liquid extraction in parallel microflows. *Ind. Eng. Chem. Res.* 53:8171–81.

Vural Gürsel, I., N. Kockmann, and V. Hessel. 2017. Fluidic separation in microstructured devices – Concepts and their Integration into process flow networks. *Chem. Eng. Sci.* 169:3–17.

Wang, K., and G. S. Luo. 2017. Microflow extraction: A review of recent development. *Chem. Eng. Sci.* 169:18–33.

Wang, K., Y. C. Lu, and G. S. Luo. 2014. Strategy for scaling-up of a microsieve dispersion reactor. *Chem. Eng. Technol.* 37:2116–22.

Wang, K., Y. C. Lu, H. W. Shao, and G. S. Luo. 2008. Heat-transfer performance of a liquid–liquid microdispersed system. *Ind. Eng. Chem. Res.* 47:9754–8.

Wannachod, T., V. Mohdee, S. Suren, P. Ramakul, U. Pancharoen, and K. Nootong. 2015. The separation of Nd(III) from mixed rare earth via hollow fiber supported liquid membrane and mass transfer analysis. *J. Ind. Eng. Chem.* 26:214–7.

Wasewar, K. L., A. B. M. Heesink, G. F. Versteeg, and V. G. Pangarkar. 2002. Reactive extraction of lactic acid using alamine 336 in MIBK: Equilibria and kinetics. *J. Biotechnol.* 97:59–68.

Wasewar, K. L., A. B. M. Heesink, G. F. Versteeg, and V. G. Pangarkar. 2004. Intensification of conversion of glucose to lactic acid: Equilibria and kinetics for back extraction of lactic acid using trimethylamine. *Chem. Eng. Sci.* 59:2315–20.

Weeranoppanant, N., A. Adamo, G. Saparbaiuly, et al. 2017. Design of multistage countercurrent liquid–liquid extraction for small-scale applications. *Ind. Eng. Chem. Res.* 56:4095–103.

Wei, J., X. J. Ju, R. Xie, C. L. Mou, X. Lin, and L. Y. Chu. 2011. Novel cationic pH-responsive poly(N,N-dimethylaminoethyl methacrylate) microcapsules prepared by a microfluidic technique. *J. Colloid.Interface Sci.* 357:101–8.

Weigl, M., A. Geist, U. Müllich, and K. Gompper. 2006. Kinetics of americium(III) extraction and back extraction with BTP. *Solvent Extr. Ion Exch.* 24:845–60.

Woitalka, A., S. Kuhn, and K. F. Jensen. 2014. Scalability of mass transfer in liquid–liquid flow. *Chem. Eng. Sci.* 116:1–8.

Xu, B. Y., S. W. Hu, X. N. Yan, X. H. Xia, J. J. Xu, and H. Y. Chen. 2012. On chip steady liquid–gas phase separation for flexible generation of dissolved gas concentration gradient. *Lab Chip* 12:1281–8.

Xu, J. H., J. Tan, S. W. Li, and G. S. Luo. 2008. Enhancement of mass transfer performance of liquid–liquid system by droplet flow in microchannels. *Chem. Eng. J.* 141:242–9.

Yang, L., M. J. Nieves-Remacha, and K. F. Jensen. 2017. Simulations and analysis of multiphase transport and reaction in segmented flow microreactors. *Chem. Eng. Sci.* 169:106–16.

Yang, L., Y. Zhao, Y. Su, and G. Chen. 2013. An experimental study of copper extraction characteristics in a T-Junction microchannel. *Chem. Eng. Technol.* 36:985–92.

Yin, S., K. Chen, C. Srinivasakannan, et al. 2018a. Microfluidic solvent extraction of Ce (III) and Pr (III) from a chloride solution using EHEHPA (P507) in a serpentine microreactor. *Hydrometallurgy* 175:266–72.

Yin, S., J. Pei, J. Peng, L. Zhang, and C. Srinivasakannan. 2018b. Study on mass transfer behavior of extracting La(III) with EHEHPA (P507) using rectangular cross-section microchannel. *Hydrometallurgy* 175:64–9.

Yue, J., L. Luo, Y. Gonthier, G. W. Chen, and Q. Yuan. 2009. An experimental study of air–water Taylor flow and mass transfer inside square microchannels. *Chem. Eng. Sci.* 64:3697–708.

Zhang, L., V. Hessel, and J. Peng. 2018. Liquid–liquid extraction for the separation of Co(II) from Ni(II) with Cyanex 272 using a pilot scale Re-entrance flow microreactor. *Chem. Eng. J.* 332:131–9.

Zhang, L., V. Hessel, J. Peng, Q. Wang, and L. Zhang. 2017. Co and Ni extraction and separation in segmented micro-flow using a coiled flow inverter. *Chem. Eng. J.* 307:1–8.

Zhao, B., J. S. Moore, and D. J. Beebe. 2001. Surface-directed liquid flow inside microchannels. *Science* 291:1023–6.

Zhao, F., Y. C. Lu, K. Wang, and G. S. Luo. 2015. Modeling of kinetics of a microfluidic reaction–extraction process for the preparation of KH_2PO_4. *Sep. Purif. Technol.* 156:108–15.

Zhao, F., Y. C. Lu, K. Wang, and G. S. Luo. 2016a. Kinetic study on selective extraction of HCl and H_3PO_4 in a microfluidic device. *Chin. J. Chem. Eng.* 24:221–5.

Zhao, F., Y. C. Lu, K. Wang, and G. S. Luo. 2016b. Back extraction of HCl from TOA dissolved in n-octanol by aqueous ammonia in a microchannel device. *Solvent Extr. Ion Exch.* 34:60–73.

Zhu, P., and L. Wang. 2016. Passive and active droplet generation with microfluidics: A review. *Lab Chip* 17:34–75.

7 Drop-Based Modeling of Extraction Equipment

David Leleu and Andreas Pfennig

CONTENTS

7.1 INTRODUCTION

The design of equipment for solvent extraction is a challenging task. Two major types of equipment shall be regarded here, namely extraction columns and mixer-settlers. The first steps of design will typically be based on a stage-based method like utilizing a McCabe–Thiele diagram or a corresponding simulation tool to determine the number of theoretical or equilibrium stages required for the separation task to be solved. Since appropriately designed mixer-settlers realize essentially one theoretical stage—that is, the streams leaving each mixer-settler unit are essentially in equilibrium—this method accurately allows one to determine the required number of units in a mixer-settler cascade. The challenge in designing a mixer-settler with respect to its major dimensions lies in the difficulty of predicting coalescence and thus settling behavior, which strongly depends on the material system due to the significant influence of trace impurities like salts and surfactants on the drop coalescence. For extraction columns, on the other hand, it turns out that the efficiency of a column with some selected internals strongly depends also on the specific material system and on the operating conditions. Thus, even knowing the number of equilibrium stages required to solve a separation task will not allow the proper design of the specific extraction equipment.

These challenges in equipment design are typically mastered by performing experiments with the original material system to characterize settling behavior and the performance of extraction columns. For the design of gravity settlers, for example, a simple lab-scale settling test allows dimensioning of the technical equipment in principle, if evaluated with sufficient detail (Henschke, Schlieper, and Pfennig 2002). The design method proposed by Henschke, which is one of the most detailed methods available, assumes a mono-dispersed drop-size distribution. For extraction columns, on the other hand, pilot-plant scale experiments are required (Pfennig, Pilhofer, and Schröter 2006). Based on the results obtained from these experiments, the technical columns can then be designed using, for example, the HTU-NTU method (height of transfer unit, number of transfer units) to adjust the experimentally obtained purities to those desired, followed by applying appropriate scale-up rules. Unfortunately, this method typically requires that the pilot-plant experiments are performed with the column internals identical to those later applied in the technical equipment and that they are performed at similar area-specific loads of the column. This apparently significantly limits the freedom in design and requires expert knowledge to select appropriate column internals prior to performing the pilot-plant scale experiments. If later, based on the first experiments, it should turn out that other internals would be better suited, the pilot-plant experiments need to be repeated with these internals. Also, for pilot-plant experiments, several hundred liters of either phase have to be available to reach steady state, which is usually not the case at least in the early stages of process design.

In the future, the challenges in designing extraction columns will possibly further increase. The cause for this trend is the shift from fossil feedstock in chemical processes to increased utilization of bio-based resources, which will lead, for example, to higher viscosities and the presence of fine particles in the streams to be treated. This will typically lead to broader drop-size distributions and further hindrance in drop coalescence.

To overcome these obstacles and to allow accurate description of extraction performance, drop-based design methods can be applied, which require only a minimal amount of the corresponding phases. How these can be realized together with exemplary examples of their application will be presented in the following. First, extraction-column design will be regarded, then the design of gravity settlers.

7.2 DROP-BASED EXTRACTION-COLUMN DESIGN

At the end of the last millennium, it was realized in the German extraction community from industry and academia that, despite all previous efforts in developing design tools for extraction equipment, there was a significant qualitative difference between the design of distillation and extraction equipment. In distillation-equipment design, knowledge of thermodynamic equilibrium together with some physical properties of the system allows one to realize an already relatively reliable design of required column dimensions. In contrast, for the design of solvent extraction columns, pilot-plant experiments were still required. In Figure 7.1, a schematic representation of an extraction column is shown, where the light phase is dispersed to form droplets, which sediment upward while undergoing mass transfer with the

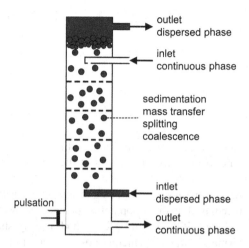

FIGURE 7.1 Scheme of an extraction column with sieve trays as internals and pulsation to ensure a narrow drop-size distribution.

continuous phase. At the top of the column, the dispersed-phase droplets coalesce to form a homogeneous phase, which can then be withdrawn from the column. For improving the drop-size distribution, energy is introduced via pulsation. For such columns, it was realized that the complexity of the equipment design is caused by the critical interplay between sedimentation and mass transfer, which is modulated by the influence of impurities especially strongly in the range of drop diameters between 1 mm and 3 mm, which corresponds exactly to the typically desired drop size in columns. At such drop diameters, traces of surface-active components may hinder internal circulation inside a sedimenting drop, which would otherwise occur in a clean system (Gross-Hardt, Amar, Stapf, Blümich, and Pfennig 2006; Wegener 2009). This influence can shift sedimentation velocity and thus residence time by a factor of up to almost 2 in the relevant diameter range. As a consequence, no general mass-transfer correlations for determination of HTU or HETS (height equivalent to the theoretical stage) are available; that is HTU, as well as HETS, always have to be determined experimentally for the specific material system. Thus, with the stage-based models and the HTU-NTU method, only approximate concentration profiles may be calculated. However, the required column height for a defined extraction task cannot be determined, nor can the operating limits be obtained. The latter depends on trace impurities as well, which influence coalescence, affecting the average drop size at given operating conditions, since this is determined by the interplay between coalescence and splitting of drops, in turn defining the holdup and thus the flooding point.

As a consequence, the question arose, if it would be possible to combine results from appropriate lab-scale experiments, which characterize these effects of the specific material system, with corresponding drop-based simulation tools to predict column performance.

The basis for such a simulation was available since on the one hand a wide variety of models describing the fundamental drop behavior exist in the literature. As indicated in Figure 7.1, these fundamental aspects for the drops to be accounted for are:

- sedimentation
- mass transfer
- coalescence
- splitting

For each of these effects, models were available, which also allow one to account for the influence of the drop swarm as well as of the interaction with the column internals.

On the other hand, preliminary studies trying to show the feasibility of evaluation of column performance based on drop-population balances showed that a consistent picture could be developed. Pilhofer and Mewes had already shown that description of column performance was in principle possible based on a drop-based picture at least for non-pulsed sieve-tray columns if an average drop size was applied (Pilhofer and Mewes 1979). As another example, Vogelpohl and coworkers showed that drop-based description allows consistent interpretation of column performance and even a prediction of holdup, drop-diameter, and concentration profiles along the column, if experimental data on drop behavior were utilized (Haverland, Vogelpohl, Gourdon, and Casamatta 1987; Mohanty and Vogelpohl 1997; Qi, Haverland, and Vogelpohl 2000). At that time, first detailed approaches for simulation of column performance had already been proposed, which intended a full prediction based on a consistent modeling of drop behavior (Kronberger, Ortner, Zulehner, and Bart 1995; Ortner, Kronberger, Zulehner, and Bart 1995; Arimont, Soika, and Henschke 1996; Henschke and Pfennig 1996; Modes 1999).

Based on these first steps and results, a joint project between three German research groups supported by industry was initiated, which was funded by the German federal ministry for economy and energy and which brought together the expertise on pulsed and stirred columns as well as on drop behavior. This project resulted in the experimental validation of the drop-based prediction of extraction-column behavior (Bart, Garthe, Grömping, Pfennig, Schmidt, and Stichlmair 2005, 2006; Kopriwa, Buchbender, Ayesterán, Kalem, and Pfennig 2012). The general idea of this drop-based approach is to apply appropriate models for the quantification of the drop aspects mentioned above. Since these models need to contain material-specific parameters, these are determined from single-drop experiments with the original material system performed at lab scale with dedicated measuring cells. The models together with the experimentally determined parameters are then used to simulate the effect of the interplay of an ensemble of drops in an extraction column.

For the simulation, two different approaches were developed based on previous experiences in each group. On the one hand, a direct solution of the drop-population balance was realized, which was implemented using the method of moments to efficiently solve the set of defining equations (Attarakih 2004; Attarakih, Bart, Steinmetz, Dietzen, and Faqir 2008; Alzyod, Attarakiha, Hasseinec, and Bart 2017). On the other hand, a Monte–Carlo approach was developed, which follows a sufficient number of individual drops along their trajectory through the column

(Henschke 2004; Kalem, Buchbender, and Pfennig 2011; Kopriwa, Buchbender, Ayesterán, Kalem, and Pfennig 2012; Ayesterán, Kopriwa, Buchbender, Kalem, and Pfennig 2015). The resulting simulation tool is called ReDrop, which stands for "representative drops."

It is believed that while the direct solution of the population-balance equations leads to faster solutions, the ReDrop approach has a certain advantage because an essentially arbitrary number of variables describing the drops can be accounted for without difficulty (Kalem and Pfennig 2007; Ayesterán, Kopriwa, Buchbender, Kalem, and Pfennig 2015). Any new variable to be considered in the direct solution of the drop-population balances leads to a new dimension, which in principle needs to be regarded as distributed by subdivision into classes or the moments taken into account. In the ReDrop approach, the dimensionality is always two, one dimension being the drop index, the other dimension collecting all drop characteristics to be considered. This allows one to account for the contact time of a drop with the continuous phase, the concentrations of reacting components inside the drop, and other variables, which may turn out to be of relevance.

The general program scheme of ReDrop for extraction columns is sketched in Figure 7.2. After initialization, the algorithm consists essentially of three loops. The outer loop is the time loop; thus, the transient behavior of the extraction column can be described. For each time step, first new drops can be fed into the column either through the distributor or at a feed. Thus, extraction with feed can also be simulated. In the following drop loop, each individual drop is regarded with respect to sedimentation, mass transfer, chemical reactions in the drop or at

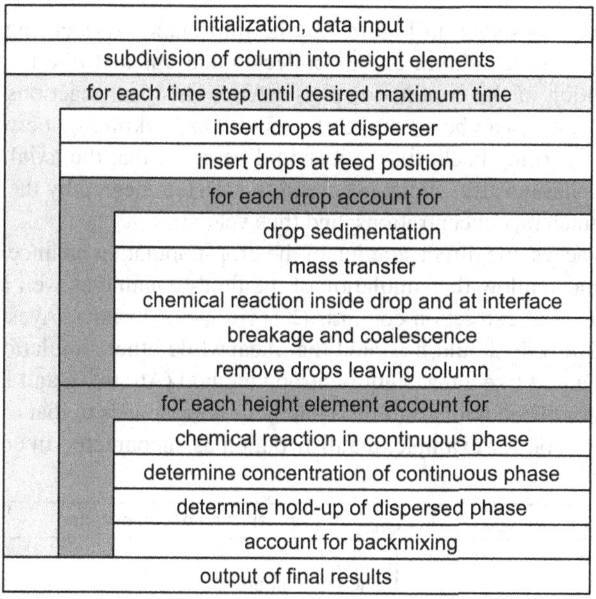

initialization, data input
subdivision of column into height elements
for each time step until desired maximum time
insert drops at disperser
insert drops at feed position
for each drop account for
drop sedimentation
mass transfer
chemical reaction inside drop and at interface
breakage and coalescence
remove drops leaving column
for each height element account for
chemical reaction in continuous phase
determine concentration of continuous phase
determine hold-up of dispersed phase
account for backmixing
output of final results

FIGURE 7.2 Schematic representation of ReDrop for the transient simulation of extraction-column performance.

its interface, drop splitting, and coalescence. Models for all of these drop phenomena also under the influence of the drop swarm as well as the internals are available in the literature for a variety of column internals (Henschke 2004, Kopriwa, Buchbender, Ayesterán, Kalem, and Pfennig 2012, and many others). For coalescence and splitting, as well as for the movement of drops between compartments in a stirred column, the models describing the corresponding probabilities have to be applied to the individual drops with a corresponding Monte–Carlo step. This is indicated in Figure 7.3 for drop splitting or coalescence. At each time step the splitting probability $p_{\text{splitting}}$ and the coalescence probability $p_{\text{coalescence}}$ are evaluated, which besides the physical properties and drop diameter depend, for example, on local holdup and energy input. These probabilities are then compared to a random number between 0 and 1 as indicated in Figure 7.3. The range to which the random number points determines if a drop will coalesce with another drop, will split, or will remain unchanged. The nearest drops, for which coalescence is indicated, will then coalesce. For splitting, on the other hand, random numbers are used in a second step to determine the number of daughter drops and their size distribution (Henschke 2004; Kopriwa and Pfennig 2016).

For stirred columns, toroidal circulation zones may occur below and above the stirring element in each compartment, in which the drops can be caught. Each compartment can thus be divided into three zones, a lower vortex zone, the stirring zone, and an upper vortex zone (Buchbender, Fischer, and Pfennig 2013). Again, a Monte–Carlo approach is used to evaluate, if an individual drop leaves either of these zones and if this occurs in the direction toward the top or the bottom of the column, since it has been observed that individual drops may sometimes move opposite the direction induced by gravity.

In the third loop shown in Figure 7.2, the information concerning the different height elements in the column is then collected and updated like the local holdup and concentration of the continuous phase. Also, chemical reactions occurring in the continuous phase can be accounted for as well as backmixing between the compartments. Concerning backmixing, it should be noted that the axial dispersion of the dispersed phase is already taken care of to a certain degree by the drops having individual diameters, concentrations, and thus velocities.

Both approaches, the direct solution of the drop-population balance as well as the ReDrop approach, allow the simulation of the fluid-dynamic as well as the separation performance of extraction columns (Kopriwa, Buchbender, Ayesterán, Kalem, and Pfennig 2012). It should be noted that meanwhile other simulation approaches have been proposed like a maximum-entropy method (Attarakih and Bart 2014).

The scale simulated with all of these methods corresponds to that of a pilot-plant. Thus, the effects of, for example, maldistribution as encountered in columns with a

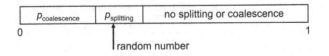

FIGURE 7.3 Using a Monte–Carlo step in ReDrop to determine, if an individual drop is coalescing, splitting, or remains unchanged.

larger diameter are not considered. If these effects can be quantified with appropriate model equations, also they can, of course, be included in the simulations.

7.2.1 SEDIMENTATION AND MASS TRANSFER

As mentioned, the drop-based simulations require experiments to characterize the behavior of the specific material system of interest, because trace components can strongly affect the drop behavior. Traces of surface-active components or solid particles can influence the drop diameter at which the internal toroidal circulation is induced by the shear forces acting at the surface of the sedimenting drop. This, in turn, may influence mass transfer, which is additionally influenced by interfacial instabilities, if the concentration gradient exceeds certain system-specific limits. To perform corresponding experiments with dedicated drops, lab-scale measuring cells have been proposed by many authors (Handlos and Baron 1957; Schügerl, Hänsel, Schlichting, and Halwachs 1986; Henschke, and Pfennig 1999; Mörters and Bart 2003; Wegener 2009). Two of the main principles are sketched schematically in Figure 7.4. As shown to the left, the drops produced in this arrangement sediment in a cylindrical cell filled with the continuous phase. The drops of defined volume can either be generated continuously with a small continuous flow through a nozzle or individually by a suitable technique, for example, with a computer-driven syringe in combination with a nozzle of appropriate diameter. The contact time in this cylindrical cell is defined by the distance between drop-production nozzle and withdrawal funnel as well as the sedimentation velocity. Using a manageable height of the cell, depending on density difference, viscosity, and drop diameter, the contact time is limited to a few 10 s. If only the sedimentation velocity is to be evaluated, no sampling is required. If mass transfer is to be quantified, the drop has to be analyzed

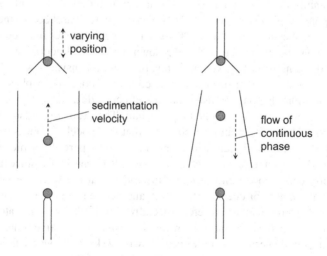

FIGURE 7.4 Two principles for single-drop experiments. Left: Drop sediments in cylindrical cell with adjustable position of sampling funnel. Right: Drop is stabilized in a conical cell with the counterflow of continuous phase, which is switched off to allow sampling. Both variants thus allow variation of contact time between phases.

after a defined contact time with the continuous phase. Either inline techniques are applied (Schügerl, Hänsel, Schlichting, and Halwachs 1986; Wegener 2009), or a sufficient number of drops is collected with a drop-collection funnel, which is again connected to a computer-driven syringe. An alternative arrangement is shown on the right side of Figure 7.4, where an individual drop is stabilized for an arbitrary time within a Venturi-type conical cell at a position, where the sedimentation velocity matches the counterflow of continuous phase. After switching off the flow of the continuous phase for some seconds, the drops can then be withdrawn again with a drop-collection funnel and a computer-driven syringe.

The advantage of these single-drop experiments is the small amount of substance required to characterize drop sedimentation and mass transfer. Also, sedimentation and mass transfer are generically characterized, independent of any specific extraction equipment. This allows in principle to utilize these data in the simulation of any extraction equipment, for which the internals influence has already been characterized.

Detailed studies on single drops have simultaneously significantly increased the knowledge about drop behavior. For example, it turns out that mass transfer is significantly faster than expected. To demonstrate this, single-drop mass transfer can be compared to one of the fastest models available neglecting mass-transfer resistance outside the drop and assuming ideal mixing along the streamlines induced inside the drop by the internal circulation (Kronig, and Brink 1950) and evaluating the mass transfer with the Newman model (Newman 1931). It turns out that this fastest model is slower than experimentally observed mass transfer by at least a factor of three for the EFCE standard test systems for solvent extraction (Míšek, Berger, and Schröter 1985; Henschke and Pfennig 1999) (European Federation of Chemical Engineering). While this effect has originally been attributed very generally to mass-transfer induced instabilities, it has later been shown that a source of this enhancement is the Marangoni effect (Wegener 2009; Wegener, Fevre, Paschedag, and Kraume 2009; Engberg, Wegener, and Kenig 2014). These mass-transfer induced interfacial instabilities consistently lead to a slowing down of drop sedimentation as shown in Figure 7.5, if the non-reacting system is regarded. Also consistent with this picture is the observation that, if a drop is contacted with a continuous phase of a constant concentration, the mass-transfer induced effects fade away as soon as the mass transfer has significantly approached equilibrium. The drops then also speed up in sedimentation. These observations also imply that the models for mass transfer and sedimentation have to be linked, which until now is only partially achieved.

Here, the Marangoni effect is only mentioned rather generically, since mass transfer apparently can induce quite different instabilities at the interface, which is of course already known for decades (Sternling and Scriven 1959; Sawistowski 1973). Marangoni effects in this context refer especially to roll cells and oscillatory motion of the interface, which can be induced by mass transfer. At the same time, it has been realized that so-called eruptions may occur (Sawistowski 1973), where fluid elements appear to be spontaneously expelled from the vicinity of the drops far into the continuous phase. While the early ideas were rather vague, it has been shown that these eruptions may be induced by a combination of mass transfer and thermodynamic instability, leading to nano-drops occurring in the very close vicinity of the interface

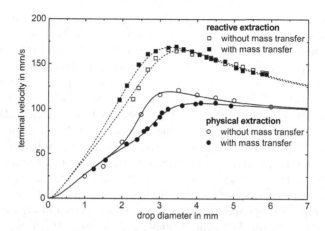

FIGURE 7.5 Influence of mass transfer on the sedimentation velocity of single drops in reactive and physical extraction systems obtained by experiment and compared with the modeling. The data are taken from Hoting (1996), Henschke and Pfennig (1999) for the non-reactive system water + acetone (transfer component) + butyl acetate, and from Kalem, Altunok, and Pfennig (2010) for the reactive system water + zinc (transfer component) + isododecane + D2EHPA, where the pH was controlled with sulfuric acid.

(Schott and Pfennig 2004). The instabilities occur due to a curved concentration profile along a diffusion path originating from significant cross-effects in the diffusion of all components including the major phase-forming solvents, which intersects the thermodynamic stability limit characterized by the binodal and spinodal. The nano-drops, of which many form in a layer covering the entire interface, can then undergo a concerted action in a domino effect. Since the nano-drops have a lower interfacial tension than the major interface of the drop, their integration into the major interface upon coalescence induces a motion of the interfacial region away from the point where these nano-drops are inserted. Due to the incompressible continuity of the fluids, the volume of these regions then has to bend away from the interface. Thus, fluid elements stemming from the vicinity of the interface, which underwent already some mass transfer, are quickly moved far into the continuous phase, which is realized as an eruption. In some cases, such eruptions do not occur, but instead a foggy appearance behind the drops is observed, for which the descriptive characterization of "smoking drops" has sometimes been used. Such spontaneous emulsification can also be explained by the nano-droplet formation, where the nano-droplets apparently grow until they become visible and move away from the drop surface. It should be noted that the identical interplay between mass transfer and thermodynamic instability at the interface leads to aerosol formation in absorption or to fog formation over a cup of hot black coffee, wherein the latter case a temperature gradient is relevant instead of a concentration gradient.

From the discussion of these effects, it is apparent that predicting such instabilities—while being possible in principle—is difficult to impossible. One reason is that the required physical data are typically lacking, especially for technical systems. To predict the different types of instabilities, the concentration-dependent diffusion-coefficient matrix in the multicomponent system would be required, which in the past has hardly been measured. It is only in recent years that feasible methods have become available, though until now they have not been systematically used. (Rutten 1992; Pertler 1996; Bollen 1999; Bardow, Göke, Koß, Lucas, and Marquardt 2005). As a consequence, the mass-transfer behavior of drops of the specific material system can only be determined experimentally and then described with appropriate models, the parameters of which are fitted to the experimental results. At the same time, it is realized that generally the interfacial instabilities enhance mass transfer and slow down the drops, so that mass transfer per column height is positively influenced by the instabilities. Thus, without these non-idealities, solvent extraction would be significantly less efficient.

In contrast to this behavior, for systems undergoing reactive extraction of zinc with di-(2-ethylhexyl)phosphoric acid (D2EHPA) diluted in isododecane, it has been shown that mass transfer speeds up the drops as also shown in Figure 7.5, if the reactive system is regarded (Kalem, Altunok, and Pfennig 2010). It was found that the influence on sedimentation velocity is most pronounced for drops between 2 mm and 3 mm diameter, that is, again exactly in the range most relevant for solvent extraction. This also is roughly the range in which internal circulation is setting in. This effect has been explained by the properties of D2EHPA, which acts as a surfactant, stabilizes the drop interface, and hinders the onset of internal circulation in the drop. As the mass transfer of zinc ions occurs, D2EHPA forms a complex, which is dissolved inside the drop, and thus D2EHPA is removed from the interface. Thus, upon mass transfer and reaction, the interface becomes more mobile, and the internal circulation sets in, leading to faster sedimentation, an effect, which fortunately is not very pronounced. Thus, also in this case, a physically consistent view of drop phenomena could be developed.

The interdependence between mass transfer and sedimentation actually has to be dealt with in even more detail, especially taking into account that the concentration profile encountered by a drop in the lab-scale measuring cells and the extraction column differ significantly. In the single-drop cell, the concentration of the continuous phase is essentially constant, while in an extraction column, the drop experiences a different concentration of the continuous phase at each moment due to its sedimentation along the concentration profile in the column. To study the interplay between concentration gradient, sedimentation, and mass transfer, Kalvolda designed a new lab-scale cell of 1 m in length and 80 mm diameter, in which a concentration gradient can be generated by pumps that layer the levels of continuous phase above each other with different concentrations (Kalvoda 2016; Ayesterán, Buchbender, Kalem, Kalvoda, and Pfennig 2017). The drop-collection funnel can be adjusted in height so that the contact time can be varied. A more detailed model is then required to model mass transfer, which applies a numerical solution of the diffusion process based on a shell model for the drops, where again an effective diffusion coefficient can be used, similar to that proposed by Henschke (Henschke and Pfennig 1999). It was found that an enhanced mass transfer and slowing down of drops sets in, if the concentration gradient at the start exceeds a certain limit. The effective diffusion coefficient in this case and for the system investigated turns out to be higher than without the instabilities by a factor of 1.5. While a basic understanding has thus been derived in a consistent way, the quantitative description still needs to be worked out.

7.2.2 Accounting for Influence of Drop Swarm and Column Internals

Since the models for the drop-based simulation of extraction-column behavior also need to take the interaction between the drop swarms and the column internals properly into account, the lab-scale measuring cells for drop behavior have been augmented correspondingly. This allows one to characterize the influence of a sieve-tray, regular and random packings, as well as a variety of stirred geometries. The individual results from these studies have at the same time led to significantly more detailed insight into drop behavior, some of which is quite surprising.

To analyze the influence of a sieve-tray on mass transfer, Henschke placed a small sieve-tray in the conical part of the mass-transfer cell and kept the produced drops for defined times below and above the sieve-tray, stabilized by the counterflow of continuous phase (Henschke and Pfennig 2002). For organic drops in aqueous continuous phase, he showed via an appropriate evaluation that the enhancement of mass transfer is not induced by the squeezing of the drops through the holes of the sieve-tray. Instead, during the phase of pulsation with an increased downward flow of the continuous phase, the drops cannot pass through the holes and stay below the tray. At that position, they experience a directed flow of continuous phase through the holes "blowing" at them, which enhances mass transfer.

For regular packings, on the other hand, it turned out that the mass transfer for a given contact time between the phases is hardly affected by the presence of the internals (Grömping 2015). Only upon the first contact of a drop meeting the first edge of the packing, some enhancement of mass transfer could be detected. This implies that the influence of the packings on mass transfer is not the direct enhancement of transfer rate but rather the slowing down of the sedimentation of the drops so that a larger residence time per column height can be realized, which effectively leads to more mass transfer per column volume.

For mass transfer in a stirred compartment, it turned out that for a light dispersed phase, the residence time below the stirring element is rather small as compared to the overall residence time in the compartment (Buchbender, Fischer, and Pfennig 2013). Thus, column volume can be saved by simply reducing the space available below the stirring element without significant loss in separation performance.

7.2.3 Drop Coalescence and Breakage

In the extraction column, coalescence and breakage of the drops determine the drop-size distribution and, therefore, the volume-specific interfacial area for mass transfer. Also, coalescence and breakage lead to a mixing of a composition of drops with different size and thus influence axial dispersion. Finally, the drop diameter is also a key property for the sedimentation velocity, namely the drops residence time in the column. Several research groups have thus investigated coalescence and breakage with rather different setups and applying different model concepts.

Coalescence can be investigated in a Venturi-type measuring cell, similar to the one used for determining mass transfer. In the conical region of the cell, a swarm of drops is produced and kept floating (Simon 2004). Each additional drop produced then either hits one of the previous drops such that coalescence takes place or is caught by the counter-current flow without coalescing and becomes a new member of the stabilized swarm. In the case of coalescence, the counter-current flow of continuous phase can no longer stabilize the resulting bigger drop, which continues to rise and can be detected leaving the drop swarm toward the top of the cell. It turns out that coalescence among drops within the drop swarm kept in the cell only very rarely occurs.

Also, the coalescence between a pair of drops created, for example, with syringe dosing systems have been investigated, such as in the groups of Bart in Kaiserslautern and Kraume in Berlin (Gebauer, Villwock, Kraume, and Bart 2016; Kamp 2017; Kamp, Villwock, and Kraume 2017). These investigations show that the time until drops coalesce may vary significantly and is influenced strongly by factors such as the presence of salts. This pronounced salt effect is induced by electrostatic repulsion of the interfaces of approaching drops (Pfennig, and Schwerin 1998; Webber, Edwards, Stevens, Grieser, Dagastine, and Chan 2008), which is determined by the ion combination present in the system. The coalescence process itself, on the other hand, is a very fast process taking place during some milliseconds. A systematic dependence on the physicochemical properties appears to be difficult to find.

Since such specific cells measuring either splitting or coalescence individually create a rather artificial environment as compared to the drop swarms occurring in extraction columns, Klinger (2008) developed a cell that allows for discriminating both phenomena under more realistic column conditions. The cell she developed mainly consists of a vertical column section, which is filled with a continuous phase at the beginning of the experiment. The dispersed phase is then introduced at the bottom of the cell through a disperser. Within the cell, several short sections of random packing are introduced, which are connected via rods with an eccentric motion motor to generate pulsation. With a set of cameras positioned at different heights in front of the cell, the drop swarms between the packing sections are photographed and the drop-size distributions evaluated with appropriate image-analysis software. Since breakage and coalescence also always occur simultaneously in this cell, two measurements are required to discriminate between the individual effects. Starting out with smaller drops generated with the help of a disperser with many small holes, coalescence is more dominant as compared to a second experiment at otherwise identical conditions but starting out with larger drops created with a disperser having few large holes. Thus, starting with the smaller drops, coalescence predominantly influences drop-size development, while for the larger starting drop diameter, splitting is more important. If these two experiments are then compared and simultaneously evaluated with appropriate models, the parameters of both models can significantly be determined (Klinger 2008). This cell has also been used by Kopriwa (2013), who demonstrated how strongly trace impurities affect coalescence, even though no physical properties of the macroscopic system like densities or viscosities are changed.

By studying coalescence behavior in extraction columns, vertical settlers, and a batch-settling experiment, Kopriwa finally demonstrated that coalescence can be described in principle as shown in Figure 7.6 (Kopriwa 2013). The major result is that the influence of fluid dynamics, which is determined by the specific type of extraction equipment, and the influence induced by the specific material system, which affects

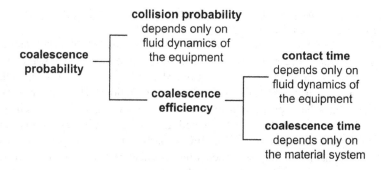

FIGURE 7.6 Schematic sketch of modeling coalescence probabilities in extraction processes in different extraction equipment. (Adapted from Kopriwa 2013.)

the coalescence of drops, can be clearly separated. This allows for the quantification of the influence of the material system such as in a lab-scale settling experiment, described in the following section in some detail, and using these results in column simulations with the appropriate models accounting for the fluid-dynamic effects. The fluid dynamics determine how often drops meet, characterized via the collision probability, and how long they stay in contact. The material system only influences how long it would take for drops to coalesce. This coalescence time together with the contact time then determine the probability that the drops, once they meet, indeed coalesce, which is expressed as coalescence efficiency. Also, for coalescence, an interrelation with mass transfer unfortunately exists, which is typically quite pronounced. If mass transfer occurs from the dispersed to the continuous phase, coalescence is enhanced leading to rather large drops or requiring significantly increased energy input to ensure sufficiently small drops. While operation in this regime is possible in principle, it may be difficult to control. Also, one has to keep in mind that one advantage of solvent extraction is that the auxiliary component, the extractant, can be chosen to have a good selectivity for the mass-transfer component. As a result, the flow rate of the extractant is usually smaller than that of the raffinate phase, so that the extract phase is often chosen to be the dispersed phase and thus mass transfer occurs from the continuous to dispersed phase consequently hindering coalescence. Thus, in the drop-based simulations, the interrelation between mass transfer and coalescence is usually neglected; that is, systems with hindered coalescence are assumed.

For the description of drop breakage, available concepts are typically applied in drop-based column simulations, which depend essentially only on the physical properties of the system and some suitable characterization of the energy introduced via the extraction process like pulsation or stirring. Nevertheless, key concepts shall be briefly mentioned. In corresponding experiments, it has been shown that for a given arrangement and operating conditions, drops below a so-called stable drop diameter do not break, while above a certain diameter essentially all drops break, such as during one passage through a stirred compartment or through a sieve-tray (Haverland 1988; Leu 1995; Henschke 2004; Grömping 2015). Between these limits, the breakage probability increases from 0 to 1 and also the number of daughter drops increases, for which a variety of models for different geometries have been proposed

(Pietzsch 1984; Henschke 2004; Garthe 2006; Schmidt 2006; Kopriwa, Buchbender, Ayesterán, Kalem, and Pfennig 2012; Grömping 2015).

Equipped with this information on drop behavior, extraction columns can be simulated. Already here it should be stressed that it is not at all necessary to operate a multitude of single-drop equipment in order to characterize the drop behavior and to obtain the basic parameters to run the simulations. These many different experiments have been performed to gain the basic understanding, which is now coded in the structure of the models applied. Before performing extraction-column simulations, in most cases a batch-settling experiment is required to characterize coalescence behavior, some experiments on sedimentation velocity, as well as few experiments on mass transfer. Overall, the experimental effort can be less than one week, and with some additional assumptions it can be minimized further, as described later in this chapter.

Since our own experience is mainly related to the simulations with the ReDrop concept, the corresponding results shall be briefly described here. For the Monte–Carlo based simulations, about 1000 drops per meter of column height need to be considered, leading to simulations which take few minutes to reach steady state on an ordinary desktop computer. More drops lead to excessive computer-time consumption, and fewer drops lead to fluctuating results or even to quantitative deviations if the number of drops is much too small. As already mentioned, the simulations depict the time-dependent behavior of the extraction column in all detail. Besides the concentration profile, also the local holdup and the drop-size distributions at any column height are available. If the flowrates are increased, the simulations even indicate at which limiting flow rate the operation becomes infeasible, namely when the flooding point is reached. This shows impressively the power of the drop-based approach: Even the limits of operability can be predicted consistently within the framework of drop-based simulations. Thus, no additional flooding correlations are required.

7.2.4 APPLICATION TO DIFFERENT COLUMN TYPES

Up to this point it has been shown that this drop-based approach can be successfully applied to pulsed as well as to stirred columns for a variety of individual geometries. For pulsed columns such as sieve trays, regular and random packings have been depicted with appropriate models (Henschke 2004; Grömping 2015). In the project mentioned in the introductory paragraphs (Bart, Garthe, Grömping, Pfennig, Schmidt, and Stichlmair 2005), it has even been shown that true predictions are possible. To that end, Ineos Phenol supplied the industrial material system, for which the column experiments had been performed at Bayer Technology Services, the results of which were not directly communicated. At RWTH Aachen, the single-drop behavior was then characterized experimentally and the ReDrop simulations performed. Since the results of the pilot-plant experiments were not yet known, this was a true prediction. When the results were later compared to the pilot-plant results, it turned out that, even for this industrial system, the accuracy of the flooding point and separation performance were better than 10% as already shown previously for EFCE standard test systems (Grömping 2015). This order of magnitude of deviation between experiment and simulation was found as typical for a variety of cases, where the average deviation is always below and sometimes significantly below 10% with respect to separation efficiency as well as flooding point (Henschke 2004; Grömping 2015).

Even challenging systems like aqueous two-phase systems and ionic liquids have been dealt with by ReDrop simulations (Buchbender, Onink, Meindersma, de Haan, and Pfennig 2012; Mohammadi Sarab Badieh, Quaresima, Pfennig, and Saien 2017; Quaresima, Schmidt, and Pfennig 2017). ReDrop has also been applied to various systems in the area of reactive extraction with comparable accuracy, where different models have been used to describe the chemical reaction. Either an effective diffusion, which corresponds to an enhancement factor, has been used, but also a variety of more complex combinations of diffusion and reaction-kinetic models have been applied (Kalem, Altunok, and Pfennig 2010; Altunok, Kalem, and Pfennig 2012; Kalem 2015). Also, for such demanding systems, the accuracy is better than 10%, for example, expressed in the overall separation efficiency of the equipment as demonstrated in Figure 7.7, in which experimental results and ReDrop simulations for the D2EHPA standard test system are compared. Currently, these models are still being developed further, since it would be desirable to deduce the drop behavior from standardized experiments on reaction kinetics as, for example, obtained in a Lewis cell with a defined horizontal interface between the phases, which until now is not possible.

Finally, since performing the required single-drop experiments is a challenging task in an industrial environment, the question arose if it would in principle be

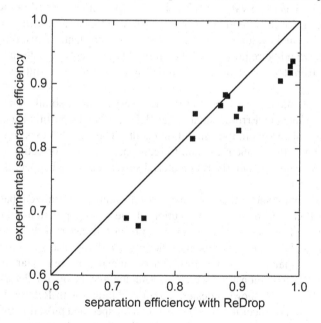

FIGURE 7.7 Parity plot for separation efficiency of a pulsed sieve-tray column with the reactive system water + zinc (transfer component) + isododecane + D2EHPA for a variety of relative flow rates and sulfuric-acid concentrations in the extractant phase fed to the column. (Data from Kalem 2015.)

possible to simulate extraction-column behavior with limited information or utilizing some standard parameters for typical systems. This has been realized in cooperation with BASF SE for a technical system (Buchbender, Schmidt, Steinmetz, and Pfennig 2012) where the parameters have partly been adjusted to the findings from pilot-plant experiments. It was possible to derive consistent simulations indicating that it may be possible to further reduce the experimental effort for characterizing the single-drop behavior. Partly, the single-drop experiments can be replaced by pilot-plant experiments to determine some characteristic parameters. Especially coalescence may be a sensible parameter, which strongly depends on the exact composition of the system used, but which also can easily be backed out from the observed drop diameter in the pilot-plant experiment.

At the same time, it has to be kept in mind that presumably one reason why the drop-based method can describe column experiments so accurately is that if mass transfer and sedimentation velocity have been determined experimentally, the quality of the models in describing these key performance parameters can directly be assessed by comparison with the experimental results. The models can then be modified, if deviations become apparent. Based on these proper models, the extraction column can then be well described. Thus, trying to reduce the single-drop experiments below some reasonable limit would significantly increase the uncertainty in column prediction. At the same time, it should be cautioned that the drop-based simulation until now has only been validated for typical situations. Under extreme conditions, such as with respect to phase ratio, concentrations, or other parameters, the description has to be validated. Thus, in the project mentioned in the introduction to this chapter, the intention was rather to have a tool available that allows much better planning and evaluation of the pilot-plant experiments. Thus, only very few experiments would be required, and the optimal type of column internals may even be deduced from the simulations by comparing the simulated results with different internals.

Finally, it should also be stressed that these very accurate simulations need accurate model parameters. Errors such as when determining the partition coefficient can lead to significant errors in the simulation results. Thus, the lab experiments should be carried out with sufficient care and precision, since the correcting tendency of pilot-plant experiments with the real material system does not happen, if only simulations are performed.

These examples show that the drop-based simulation realized based on the ReDrop methodology or by direct solution of the drop-population balance allows the prediction of hydrodynamic behavior and mass transfer of extraction column solely based on lab-scale experiments. Therefore, ReDrop offers a better and more reliable method than the design methods based on expensive pilot-plant experiments used previously. To demonstrate which information may be obtained via drop-based simulation, Figure 7.8 shows a design diagram obtained with ReDrop for a pulsed sieve-tray column of specified geometry and under specified pulsation intensity. The design point lies typically at 80% of the flooding limit and can be chosen so as to maximize separation performance, that is, the number of theoretical stages realized in the given column. Of course, all of these parameters can be varied and depending on the simulation results optimized.

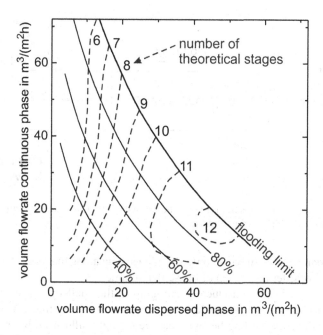

FIGURE 7.8 Design diagram for a sieve-tray column of 3 m active height, on open-area fraction of 39%, a hole diameter of 4 mm and a pulsation intensity of 11.7 mm/s operated with the system water (continuous) + butyl acetate (dispersed) + acetone (transfer component).

From this discussion, it is also obvious that the simulations go far beyond the stage-based modeling or even application of the HTU-NTU model. Instead, a real design simulation is performed that takes all details of the column with its internals into account, including aspects like the hole diameter of sieve-tray columns.

7.3   DROP-BASED DESIGN OF GRAVITY SETTLERS

A schematic horizontal gravity settler is shown in Figure 7.9. The liquid–liquid dispersion is entering at the left, where in this sketch the situation is shown for a dispersed light phase. The momentum introduced via the inlet stream is typically removed by inserting a hole plate or some other suitable device. Typically, a dispersion wedge forms, in which the majority of the coalescence events takes place for systems with hindered coalescence. The drops introduced with the feed stream in this case sediment toward the dispersion wedge from below. If properly designed, coalescence occurs until two clear phases are obtained, which are then exiting the settler, where also here different geometries can be used including a set of weirs to separate the phases. With the flow rates of the leaving phases, the level of the major interface can be adjusted, which is typically chosen such that the phase volumes inside the settler roughly correspond to the phase ratio of the feed, because then the average velocity of the phases is similar, minimizing the relative flow of the phases at the interface. The wedge height not necessarily has to decrease to zero along the settler for proper operation. If the settler length is reduced below the wedge length, a

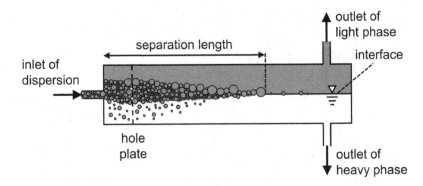

FIGURE 7.9 Schematic sketch of a horizontal gravity settler, in which a liquid–liquid dispersion is separated into two clear phases.

dispersion band with a thickness only slightly varying along the settler length forms, which leads to better utilization of the settler volume.

As mentioned in the introductory paragraphs, the challenge in designing such a settler is to account for the coalescence behavior of the specific material system of interest. Since the settling behavior can be strongly influenced by trace impurities (Soika and Pfennig 2005; Kopriwa 2013), settling experiments are mandatory for settler design. Typically, the result of simple settling experiments performed for example in a sufficiently large beaker after generating the dispersion by stirring is converted into an area-specific coalescence rate, which is then used for designing the gravity settler. This design method is apparently limited, which can easily be realized, if settlers with different height-to-width ratio but identical cross-sectional area; that is, identical average flowrates are regarded. The higher the settler, the stronger will be the influence of the drops sedimenting toward the interface, while for a wide settler with a low height, settling is comparably fast, and the performance is determined by the coalescence rate in the close-packed zone. Similar considerations apply if the phase ratio is changed for a given direction of dispersion, because this also shifts the relative importance of sedimentation and coalescence.

To overcome these limitations for the design of gravity settlers, the drop-based approach can be applied as well. This has been realized by Henschke, who designed a standardized settling experiment to characterize the settling behavior of the specific material system (Henschke 1995; Henschke, Schlieper, and Pfennig 2002; Leleu and Pfennig 2017). The experimental setup consists of a glass vessel with at least 80 mm diameter and a recommended height of at least 15 cm, better 20 to 25 cm so that roughly 1 liter of two-phase system is required. The large diameter ensures that wall effects can be excluded, while a sufficient height is required to ensure acceptable accuracy in evaluating the different time-dependent heights observed during settling. This cell, which may be equipped with a glass jacket linked to a thermostatic bath for temperature control, is first filled with the two-phase system of interest. A dispersion is then created by stirring with stirrers, four of which are mounted on two rotation axes, which are rotating at 800 min⁻¹ for 30 s. The axes rotate in an opposite direction, which leads to reliable production of dispersion without the need of baffles and at the same time does not introduce an overall rotation of the fluids. Thus, after the stirring motor is stopped, the liquid comes to complete rest within very few seconds so that the initial conditions of the settling process are well defined. The results obtained with this arrangement appear to be relatively insensitive to the stirring intensity and duration (Henschke 1995; Leleu and Pfennig 2017). The settling cell is homogeneously lighted from behind and a video of the settling recorded.

A principal sketch of the settling process as it can then be observed for a light phase dispersed and assuming mono-sized drops is shown in Figure 7.10 (Henschke 1995; Jeelani and Hartland 1998). In the sedimentation zone, the mono-sized drops are all moving with identical velocity. The sedimentation velocity of each drop also corresponds to the slope of the sedimentation curve. The sedimentation curve for a mono-dispersed drop-size distribution is thus marked by the last sedimenting drop. When the drops approach the interface, typically a close-packed layer is formed, because coalescence is the rate-determining step. Coalescence occurs among the drops in the close-packed zone as well as between the drops and the major interface if they are in contact. Typically, the limit between close-packed and sedimentation zone cannot be observed experimentally, since both regions are rather turbid. The time-dependent position of the major interface marks the coalescence curve. This basic picture is regularly applied when the batch-settling process is regarded, evaluated, and modeled.

7.3.1 Description with Mono-Sized Drops

Drop sedimentation is determined by physical properties like viscosities and densities as well as holdup, which are all known or can easily be determined for a system of interest. Since the drops in this evaluation are assumed to be of identical size, the holdup is constant in the sedimentation zone and corresponds to the overall holdup of the settling experiment. Then, the only unknown determining the drop sedimentation

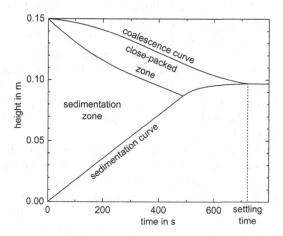

FIGURE 7.10 Sketch of behavior observed in a settling experiment. (Data from Jeelani and Hartland 1998, Henschke 1995.)

is the drop diameter, which can correspondingly be backed out from the slope of the sedimentation curve with appropriate models. Since the drops are typically rather small, almost any suitable sedimentation model for single drops valid in that region can be applied, while the swarm effect can be accounted for, e.g. with the model of Richardson and Zaki (1954).

For describing coalescence, Henschke (1995; Henschke, Schlieper, and Pfennig 2002) suggested a model accounting for an asymmetric dimple forming at the contact region between two drops as well as between drop and major interface. This model contains a single adjustable parameter, which characterizes the asymmetry of the dimple, which is to be fitted to the coalescence curve. The model proposed by Henschke then allows one to also calculate the limit between sedimentation zone and close-packed region.

Henschke showed that the coalescence parameter does not depend on stirring rate and phase ratio for a given direction of dispersion. The intensity of stirring only affects the starting drop diameter, but this influence is typically minor for many systems. It thus appears as if the material system would have an intrinsic drop diameter, which is obtained by stirring irrespective of stirring intensity and duration.

Based on the detailed evaluation of the settling experiment, technical batch settlers can also directly be described, because the underlying phenomena in the technical equipment are identical to those of the lab-scale settling test. The only difference is the difference in equipment size, which leads to longer sedimentation paths and larger volumes to be coalesced. Thus, any proper model used to evaluate the lab-scale settling experiment can directly be applied to design a technical batch settler. Of course, the underlying assumptions such as on mono-sized drops have to be verified, and possibly some additional time needs to be accounted for, which characterizes the decay of initial turbulences resulting from previous stirring or the filling process.

For horizontal and vertical gravity settlers, Henschke also developed a design method, where for horizontal settlers a wedge of close-packed dispersion and a dispersion band of constant height throughout the settler can be accounted for (Henschke 1995). In these simulations, again a mono-disperse drop-size distribution is assumed for the sedimenting drops, whereas inside the close-packed zone average drop diameters result, which depend on height and horizontal position. This approach by Henschke has been recognized in the literature as one of promising approach for gravity-settler design (Frising, Noïk, and Dalmazzone 2006; Kamp, Villwock, and Kraume 2017). The accuracy as compared to pilot-plant settlers is again of the order of 10% as shown in the comparison between calculated and measured wedge lengths in Figure 7.11.

Similar approaches to describe batch settling, solving the different rates at which the limiting curves in Figure 7.10 develop, have been proposed, which take slightly different assumptions and basic models into account and especially relax the assumptions made in the Henschke model. Notably, Noïk et al. (Noïk, Palermo, and Dalmazzone 2013) account for the coalescence also during sedimentation, which leads to a curved sedimentation curve, since as the drops coalesce and increase in diameter their sedimentation velocity increases. Such a curved sedimentation curve is also frequently observed in the experiments under certain conditions. Additionally,

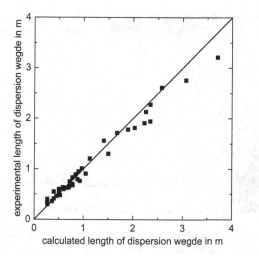

FIGURE 7.11 Comparison of experimental and calculated length of the dispersion wedge in a horizontal gravity settler for the systems water + butyl acetate and water + n-butanol with a phase ratio of 1:1 for settlers between 0.1 m and 0.6 m diameter. (Data from Henschke 1995.)

Grimes sets up and solves a population balance for drop sedimentation and coalescence, which is then solved and compared to experimental data (Grimes 2012; Grimes, Dorao, Opedal, Kralova, Sørland, and Sjöblom 2012). Also, these authors find curved sedimentation curves. In both cases, the evaluations compare reasonably well with the experimental results on the time-dependent local holdup obtained either with a capacity-method or via NMR (nuclear magnetic resonance).

Apparently, the simple drop-based approach of Henschke was rather limited due to the assumption of mono-sized drops and the neglecting of coalescence during sedimentation. This is obvious when the result of a settling experiment with a more challenging system is observed as shown in Figure 7.12. The settling diagram is produced from the video by cutting a slice from the video frames every 3 s and combining these side by side. Such a representation is always advised, if the boundaries between the different zones are not clearly visible because this representation allows one to nevertheless obtain good estimates of their position. It becomes obvious that the system is relatively turbid. After the major sedimentation front has passed, significant turbidity remains that gradually decreases. The sedimentation curve appears to be curved upward, and upon closer inspection, it is apparent that there is a lag time for the onset of sedimentation of several seconds. Such behavior can obviously not be represented with a mono-dispersed model.

7.3.2 Accounting for Drop-Based Behavior with Drop-Size Distribution

This can also be overcome with a ReDrop approach, namely by accounting for a sufficiently large ensemble of individual drops with a drop-size distribution, which are traced during their lifetime in the settling cell while accounting for the different effects which can occur. This has recently been realized and implemented in a

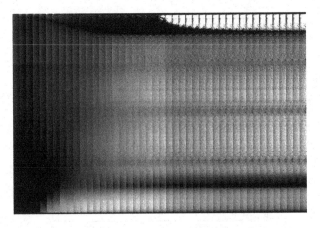

FIGURE 7.12 Result of settling experiment with 10% paraffin oil (FC1013, Fauth GmbH + Co.KG, Mannheim, viscosity 1.3 mPas, density 748 kg/m³) and water.

corresponding simulation tool. While the simulations based on the ReDrop concept may take longer than the direct solution of the drop-population balance, the advantage of ReDrop is that any effect can directly be implemented based on a physical picture as it acts during a time step. Thus, modification of drop-based models and testing various assumptions is facilitated. The individual drop models can be chosen from a selection of models described in the literature. In the simulations presented here, Stokes law has been used for drop sedimentation together with the extension of Richardson and Zaki (1954) to account for the swarm effect. For coalescence, very different model types are available in the literature (Kopriwa, Buchbender, Ayesterán, Kalem, and Pfennig 2012; Kopriwa 2013; Kopriwa and Pfennig 2016; Kamp, Villwock, and Kraume 2017).

In one class of coalescence models, the energy of the collision is accounted for, and it is assumed that the available energy has to overcome effectively repelling forces between two drops as they approach in order to induce coalescence. It turns out that this behavior cannot be found in experiments. For example, Kamp (2017) clearly showed that the high velocity of a drop hitting a drop hanging at a tip of a nozzle leads to a bouncing effect; that is, the drops do not coalesce. A similar bouncing effect has also been observed when a drop impacts a planar interface (Scheele and Clark 1980).

An alternative approach characterizes coalescence probability via contact time in relation to the coalescence time. Several authors apply the model of Coulaloglou and Tavlarides (1977) to characterize the coalescence probability $p_{coalescence}$, which realizes this approach:

$$p_{coalescence} = p_{collision}\lambda,$$

where $p_{collision}$ is the collision probability and λ the coalescence efficiency. This model corresponds to the general scheme for describing coalescence as discussed for extraction columns and shown in Figure 7.6. The collision probability characterizes how frequently the drops meet, and the coalescence efficiency describes the efficiency with which such a collision leads to coalescence. For the coalescence efficiency it is typically written:

$$\lambda = \exp\left(-\frac{t_{\text{coalescence}}}{t_{\text{contact}}}\right),$$

where the contact time describes for how long the drops stay in contact once they meet and the coalescence time how long it characteristically takes for coalescence to occur. Both times are to be understood as average values.

The coalescence time can again be described by various models, which account for the details of the approach of the drops or a drop and an interface until the rupture of the dimple-shaped film of continuous phase between the approaching entities occurs (Henschke 1995; Kopriwa 2013; Kopriwa and Pfennig 2016). After evaluation of various models in the literature, which could not quantitatively describe experimental results, Henschke (1995) proposed an asymmetric-dimple model, where the asymmetry is a system-specific parameter, which turns out to be independent of drop size and phase ratio of the experiment.

Upon closer inspection, the contact-time approach is promising, but its mathematical form appears not to be appropriate. To show this, it can be assumed that in each time interval Δt, during which the drops stay in contact, the probability of coalescence is comparable. Then the following should hold for the probability that the drops do not coalesce:

$$p_{\text{non-coalescence},2\Delta t} = p_{\text{non-coalescence},\Delta t}^{2}.$$

This condition can only be fulfilled by the following ansatz:

$$p_{\text{non-coalescence},\Delta t} = \exp\left(-\frac{\Delta t}{t_{\text{coalescence}}}\right).$$

Summed up over all time steps during the contact this leads to:

$$p_{\text{coalescence}} = 1 - \exp\left(-\frac{t_{\text{contact}}}{t_{\text{coalescence}}}\right),$$

where the coalescence time is again characteristic for the specific material system investigated and can be described by the asymmetric-dimple model, which is also used in the results presented here. This final equation apparently has a fundamentally different mathematical form as that typically applied.

Even if for each Δt of contact between the drops the probability of coalescence is slightly different, the principal form would remain identical, the argument of the exponential function would only need to be understood as the summation over all contributions from the individual time steps. A similar ansatz has already been derived from experiment but applied after modification with a slightly different physical picture (Cockbain and McRoberts 1953; Gillespie and Rideal 1956; Yu and Mao 2004). The advantage of this model is that it is independent of how coalescence is treated in the simulations. The probability of non-coalescence between two drops can either be evaluated between two possibly interacting individual drops for each time step Δt or for the integral contact time between two drops $t_{contact}$, where each drop encounter is only evaluated once. For both simulation schemes, consistent results are obtained, while this is not the case with the Coualaloglou and Tavlarides model.

If the ReDrop approach is used to evaluate the settling experiment, it turns out that a rather large number of drops has to be accounted for at the start of the simulation in order to ensure a sufficiently smooth result, which is not determined by the statistics of the simulation. This is a result of the big drops quickly sedimenting leaving small drops behind. Due to the transient nature of the settling process, also time-averaging cannot be applied to smooth the results. If the sedimentation of the big drops as well as the behavior of drops left behind are to be smoothly characterized, of the order of some 100,000 drops are required. Since the simulation time scales with the number of drops to an exponent significantly larger than 1, simulation time becomes excessive, especially if model parameters are to be fitted. To overcome this problem, the concept of fractional drops has been implemented. During each time step, it is checked, if the number of drops within a height element of the simulation lies within a predefined range. To give an impression of the order of magnitude, 100 height elements and between 100 and 150 drops per height element can be considered leading already to relatively smooth results. This corresponds to between 10,000 and 15,000 drops, which can be handled in sufficiently fast simulations of 3 minutes in a typical case. Of course, larger drop numbers will lead to still smoother results, for example when used to prepare final diagrams for presentation.

If the drop number in a height element does not lie within the predefined limits, drops are removed, or new drops are created. The drop to be removed in a height element is selected randomly. In order to obtain stable results, the remaining drops in the height element are attributed a higher weight, which ensures that the overall drop volume remains constant. The drop volume is then obtained via summation over the individual drop volumes times their weight. For additional drops to be generated, a drop within the regarded height element is selected randomly to be copied, and the copy is placed at a random position in the height element. The weight of all drops in the height element is then decreased so as to again ensure the constant volume of the dispersed phase.

The weights attributed to the drops need to be considered consistently in the various calculations. For determining local volume and surface area of the dispersed phase, the weights have to be accounted for. These values are then used to determine the local holdup of the dispersed phase, which is used for example in the calculation of drop sedimentation and the local Sauter-mean diameter. On the other hand, the stochastic coalescence probabilities need to be applied to all individual drops including the fractional ones without accounting for their weight. When coalescence occurs, the weight of the coalesced drop has again to be determined such that the volume of the dispersed phase remains constant. This ReDrop approach for describing settling behavior is currently under further development, especially including more realistic coalescence models, so that comparison with experimental results we become available in the future.

7.3.3 APPLICATION TO DESCRIBE BATCH-SETTLER PERFORMANCE

Based on such a ReDrop simulation, the results shown in Figure 7.13 have been obtained. The starting drop-size distribution has been assumed to be log-normal with an average drop size of 0.05 mm, which has also been found in some experiments. The standard deviation of the distribution is 20% of the average diameter in all simulations shown. The starting holdup of dispersed light phase is 30%, which is assumed to have physical properties of low-viscosity paraffin oil. The curves shown in white are the position of the major interface as well as the lower limit of the close-packed zone, in which the drops are in direct contact, and in black the line of 10% holdup, which corresponds roughly to what might be evaluated as sedimentation curve from the videos of the experiment. Other choices of the latter limit lead to comparable principal results. If the time- and height-dependent holdup profile shown in Figure 7.13 is compared to the simple picture presented in Figure 7.10, at least two aspects are worth noting. First of all, there appears to be a lag time; that is, during the first 20 seconds the sedimentation and coalescence curves are essentially horizontal. Such a lag time is frequently observed in settling experiments with technical systems and is also visible in Figure 7.12. It is obvious also from the curved sedimentation curve that during these first seconds considerable coalescence occurs, leading to an increase in drop size until the drops get so large that their sedimentation velocity leads to a visible sedimentation curve. Also, this corresponds to the behavior observed in real systems as seen in Figure 7.12. In Figure 7.14, the corresponding average drop size is plotted vs. height and time. Comparing this drop-size profile to the holdup profile in Figure 7.13, it becomes apparent that the drops for which sedimentation becomes obvious have reached a diameter of around 0.25 mm. Thus, of the order of 100 original drops have coalesced during this "ripening" time until the drops have become large enough to sediment so quickly that phase

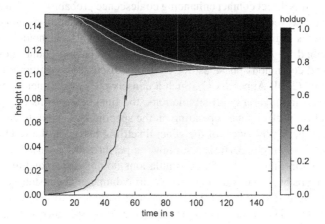

FIGURE 7.13 Holdup profile from ReDrop simulation of settling experiment with starting diameter distribution around 0.05 mm. The white lines separate the close-packed layer from the coalesced previously dispersed phase and the sedimentation zone. The black line corresponds to 10% holdup. A lag time as well as a densely packed zone within the sedimentation zone are clearly visible.

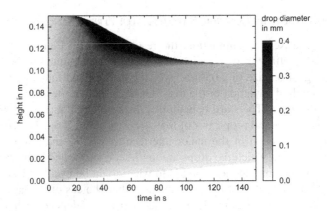

FIGURE 7.14 Sauter-mean diameter for the simulation of Figure 7.13. In the sedimentation zone relevant for the major settling the average drop size has increased from the starting value of 0.05 mm to around 0.25 mm.

separation becomes apparent. Interestingly, if corresponding settling experiments are performed for technical systems and evaluated with the Henschke model assuming mono-sized droplets while shifting the evaluation by the lag time, frequently drop sizes in the range between 0.2 and 0.4 mm are obtained. Only for clean systems, for example, as recommended as standard test systems for solvent extraction by EFCE (Míšek, Berger, and Schröter 1985), no lag time occurs, and the drops apparently start out already relatively large at diameters above typically 0.5 mm (Henschke 1995).

The second observation of Figure 7.13 is that besides the close-packed zone, where the drops are assumed to be in contact so that they can transmit their buoyancy force by this direct contact enhancing coalescence probability, a densely packed zone can be observed. In this zone, the relative flow between drops and continuous phase stabilizes a dispersion which has a relatively high local holdup. This situation corresponds to a second stable fluid-dynamic point of a counterflow between dispersed and continuous phase as, for example, discussed by Pilhofer and Mewes (1979) or Bart (2001, Appendix C). Such a densely packed sedimentation zone has been observed in essentially all simulations, the thickness of this zone relative to that of the close-packed zone depending on the specific system parameters chosen.

In order to get an overview of the general settling behavior, the ReDrop simulation of a batch-settling experiment can now be performed for varied major parameters as shown in Figure 7.15. These simulations do not correspond to any specific material systems but rather vary the average drop diameter at constant relative width of the drop-size distribution and the coalescence parameter for a system with fixed other parameters. The variation of starting average drop size corresponds to a variation of stirring intensity during the preparation of dispersion; the varied coalescence parameter mimics the influence of trace impurities like surfactants and salts on the drop coalescence (Soika and Pfennig 2005; Kopriwa 2013). The variations show that even though the initial drop size is varied in a wide range, the sedimentation curve has a comparable slope after the initial ripening of the dispersion for each value of

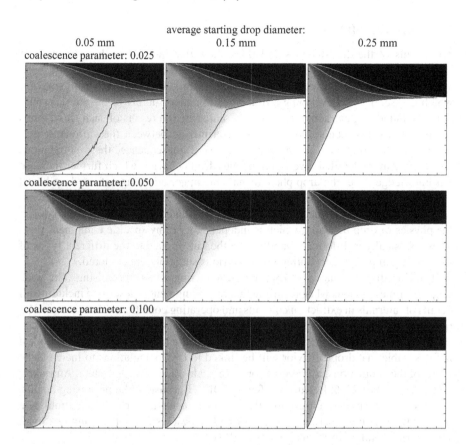

average starting drop diameter:

| 0.05 mm | 0.15 mm | 0.25 mm |

coalescence parameter: 0.025

coalescence parameter: 0.050

coalescence parameter: 0.100

FIGURE 7.15 Drop-based simulations with varied average starting drop diameter and coalescence parameter. The results indicate that stirring intensity has an effect mainly on lag time while beyond that lag time coalescence is little influenced by starting drop diameter.

coalescence parameter, namely for each material system. This means that the general observation mentioned above that a system has a typical drop diameter, which is obtained independent of the intensity with which the dispersion is generated over a wide range of this intensity, is a result of the interplay between coalescence during sedimentation and sedimentation of the drop ensemble.

Similarly, the time until complete phase separation is reached depends essentially only on the coalescence parameter, if the different lag times are disregarded. The coalescence time is thus mainly affected by the coalescence parameter as is to be expected. The observed decoupling of overall coalescence behavior after ripening and the initial drop size is noteworthy.

In all the cases shown, the densely packed region can be observed below the close-packed zone. This has to be accounted for appropriately when performing a drop-based simulation of the settling process, which can in principle be performed with the Monte–Carlo approach as in ReDrop or by direct solution of the drop-population balances.

7.4 CONCLUSIONS

The results of the drop-based simulations show that they allow the prediction of the behavior of extraction equipment on a pilot-plant and technical scale with good accuracy. At the same time, they help to gain detailed insight into drop behavior, which is essential for developing a fundamental understanding of the physical phenomena underlying extraction processes. Comparing the results of such simulations to dedicated experiments allows one to discriminate between the various modeling approaches presented in the literature. As a consequence, these simulations also trigger more detailed investigations into drop behavior, which further increase our knowledge about the drop phenomena. One major advantage of the drop-based approach is that drop behavior can be investigated on lab scale with a little amount of substance. The experimental results can be evaluated with models, which depict the physics of drop behavior, which is independent of any specific equipment type. These physically well-founded models can then be applied to the different types of extraction equipment, accounting for the various situations encountered.

The overall performance of drop-based modeling is very promising. The wide scope becomes obvious when recalling that real equipment is described including all details of internals in extraction columns and operating conditions. Thus, the simulation happens not just on the level of theoretical stages but on the equipment level.

For the next steps in applying the drop-based method for solvent extraction, it is foreseeable that drop behavior will be linked to CFD simulations to increase the scope of the simulations, accounting for large-scale effects (Hlawitschka, Attarakih, Alzyod, and Bart 2016; Leleu and Pfennig 2018). Because of the promising results, activities in other research groups also try to transfer the drop-based simulation method to other unit operations like fiber-bed coalescers or bubble columns (Speth 2004; Grünewald, Lausch, and Schlüter 2013).

REFERENCES

Altunok, Mehmet Yücel, Murat Kalem, and Andreas Pfennig. 2012. "Investigation of mass transfer on single droplets for the reactive extraction of zinc with D2EHPA." *AIChE Journal* 58(5):1346–1355.

Alzyod, Samer, Menwer Attarakih, Abdelmalek Hasseine, and Hans-Jörg Bart. 2017. "Steady state modeling of Kühni liquid extraction column using the Spatially Mixed Sectional Quadrature Method of Moments (SM-SQMOM)." *Chemical Engineering Research and Design* 117:549–556.

Arimont, Klaus, Michael Soika, and Martin Henschke. 1996. "Simulation einer pulsierten Füllkörperextraktionskolonne." *Chemie Ingenieur Technik* 68(3):276–279.

Attarakih, Menwer. 2004. *Solution Methodologies for the Population Balance Equations Describing the Hydrodynamics of Liquid–Liquid Extraction Contactors*. Ph.D. thesis, University of Kaiserslautern.

Attarakih, Menwer, and Hans-Jörg Bart. 2014. "Solution of the population balance equation using the differential maximum entropy method (DMaxEntM): An application to liquid extraction columns." *Chemical Engineering Science* 108:123–133.

Attarakih, Menwer M., Hans-Jörg Bart, Tilmann Steinmetz, Markus Dietzen, and Naim M. Faqir. 2008. LLECMOD: "A bivariate population balance simulation tool for liquid–liquid extraction columns." *The Open Chemical Engineering Journal* 2(1):10–34.

Ayesterán, José, Nicole Kopriwa, Florian Buchbender, Murat Kalem, and Andreas Pfennig. 2015. "ReDrop – A simulation tool for the design of extraction columns based on single-drop experiments." *Chemical Engineering and Technology* 38(10):1894–1900.

Bardow, André, Volker Göke, Hans-Jürgen Koß, Klaus Lucas, and Wolfgang Marquardt. 2005. "Concentration-dependent diffusion coefficients from a single experiment using model-based Raman spectroscopy." *Fluid Phase Equilibria* 228–229:357–366.

Bart, Hans-Jörg. 2001. *Reactive Extraction*. Berlin: Springer.

Bart, Hans-Jörg, Daniel Garthe, Tobias Grömping, Andreas Pfennig, Stephan Schmidt, and Johann Stichlmair. 2005. "Vom Einzeltropfenexperiment zur Extraktionskolonne." Final project report, AiF project 40 ZN/1+2+3. http://hdl.handle.net/2268/182578.

Bart, Hans-Jörg, Daniel Garthe, Tobias Grömping, Andreas Pfennig, Stephan Schmidt, and Johann Stichlmair. 2006. "Vom Einzeltropfen zur Extraktionskolonne." *Chemie Ingenieur Technik* 78(5):543–547.

Bollen, Arnoud Maurits. 1999. *Collected Tales on Mass Transfer in Liquids*. Ph.D. thesis, Rijksuniversiteit Groningen.

Buchbender, Florian, Armin Fischer, and Andreas Pfennig. 2013. "Influence of compartment geometry on the residence time of single drops in Kühni extraction columns." *Chemical Engineering Science* 104:701–716.

Buchbender, Florian, Ferdy Onink, Wytze Meindersma, André de Haan, and Andreas Pfennig. 2012. "Simulation of aromatics extraction with an ionic liquid in a pilot-plant Kühni extractor based on single-drop experiments." *Chemical Engineering Science* 82:167–176.

Buchbender, Florian, Markus Schmidt, Tilmann Steinmetz, and Andreas Pfennig. 2012. "Simulation von Extraktionskolonnen in der industriellen Praxis." *Chemie Ingenieur Technik* 84(4):540–546.

Cockbain, E.G., and T.S. McRoberts. 1953. "The stability of elementary emulsion drops and emulsions." *Journal of Colloid Science* 8(4):440–451.

Coulaloglou, C. A., and L. L. Tavlarides. 1977. "Description of interaction processes in agitated liquid–liquid dispersions." *Chemical Engineering Science* 32(11):1289–1297.

Engberg, Roland F., Mirco Wegener, and Eugeny Y. Kenig. 2014. "The impact of Marangoni convection on fluid dynamics and mass transfer at deformable single rising droplets – A numerical study." *Chemical Engineering Science* 116:208–222.

Frising, Tom, Christine Noïk, and Christine Dalmazzone. 2006. "The liquid/liquid sedimentation process: From droplet coalescence to technologically enhanced water/oil emulsion gravity separators: A review." *Journal of Dispersion Science and Technology* 27(7):1035–1057.

Garthe, Daniel. 2006. *Fluid dynamics and Mass Transfer of Single Particles and Swarms of Particles in Extraction Columns*. Ph.D. thesis, TU München.

Gebauer, Felix, Jörn Villwock, Matthias Kraume, and Hans-Jörg Bart. 2016. "Detailed analysis of single drop coalescence – Influence of ions on film drainage and coalescence time." *Chemical Engineering Research and Design* 115(B):282–291.

Gillespie, T., and Eric K. Rideal. 1956. "The coalescence of drops at an oil–water interface." *Transactions of the Faraday Society* 52:173–183.

Grimes, B.A. 2012. "Population balance model for batch gravity separation of crude oil and water emulsions. Part I: Model formulation." *Journal of Dispersion Science and Technology* 33(4):578–590.

Grimes, B. A., C. A. Dorao, N. V. D. T. Opedal, I. Kralova, G. H. Sørland, and J. Sjöblom. 2012. "Population balance model for batch gravity separation of crude oil and water emulsions. Part II: Comparison to experimental crude oil separation data." *Journal of Dispersion Science and Technology* 33(4):591–598.

Grömping, Tobias Christoph. 2015. *Auslegung von Extraktionskolonnen mit geordneten Packungen*. Ph.D. thesis, RWTH Aachen, Aachen: Shaker Verlag.

Gross-Hardt, Edwin, Andrea Amar, Siegfried Stapf, Bernhard Blümich, and Andreas Pfennig. 2006. "Flow dynamics measured and simulated inside a single levitated droplet." *Industrial and Engineering Chemistry Research* 45(1):416–423.

Grünewald, Marcus, Hans-Rolf Lausch, and Michael Schlüter. 2013. "'Campus Blasensäulen' – nachhaltige Entwicklung von Messtechniken und Auslegungswerkzeugen für Mehrphasenströmungen." *Chemie Ingenieur Technik* 85(7):967.

Handlos, A. E., and T. Baron. 1957. "Mass and heat transfer from drops in liquid–liquid extraction." *AIChE Journal* 3(1):127–136.

Haverland, Hartmut. 1988. "Untersuchungen zur Tropfendispergierung in flüssigkeitspulsierten Siebboden-Extraktionskolonnen." Ph.D. thesis, TU Clausthal.

Haverland, Hartmut, Alfons Vogelpohl, Christophe Gourdon, and Gilbert Casamatta. 1987. "Simulation of fluid dynamics in a pulsed sieve plate column." *Chemical Engineering and Technology* 10(1):151–157.

Henschke, Martin. 1995. *Dimensionierung liegender Flüssig-flüssig-Abscheider anhand diskontinuierlicher Absetzversuche.* Reihe 3 Verfahrenstechnik, No. 279. Düsseldorf: VDI-Verlag.

Henschke, Martin. 2004. *Auslegung pulsierender Siebboden-Extraktionskolonnen. Berichte aus der Verfahrenstechnik.* Aachen: Shaker Verlag.

Henschke, Martin, and Andreas Pfennig. 1996. "Simulation of packed extraction columns with the REDROP model." In *12th International Congress of Chemical and Process Engineering*, CHISA, Praha, Czech Republic, August 25–30. http://hdl.handle.net/2268/180407.

Henschke, Martin, and Andreas Pfennig. 1999. "Mass-transfer enhancement in single-drop extraction experiments." *AIChE Journal* 45(10):2079–2086.

Henschke, Martin, and Andreas Pfennig. 2002. "Influence of sieve trays on the mass transfer of single drops." *AIChE Journal* 48(2):227–234.

Henschke, Martin, Lars Holger Schlieper, and Andreas Pfennig. 2002. "Determination of a coalescence parameter from batch-settling experiments." *Chemical Engineering Journal* 85(2–3):369–378.

Hlawitschka, Mark W., Menwer M. Attarakih, Samer S. Alzyod, and Hans-Jörg Bart. 2016. "CFD based extraction column design – Chances and challenges." *Chinese Journal of Chemical Engineering* 24(2):259–263.

Hoting, Björn. 1996. "Untersuchung zur Fluiddynamik und Stoffübertragung in Extraktionskolonnen mit strukturierten Packungen." Ph.D. thesis, Technical University Clausthal, Düsseldorf: VDI Verlag.

Jeelani, Shaik Abdul Khadar, and Stanley Hartland. 1998. "Effect of dispersion properties on the separation of batch liquid–liquid dispersions." *Industrial and Engineering Chemistry Research* 37(2):547–554.

Jerez, Ayesterán, José Manuel, Florian Buchbender, Murat Kalem, Eva Kalvoda, and Andreas Pfennig. 2017. "Single-drop experiments for challenging conditions as basis for extraction-column simulations." In *Proceedings of the 21st International Solvent Extraction Conference ISEC2017*. 5.11. to 10.11.2017. Miyazaki, Japan.

Kalem, Murat. 2015. "Einzeltropfenbasierte Simulation von Pulsierten Siebbodenkolonnen für die Reaktivextraktion." Ph.D. thesis, RWTH Aachen, Aachen: Shaker Verlag.

Kalem, Murat, and Andreas Pfennig. 2007. "ReDrop – A general method for solving drop-population balances with an arbitrary number of property variables." Presented at 17th European Symposium on Computer Aided Process Engineering – ESCAPE17. Bucharest, Romania.

Kalem, Murat, Mehmet Yücel Altunok, and Andreas Pfennig. 2010. "Sedimentation behavior of droplets for the reactive extraction of zinc with D2EHPA." *AIChE Journal* 56(1):160–167.

Kalem, Murat, Florian Buchbender, and Andreas Pfennig. 2011. "Simulation of hydrodynamics in RDC extraction columns using the simulation tool 'ReDrop.'" *Chemical Engineering Research and Design* 89(1):1–9.

Kalvoda, Eva. 2016. "Einfluss von Konzentrationsgradienten auf das Verhalten von Einzeltropfen in Extraktionskolonnen." Ph.D. thesis, Technischen Universität, Graz. https://diglib.tugraz.at/einfluss-von-konzentrationsgradienten-auf-das-verhalten-von-einzeltropfen-in-extraktionskolonnen-2016.

Kamp, Johannes. 2017. "Systematic coalescence investigations in liquid/liquid systems – From single drops to technical applications." Ph.D. thesis, TU Berlin. https://deposit once.tu-berlin.de/handle/11303/7252.

Kamp, Johannes, Jörn Villwock, and Matthias Kraume. 2017. "Drop coalescence in technical liquid/liquid applications: A review on experimental techniques and modeling approaches." *Reviews in Chemical Engineering* 33(1):1–47.

Klinger, Sigrid. 2008. "Messung und Modellierung des Spaltungs- und Koaleszenzverhaltens von Tropfen bei der Extraktion." Ph.D. thesis, RWTH Aachen. http://darwin.bth.rwth-aachen.de/opus3/volltexte/2008/2579/index.html.

Kopriwa, Nicole Sabine. 2013. "Quantitative Beschreibung von Koaleszenzvorgängen in Extraktionskolonnen." Ph.D. thesis, RWTH Aachen. https://publications.rwth-aach en.de/record/229074.

Kopriwa, Nicole, and Andreas Pfennig. 2016. "Characterization of coalescence in extraction equipment based on lab-scale experiments." *Solvent Extraction and Ion Exchange* 34(7):622–642.

Kopriwa, Nicole, Florian Buchbender, José Ayesterán, Murat Kalem, and Andreas Pfennig. 2012. "A critical review of the application of drop-population balances for the design of solvent extraction columns: I. Concept of solving drop-population balances and modeling breakage and coalescence." *Solvent Extraction and Ion Exchange* 30(7):683–723.

Kronberger, Thomas, Andreas Ortner, Walter Zulehner, and Hans-Jörg Bart. 1995. "Numerical simulation of extraction columns using a drop population model." *Computers and Chemical Engineering* 19:639.

Kronig, R., and J.C. Brink. 1950. "On the theory of extraction from falling droplets." *Applied Scientific Research* 2:142–154.

Leleu, David, and Andreas Pfennig. 2017. "Standardizes settling cell to characterize liquid–liquid dispersion." In *Proceedings of the 21st International Solvent Extraction Conference ISEC2017.* 5.11. to 10.11.2017. Miyazaki, Japan.

Leleu, David, and Andreas Pfennig. 2018. "ERICAA, liquid–liquid gravity settlers." https://www.chemeng.uliege.be/ericaa.

Leu, Jan Thomas. 1995. "Beitrag zur Fluiddynamik von Extraktionskolonnen mit geordneten Packungen." Ph.D. thesis, TU Clausthal.

Míšek, T., R. Berger, and J. Schröter. 1985. *Standard Test Systems for Liquid Extraction*, 2nd edition. Warwickshire: The Institution of Chemical Engineers.

Modes, Gerd. 1999. "Grundsätzliche Studie zur Populationsdynamik einer Extraktionskolonne auf Basis von Einzeltropfenuntersuchungen." Ph.D. thesis, TU Kaiserslautern.

Mohammadi Sarab Badieh, Marjan, Maria Chiara Quaresima, Andreas Pfennig, and Javad Saien. 2017. "Performance study of ionic liquid in extraction based on single-drop experiments." *Solvent Extraction and Ion Exchange* 35(7):563–572.

Mohanty, Swati, and Alfons Vogelpohl. 1997. "A simplified hydrodynamic model for a pulsed sieve-plate extraction column." *Chemical Engineering and Processing: Process Intensification* 36(5):385–395.

Mörters, Martin, and Hans-Jörg Bart. 2003. "Mass transfer into droplets undergoing reactive extraction." *Chemical Engineering and Processing: Process Intensification* 42(10):801–809.

Newman, Albert B. 1931. "The drying of porous solids: Diffusion and surface emission equations." *Transactions of the American Institute of Chemical Engineers* 27:203–220.

Noïk, Christine, Thierry Palermo, and Christine Dalmazzone. 2013. "Modeling of liquid/liquid phase separation: Application to petroleum emulsions." *Journal of Dispersion Science and Technology* 34(8):1029–1042.

Ortner, Andreas, Thomas Kronberger, Walter Zulehner, and Hans-Jörg Bart. 1995. "Tropfenpopulationsmodell am Beispiel einer gerührten Extraktionskolonne." *Chemie Ingenieur Technik* 67(8):984–988.

Pertler, Manfred. 1996. "Die Mehrkomponenten-Diffusion in nicht vollständig mischbaren Flüssigkeiten." Ph.D. thesis, Technische Universität München.

Pfennig, Andreas, and Albrecht Schwerin. 1998. "Influence of electrolytes on liquid–liquid extraction." *Industrial and Engineering Chemistry Research* 37(8):3180–3188.

Pfennig, Andreas, Theo Pilhofer, and Jürgen Schröter. 2006. "Flüssig-Flüssig-Extraktion." In *Fluid-Verfahrenstechnik*, Volume 2, edited by Ralf Goedecke. Weinheim: Wiley–VCH, 907–992.

Pietzsch, Wolfgang. 1984. "Beitrag zur Auslegung pulsierter Siebbodenextraktoren." Ph.D. thesis, TU Munich.

Pilhofer, Theo, and Dieter Mewes. 1979. *Siebboden-Extraktionskolonnen. Vorausberechnung unpulsierter Kolonnen*. Weinheim: Verlag Chemie.

Qi, Mingzhai, Hartmut Haverland, and Alfons Vogelpohl. 2000. "Auslegung von pulsierten Siebboden- und Sprühkolonnen für die Extraktion auf der Basis von Einzeltropfenuntersuchungen." *Chemie Ingenieur Technik* 72(3):203–214.

Quaresima, Maria Chiara, Markus Schmidt, and Andreas Pfennig. 2017. "Solvent extraction design for highly viscous systems." In *Proceedings of the 21st International Solvent Extraction Conference ISEC2017*. 5.11. to 10.11.2017. Miyazaki, Japan.

Richardson, J. F., and W. N. Zaki. 1954. "Sedimentation and fluidisation: Part 1." *Transactions of the Institution of Chemical Engineers* 32:35–53.

Rutten, Philippus Willibrordus Maria. 1992. "Diffusion in Liquids." Ph.D. thesis, Technische Universiteit, Delft, Delft: Delft University Press.

Sawistowski, Henryk. 1973. "Surface-tension-induced interfacial convection and its effect on rates of mass transfer." *Chemie Ingenieur Technik – CIT* 45(18):1093–1098.

Scheele, G. F., and D. B. Clark. 1980. "The effect of impact velocity on the coalescence of tetrachloroethylene drops at a planar tetrachloroethylene–water interface." *Chemical Engineering Communications* 5(1–4):23–35.

Schmidt, Stephan Anton. 2006. "Populationsdynamische Modellierung gerührter Extraktionskolonnen auf der Basis von Einzeltropfen- und Tropfenschwarmuntersuchungen." Ph.D. thesis, TU Kaiserslautern, Aachen: Shaker Verlag.

Schott, Robin, and Andreas Pfennig. 2004. "Modelling of mass-transfer induced instabilities at liquid–liquid interfaces based on molecular simulations." *Molecular Physics* 102(4):331–339.

Schügerl, Karl, Rolf Hänsel, Eberhard Schlichting, and Werner Halwachs. 1986. "Reaktivextraktion." *Chemie Ingenieur Technik* 58(4):308–317.

Simon, Martin. 2004. "Koaleszenz von Tropfen und Tropfenschwärmen." Ph.D. thesis, TU Kaiserslautern.

Soika, Michael, and Andreas Pfennig. 2005. "Extraktion – Eine Frage des Wassers?" *Chemie Ingenieur Technik* 77(7):905–911.

Speth, Hauke. 2004. "Ein neues Modell zur Auslegung von Faserbett-Koaleszenzabscheidern." Ph.D. thesis, RTWH Aachen, Aachen: Shaker Verlag.

Sternling, C. V., and L. E. Scriven. 1959. "Interfacial turbulence: Hydrodynamic instability and the Marangoni effect." *AIChE Journal* 5(4):514–523.

Webber, Grant B., Scott A. Edwards, Geoffrey W. Stevens, Franz Grieser, Raymond R. Dagastine, and Derek Y.C. Chan. 2008. "Measurements of dynamic forces between drops with the AFM: Novel considerations in comparisons between experiment and theory." *Soft Matter* 4(6):1270–1278.

Wegener, Mirco. 2009. *"Der Einfluss Der Konzentrationsinduzierten Marangonikonvektion Auf Den Instationären Impuls- Und Stofftransport an Einzeltropfen."* Ph.D. dissertation, Technical University Berlin.

Wegener, M., M. Fevre, A. R. Paschedag, and M. Kraume. 2009. "Impact of Marangoni instabilities on the fluid dynamic behaviour of organic droplets." *International Journal of Heat and Mass Transfer* 52(11–12):2543–2551.

Yu, Geng-Zhi, and Zai-Sha Mao. 2004. "Sedimentation and coalescence profiles in liquid–liquid batch settling experiments." *Chemical Engineering and Technology* 27(4): 407–413.

Index

Printed in the United States
by Baker & Taylor Publisher Services